Proceedings of the 12th Reinventing Space Conference

Scott Hatton
Editor

Proceedings of the 12th Reinventing Space Conference

Springer

Editor
Scott Hatton
The British Interplanetary Society
London, United Kingdom

ISBN 978-3-319-34023-4 ISBN 978-3-319-34024-1 (ebook)
DOI 10.1007/978-3-319-34024-1

Library of Congress Control Number: 2016960756

Printed on acid-free paper

This Springer imprint is published by Springer Nature
The registered company is Springer International Publishing AG
The registered company address is: Gewerbestrasse 11, 6330 Cham, Switzerland

Contents

Skimsats: bringing down the cost of Earth Observation

Andrew Bacon
Thales Alenia Space UK Ltd
660 Bristol Business Park, Coldharbour Lane, Bristol, BS16 1EJ; +44 (0) 1179 375231
andrew.bacon@thalesaleniaspace.com

Ben Olivier
Thales Alenia Space UK Ltd
660 Bristol Business Park, Coldharbour Lane, Bristol, BS16 1EJ; +44 (0) 1179 375179
ben.olivier@thalesaleniaspace.com

ABSTRACT

The Skimsat concept is based upon the central idea of that the closer you are to a target, the smaller an imaging payload can be; leading to reductions in overall satellite size and cost. Skimsats are small satellites designed to operate in Very Low Earth Orbits (VLEO), with perigees as low as 160km, which would place them closer to the ground than all past and present operational satellites.

The Skimsat concept has been developed to meet the challenge of Unmanned Ariel Vehicles (UAVs) to the satellite Earth Observation (EO) industry. UAVs have very high persistence over a target and can achieve spatial resolutions better than all but the very largest instruments found on current EO satellites. EO satellites have several advantages over UAVs including lower vulnerability, intrusiveness and a significantly larger daily area coverage. However to match the persistence and resolution of a UAV requires constellations of expensive satellites; out of reach of all but the largest budgets.

High resolution EO satellites can be provided at costs at least an order of magnitude lower than the current state of the art by a single change; significant reduction in orbital altitude. By operating at down to 160km altitude the Skimsat platform can provide SAR and optical imagery at 1m Ground Sample Distance (GSD) with a launch mass of <75kg, which is more than four times less than the current smallest 1m GSD capable EO satellite (SSTL 300 S1).

The decrease of altitude, with respect to a 650km orbit, by a factor of four leads to a 64x reduction in radar RF power, 16x reduction in communications RF power and 4x reduction in optical aperture diameter to achieve the same performance.

The comparatively large drag forces due to increased air density at these low altitudes would normally cause a small satellite (<100Kg) to de-orbit within 1-2 months. However, Skimsats are intended to use a combination of low cross-sectional drag area and a novel air-breathing drag-compensating propulsion system to increase the operational lifetime to 24 months or more. This has the added benefit of a guaranteed propellantless de-orbit shortly after the end of the operational mission, generating no debris and clearing the orbit for follow-on missions.

Additional challenges for Skimsats include increased damage on optical surfaces from higher atomic oxygen densities found at low altitudes and the higher rates of orbital drift. Both of these can be somewhat countered by improved materials, positioning and maneuvering technology but will ultimately limit the lifetime to approximately 24 months. This will give Skimsats a higher turnaround enabling each successive generation to make use of the latest payload technology; improving performance and services.

The overall aim of developing the Skimsat EO platform is to enhance the capability of the EO sector by providing low-cost access to VLEO and to benefit from the new and enhanced applications that this will offer. The reduced cost of EO will also help bring entities that are currently regarded as non-space (e.g. UAV and aircraft operators) together with the space sector to enhance their capability.

S. Hatton (ed.), *Proceedings of the 12th Reinventing Space Conference*, DOI 10.1007/978-3-319-34024-1_1

KEYWORDS: Skimsat, Very Low Earth Orbit, Earth Observation, Small Satellite, Electric Propulsion, SAR, Telecommunications, Low Cost, Space Debris, UAV, Air Breathing

1 THE SKIMSAT CONCEPT

Thales Alenia Space UK Ltd (TAS UK) have developed the "Skimsat" concept of Very Low Earth Orbiting (VLEO) small satellites that fly at an altitude that can be considered as 'skimming' the top of Earth's atmosphere and obtain significant EO payload miniaturisation whilst continuing to improve performance.

A Skimsat EO imaging platform offers a solution to the issues with the expense of current EO platforms due to the benefits of orbiting at an unprecedented low altitude of 160km, including:

- Operation at 160km leads to an approximate 64x reduction in the required RF power of an Earth scanning radar, for the same Signal to Noise Ratio (SNR), when compared to a radar imaging satellite at 650km altitude.
- Operation at 160km leads to an approximate 4x reduction in the required aperture diameter and focal length, for the same Ground Sample Distance (GSD), when compared to an optical imaging satellite at 650km altitude.
- A reduction in the overall cost by at least an order of magnitude of building and launching high resolution radar and optical imaging systems through miniaturisation of the platform, without compromising performance. This enables many Skimsats to be launched into different orbits for the price of a single traditional high resolution imaging satellite, improving the temporal resolution and redundancy by an order of magnitude or more for the same cost.
- Secondary benefits that make Skimsats better platforms for EO imaging include:
    - Operation at 160km leads to an approximate 10x reduction in the required RF power for downlinking EO data, for the same SNR, when compared to a satellite at 500km altitude.
    - Elimination of orbital debris issues as operation at 160km guarantees total disintegration of the space segment by re-entry within approximately 30 days from mission end.

The Skimsat concept is based upon the idea of that the closer you are to a target, the smaller an imaging payload can be; leading to reductions in overall satellite size and cost. Skimsats, as shown in Figure 1 below, are small satellites designed to operate in Very Low Earth Orbits (VLEO) with perigees as low as 160km which would place them closer to the ground than all past and present operational satellites.

Figure 1 Baseline Skimsat space segment concept

The comparatively large drag forces due to increased air density at these low altitudes would normally cause a small satellite (<100Kg) to de-orbit within 1-2 months. However, Skimsats are intended to use a combination of low cross-sectional drag area and air breathing drag-compensating thrusters to increase the operational lifetime to 24 months or more. This has the added benefit of a guaranteed propellantless de-orbit shortly after the end of the operational mission, generating no debris and preserving the orbit for follow-on missions.

The overall aim of developing the Skimsat imaging platform is to enhance the UK capability by providing low-cost access to VLEO and to benefit from the new and enhanced applications that this will offer. The reduced cost of EO will also help bring entities that are currently regarded as non-space (e.g. UAV and aircraft operators) together with the space sector to enhance their capability.

2 BENEFITS

The primary benefits of developing Skimsat high resolution EO platforms brought to the providers and users of EO data, include:

- At least an order of magnitude reduction in the cost per satellite when compared to traditional EO platforms, due to the significantly reduced payload size. This means that constellation sizes can be increased by an order of magnitude or more for the same cost, leading to significantly reduced target revisit times. The availability of 1m GSD imagery with revisit times of a few hours or less will greatly enhance the quality of existing user applications and lead to the development of new ones.
- Shorter lead times from application idea to launch by the use of satellites which are considerably smaller than traditional EO platforms, for the same performance. Smaller satellites require less time to build, test and obtain launch slots for, enabling EO data providers to exploit new applications faster.
- The lower altitude will also greatly improve the link budget of Skimsat to ground communications; potentially allowing the size of the ground segment receiver to be reduced, in size or complexity, to achieve the same downlink rate. This will lead to ground segment build and operational savings to EO data providers.

There are also direct benefits from the developments of a multi-application Skimsat platform to EO payload designers:

- An equivalent EO payload on Skimsats, when compared to traditional EO satellites in higher orbits, can be greatly miniaturised with respect to optics size and active sensor power which will reduce development costs, times and risks. This is shown to-scale in Figure 2 below:

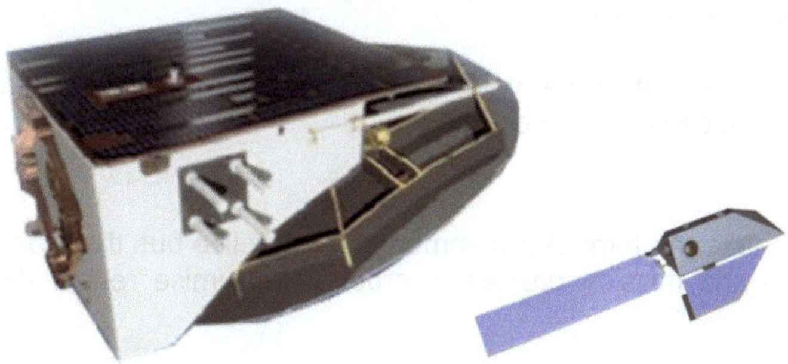

Figure 2 To-scale size comparison of a SAR-Lupe (left, image credit OHB-System AG) radar satellite and the current baseline Skimsat concept. By orbiting at an altitude three times lower Skimsats can achieve the same radar performance as SAR-Lupe on a satellite that has approximately 1/10th of the mass and 1/50th of the internal volume.

- The intended standardized payload interface (that will take the requirements for optical, hyperspectral and radar into account) will reduce the development time for payload developers and encourage entities with UAV instrument experience to create payloads.

- The reduced lifetime and lower cost of Skimsats will allow for a much faster turnaround time and more in-orbit demonstration opportunities of new EO payload developments. This will allow the industry to expand and retain skills between successive missions whilst encouraging incremental improvements to EO payloads over subsequent missions (as opposed to developing a single payload that is still required to meet user needs >5 years after launch).
- The peak of the ionosphere (the F2 layer) varies daily between 250km and 350km, as Skimsats are intended to operate lower than this radar signals will experience significantly less attenuation and propagation uncertainty than those that have to travel through the F2 layer. This will improve the SNR and accuracy of radar EO payloads and will allow the use of frequencies <10MHz which are usually reflected by the ionosphere.

Skimsats also provide non-EO benefits that are useful to the Space sector, for example:

- The improvement of communication link budgets due to the reduced altitude can also be used to reduce the power consumption and/or antenna size of transceivers on the ground, reducing the size of handheld satellite communication terminals and unmanned sensors.
- Skimsats will offer long term access to altitudes down to 160km which allows for in-situ measurements of the lower thermosphere, an area of great scientific interest for the transition between the neutral and ionised atmosphere but currently only accessible by <10 minute sounding rocket flights.
- The reduction of orbital debris issues by the widespread use of Skimsats for EO applications will help to preserve important, but currently congested, orbits (such as 600-800km sun-synchronous) for future generations. Eventually these orbits could only be used for specialist scientific missions that require them, such as astronomical observatories. Skimsats are a solution that could go a long way towards the reduction of unwanted bodies in LEO, without the cost of active debris removal missions.

3 PRECEDENT

The use of VLEO satellites has some precedent; the US CORONA 'Keyhole' satellites, launched regularly from the late 1950's until the early 1970's, operated in an orbit of 185x278km to provide high resolution imagery. This low altitude was sustained by large rocket engines with mission durations measured in tens of days.

TAS UK's concept updates and broadens this approach using 21st century technologies, most notably: miniaturised EO payloads, electric propulsion and satellite autonomy. In the past 40 years the enormous improvements in performance and miniaturisation, the concern for the prolonged life of space debris in higher orbits, and a trajectory of decreasing cost for launch into LEO combine to make a compelling case for revisiting and exploiting these orbits.

Skimsats are, in effect, a 'UAV at orbital velocity' which provides the coverage of a global UAV, with greater repeatability and reduced risk of interception.

4 PHILOSOPHY

Skimsats are intended to take the form of a common small satellite bus that can be adapted to a number of different EO payloads and mission profiles in order to maximise reuse, decrease risk and reduce deployment costs.

The comparatively short lifetime and more benign radiation environment (found when orbiting at very low altitudes) mean that there can be a lowered reliance on rad-hard parts for new systems, further reducing costs. The level of testing can also be reduced, most prominently in components, materials and mechanisms, when compared to traditional satellites designed to last for 5+ years in orbit; further reducing costs.

As the Skimsat dimensions are limited, to reduce the atmospheric drag to a minimum, internal volume will need to be highly prioritised. To maximise the use of this volume, redundancy will be limited to the most critical or vulnerable systems only or may not be used at all. The lack of internal redundancy can be offset by adding spare Skimsats to a constellation that are held in storage orbits until needed.

To overcome this minimisation of redundancy and maximise the potential of Skimsats they are best deployed into constellations or formations. When assigning multiple Skimsats to a single mission the disadvantages of small size and redundancy are significantly reduced and they can operate in the same way as a much larger satellite in an orbit much lower than could be achieved by a single large structure. The concept of payload fractionation can be used to distribute the different payloads between a formation of Skimsats so they can all observe the same point with different instruments in a similar manner to the A-Train; the CNES/NASA/JAXA formation of complementary EO satellites. Alternatively, the same payload could be put onto a number of Skimsats, orbiting in different planes, to create a constellation. Therefore global views can be built up by measuring multiple points simultaneously with the same instruments.

5 KEY CHALLENGES

The primary challenge for Skimsats is the increased air density in VLEO that leads to a greater level of atmospheric drag. Without compensating propulsion this would lead to re-entry of a small satellite within a few months. There are three past and future missions that have encountered or will soon encounter this issue:

- The European Space Agency's (ESA) Gravity Field and Steady-State Ocean Circulation Explorer (GOCE) has demonstrated sustained operation at ≈260km using a drag compensating ion engine combined with a highly sensitive accelerometer (which also acts as the payload) to conduct autonomous drag compensation. GOCE managed to operate for 55 months before running out of fuel and re-entered within 3 weeks.
- The FP7 funded QB50 project, led by the von Karman Institute, is a planned constellation of 50 CubeSats to explore the lower thermosphere. As none of these CubeSats have significant propulsion capabilities their lifetimes will be measured in months.
- The planned JAXA SLATS mission is also intending to demonstrate sustained operation at 250km altitude, descending to 180km, but only carries enough fuel for approximately 100 days below 250km [1].

The lifetime of all three of these missions is entirely dependant upon the amount of fuel that can be carried and it has been determined that using, for example, a gridded ion engine and an elliptical 'atmospheric dipping orbit', a Skimsat could operate for six months [2]. This is a limiting factor for the majority of missions and it is unlikely that conventional electric propulsion could be sufficiently optimised to achieve greater than a year's operation for a <100kg satellite.

TAS UK is therefore proposing the development of Air-Breathing Electric Propulsion (ABEP) which will gather the ambient ions and neutrals available in the lower thermosphere, ionise and accelerate them to achieve thrust. This removes the fuel restrictions on lifetime and also offers large mass reductions due to the elimination of propellant tanks and management systems. ABEP will be a game-changing technology for VLEO satellites, removing the primary lifetime limitation and allowing for much lower orbits to be accessed, with correspondingly greater benefits for EO applications.

The secondary challenges for Skimsat EO platforms and payloads are:

- At altitudes below 250km the ambient atomic Oxygen density is orders of magnitude higher than that experienced by traditional EO satellites. Atomic Oxygen erodes exposed surfaces over time; with telescope optics and solar arrays being particularly vulnerable. The degradation of optical quality and solar array power over time will become an important restriction on the lifetime of Skimsat EO payloads and platforms.
- The increased air density in VLEO has an effect on the aerostability of an EO platform, providing a resistance to attitude changes not normally encountered on satellites. This will have an impact on the ability of Skimsats to point payloads at ground targets with a sufficient accuracy and stability to achieve 1m GSD imagery. There are also issues with day/night cross-winds and

density variation in the upper atmosphere that could have undesirable effects on the Skimsat orbit.

- The orbital velocity of a satellite in a 160km circular orbit is approximately 10% faster than that in a 500km orbit which will have impacts on the design of EO payloads and the ground segment. A higher orbital velocity leads to a faster ground speed which puts limitations on the integration time of radar and optical sensors. It also leads to shorter ground station passes and faster tracking rates of steered antennae, which may prevent the use of some existing ground hardware.

6 NEW AND EMERGING MISSIONS

Whilst improvements in SNR, and potentially spatial resolution, are possible with Skimsats, the greatest benefit to applications will be the ability to launch large constellations at a lower cost for significantly improved temporal resolution. A number of existing and new applications can benefit from Skimsats, with some examples below:

- High resolution security & defence reconnaissance and surveillance can make use of a large constellation of multiple groups (in different orbital planes) of SAR, panchromatic, and multispectral equipped Skimsats flying in close (<1km) formations. With high resolution mean revisit times anywhere on the Earth of <3 hours may be achieved and the use of wider networks could allow images to be distributed in near-real time (<1 hour delay). In many cases this could eliminate the need to deploy UAVs which are vulnerable to interception, require a launch site within range and have a limited endurance.
- Obtaining high resolution imagery after natural or manmade disasters could be accomplished by the same or similar constellation described above. This can also overcome delays caused by the need to obtain air-space entry permission and flight time for disaster monitoring UAVs.
- A constellation of 20 Skimsats equipped with a 40km swath SAR payload and an AIS receiver would be able to scan Earth's oceans daily for tracking the position of maritime assets. The combination of SAR and AIS data could be used to detect vessels with no or disabled AIS transponders, potentially indicating power failure or piracy. A similar concept using MTI radar and ADS-B receivers could be used for tracking civilian aircraft, without a minimum radar height, but would require a much larger constellation or larger swath to provide a useful update interval.
- Skimsats with radar altimetry payloads could be used in a constellation of leader – follower formations with small separations (100m to 1km) to perform continuous interferometry measurements for daily high resolution and accuracy elevation maps that could be used to predict landslides, earthquakes and volcanic eruptions.
- Skimsats with scatterometer payloads could be used in a constellation to create high resolution (<1km) global wind maps daily to evaluate the effectiveness of potential sites for offshore wind turbines.
- Skimsats with thermal infrared imaging payloads could be combined with those with miniaturized SAR to detect wildfires and quantify the expected damage and monitor the loss of biomass in near-real time.
- A constellation of Skimsats with miniaturized multispectral payloads for imaging surface vegetation (as demonstrated on PROBA-V) could create daily global vegetation maps at a resolution high enough for assessment of crop stress levels and monitor deforestation.
- Skimsat EO platforms could also serve a number of non-EO applications, including:
    - Skimsats could be an ideal platform for low cost microgravity and life science experiments, without the human contact and biohazard concerns found on the ISS.
    - Store and forward communications could take advantage of the improved link budget and reduction in ionospheric effects (including attenuation and faraday rotation of linearly polarised signals) by reducing the size and power of ground terminals.

7 SUMMARY

The Skimsat concept benefits from the significant reduction in size of instruments when operating at low altitudes, driving down project and acquisition costs as well as the mass of the spacecraft. In turn, this means that Skimsats can are potentially easier to deploy into global constellations, as shown in Figure 3 below, thereby providing very high resolution optical, hyperspectral and radar imagery at a fraction of the cost of existing platforms and with reduced time delays.

Skimsat bridges the gap between conventional observation satellites and high endurance UAVs, via its superior performance relative to the former and its ability to rapidly reach targets beyond the range of UAVs within hours. But rather than rendering these systems redundant, Skimsat enhances their capabilities by improving their operational efficiency, whether by promoting more accurate target definition or by wider coverage context imagery.

The Skimsat platform's versatility extends its usefulness to a wide range of applications, from observation to scientific or communications roles, as well as offering a low risk solution to in-orbit flight tests for new technology.

Through its huge potential, Skimsat could become a showcase for the ingenuity and the ability of the UK's space industry and reaffirm the country's reputation as a leader in innovative, effective satellite systems.

Figure 3 Skimsats will be designed for versatile operation in isolation, formations and constellations

8 REFERENCES

1. https://directory.eoportal.org/web/eoportal/satellite-missions/s/slats

2. 40kg Skimsat in 160x400km orbit with 0.66m^2 drag area and 3kg of Xenon fuel

BIS-RS-2014-01

RETHINKING THE PRIVATE PROPERTY IN OUTER SPACE

Prof. Jairo Becerra
Research Professor / Catholic University of Colombia
Calle 12 No. 10 – 58 A.A. 179, Chía - Colombia; (+57) 320 862 4372
jabecerrao@ucatolica.edu.co jairoandresb@hotmail.com

ABSTRACT

This paper presents an analysis of Article II of the Outer Space Treaty - OST in its juridical and political context (policies); and how the prohibition of appropriation of outer space may be an obstacle for the development of lower cost and more responsive space systems.

The non-appropriation of outer space by any State or natural persons is a principle that has no discussion in International Law; it has become a norm of *Ius cogens* or mandatory commitment. Since the 1963 Declaration of Legal Principles Governing the Activities of States in the Exploration and Uses of Outer Space, followed by the Article II of the Outer Space Treaty and finally, Article XI of the Agreement Governing the Activities of States on the Moon and Other Celestial Bodies, the idea of not change the norm has been consistent. But, if main driver of human exploration has been the desire by acquisition of real state or private property, is difficult not to do a question: How will be the last frontier (outer space) conquered without implementing this right? And then, is it necessary to rephrase Article II of the OST, to encourage the private exploration of outer space in the next decades?

We try to establish with this paper, that to achieve more participation of private industry, it is necessary to amend Article II of the OST, and allow the appropriation of some areas of outer space like asteroids, some sections of the moon, and even areas of empty space, reinventing the space for future generations.

KEYWORDS: Space law, private property, national appropriation, equitable access, real estate.

INTRODUCTION*

Since the declaration regarding the Legal Principles Governing the Activities of States in the Exploration and Uses of Outer Space issued by the UN General Assembly in 1963[1], and later on through the Treaty of Principles Governing the Activities of States in the Exploration and Use of Outer Space, including the Moon and Other Celestial Bodies (the Outer Space Treaty, OST) [2], and finally, the Agreement Governing the Activities of States on the Moon and Other Celestial Bodies [3], space law established the principle of non-appropriation of outer space. This guiding principle of Space Law was enacted within the context of the Cold War, when the two superpowers at that time, the US and the USSR, had an open dispute regarding space conquest. The proposal to establish as a principle of space law the prohibition to appropriate space was filed by the Soviet Union in the document A/AC.105/C.2/l.1 [4] in 1962, (accepted by the United Nations Committee on the Peaceful Uses of Outer Space in 1963). Based on this prohibition, the Space countries and private entities have been carrying out their activities for more than 50 years.

S. Hatton (ed.), *Proceedings of the 12th Reinventing Space Conference*, DOI 10.1007/978-3-319-34024-1_2

Still, this proposal has been contradicted by some activities being carried out by several persons and private entities (Lunar Embassy) [5] by selling the moon and other celestial bodies, without there being any definite action by the signatory States, or at least some action together to prevent it. Is it feasible that space legislation only prevents the States from appropriating space, or does it include their citizens? Presently, we are evidencing ambiguity regarding how to proceed in this matter, since the states signing international treaties on outer space respect the principle, yet those violating it are not persecuted or sanctioned in many of these countries.

On the other hand, it is important to analyze the implications inherent in this prohibition and their impact on space exploration. Is this principle preventing the human being's permanence in space? Would some legislation allowing private property help conquering outer space?

PRIVATE PROPERTY

One of the basic human rights beings have is the right to private property, part of national orderings since the French Revolution until now, with very few exceptions. Even international law incorporates the right all individuals have to own private property. The right to private property is stated in article 17 of the Universal Declaration of Human Rights, which states: "Everyone has the right to own property individually as well as in association with others," a document adopted by the UN General Assembly in its Resolution 217 A (III), which has become one of the most important documents at an international level. Likewise, the right to own private property is acknowledged and respected in almost all democracies. In France, the 1791 Constitution included it in its article 2: "The purpose of any political association is to preserve the natural and inalienable rights of men." such rights are "freedom, property, safety, and resistance to oppression." And it is in force in its civil code and its later development, including the 1958 Constitution. Similarly, the US Constitution, even though it does not expressly mention it, does acknowledge its existence in its fifth amendment: "(…) nor shall private property be taken for public use, without just compensation (…);" likewise, it is stipulated in tens of other national legislations, from Colombia in Latin America, as stated in article 58 of its 1991 constitution, which says: "Private property and all the rights acquired according to civil laws are thus guaranteed (…)." Even in Asia, Japan's Constitution (1947), in article 29 also states: "The right to own or keep property is inviolable." As for international law, private property is acknowledged and protected in the same manner as are Human Rights and, thus, its existence is not questioned.

The said property can be of different kinds: on material or immaterial goods; on natural or real estate property. Material goods are those that can be perceived by the senses and have a real value; it is something material, such as a house or a book; immaterial goods are mere rights, such as credits or patents. Furniture is goods that can be transported from one place to another, be that through their own force or an external one; real estate is goods that cannot be moved, such as in the case of farms or pieces of land; they are the ones that remain fixed [6].

The non-appropriation principle in outer space refers indeed to the real estate on the lot or piece of land, as a material good or real estate since all the other rights in fact are present and are regulated by international and space legislation [7]. Thus, the objects we launch into space continue being under the property of their owner on Earth, and jurisdiction is kept by the State that has registered them, as well as the rights to inventions or patents, which are protected under the national legislation of the States that have jurisdiction over the spaceship or her crew, in the place where those rights have been granted [8].

Therefore, we may be able to talk about, think over, and analyze the prohibition to physically appropriate of outer space (real estate) since private property in the outer space is accepted. We don't lost the property rights over the space object just because the material goods are sent into outer space (Rights of enjoy and dispose of the thing we own)

OST, ARTICLE II

We really cannot take possession of parts of outer space? Article II of the Outer Space Treaty states: "Outer space, including the Moon and other celestial bodies, is not subject to national appropriation by claim of sovereignty, by means of use or occupation, or by any other means." And this implies the issue of national appropriation [9]. Private companies that sell parts from outer space say that this prohibition is only for the states and not for them, because this article does not explicitly mention them.

National Appropriation

It is true that article II does not explicitly refer to private entities or persons, yet it has been clearly established that such prohibition is extensive to all these entities that depend on the States. The said states are the ones responsible for such activities, and they would be the ones that are internationally accountable for the activities of private entities. In addition, private individuals cannot act in outer space without supervision from their State; moreover, they are required to have a license to carry out space activities, without which they would be violating the law [10]. This implies a clear dependence, subordination, and extension of the norm applied to the States.

On the other hand, the recognition of property rights over outer space cannot be considered, as these companies pretend it, due mainly to consuetudinary law [11] and as well as the interdependence between the prohibition and the law itself. The consuetudinary law is clearly defined by the International Court of Justice as "evidence of a general practice accepted as law" in its article 38.1 (b) [12]. That is, they are widely accepted norms that become law. They are not in writing but the actions in that sense by all the States make them so, with all the force that this implies. Since the States do not possess nor do they claim property rights, this generates an international norm of imperative enforcement. Another reason for this is that they are interdependent [13]; if there is the prohibition to claim sovereignty, there exists the prohibition of property right on the outer space that we cannot appropriate.

Virgiliu Pop, a University of Glasgow researcher, has referred to this several times when stating that even if the legal gap argued by Mr. Hope (owner of the lunar embassy) were to be accepted – that indeed private persons can appropriate themselves of physical real estate located in outer space – these companies would be the owners of nothing since no one can be the proprietor of something just because they say so [14]. We would add, in addition, that such a statement ought to be accompanied by a valid and acknowledged property deed, which cannot happen because at this time no deed is capable of fulfilling the said requirements because do not exist. Simply stated, since the requirements do not exist, nor are they defined, they cannot be met [15].

by any others means

The prohibition to appropriate outer space by claim of sovereignty or assertion, by means of use or occupation, or by any other means, we think is quite clear, since it pretended to encompass any form of appropriation of outer space as a whole.

The most common ones were mentioned, such as the claim for sovereignty by a State over some territory where it exerts effective domain; that is, its jurisdiction, the force to legally enforce – what is known as the empire of the law and, thus, considers it its own. This includes the use or occupation, which are complementary means through which there is the pretense to claim sovereignty over a space or territory, as it is understood that rights are acquired when we have used or occupied a good for enough time for it to be considered as our own, and which it is characterized by our capacity to act as its lords and masters. Thus, States could not, due to their repeated activities or their extended permanence somewhere or in outer space, assert that rights over such a territory have been acquired.

Finally, the term "by any other means" which for some authors is controversial and of uncertain scope, for us it is quite clear. Although Lanch suggests, for example, three possibilities for the scope of this last norm, "namely discovery, contiguity, and parts of outer space immediately bordering air space," or which for Christol refers to the prohibition to appropriation through private individuals or enterprises. For us it does imply a clear-cut willingness of the States to include any other means, be it already invented or to be invented [16]. They had the intension to prohibit any way of appropriation.

The goal is to fully limit appropriation of outer space, a concept which would include Lanch and Christol's conceptions, and beyond. It is quite common that at an international level when international treaties are signed, the signatory States (subject to international law) look to include fully all the problems to be solved or the obligations to be created; that is, to obtain the maximum extent of the protection, prohibition or obligation proposed [17]. That is why we believe that through such a statement the whole wide spectrum of the prohibition to appropriate could have been included even without the need to previously specifying the claims of sovereignty by use or occupation.

The desire of the States or of those drafting an international treaty is a basic norm of interpretation under International Law; that is, to determine which the intention the drafters had when they conceived the norm. As far as we are concerned, the intention to prohibit appropriation of space under any way or by any (States, individuals, private enterprises, international organizations, NGOs, belligerent subjects, etc.) is quite clear without any trace of doubt. If under a State's jurisdiction or under International Law (that is all of them) such premise is valid, it is an absolute imperative: "By any other means" is the same as "By any means."

STATES AND THEIR ACTION

The action of signatory States of treaties regarding outer space law is unequal or even contradictory. The referent as to the action against private enterprises that sold heavenly bodies takes place in China's jurisdiction. Quoting the case of the Lunar Embassy Company, China branch, which was sanctioned for speculation and usury, and its license suspended by Beijing's industrial and commercial authorities, the company sued the administration for its decision, and the legal suit was taken to court, which ruled in two instances against the company. In its last ruling in 2007, Beijing's First Intermediate People's Court used as the basis for its argument that neither private citizens nor states may claim property over the Moon, as it is established in the Outer space Treaty, of which China is a signatory state.[18]

But, what does happen in western countries? We could even state that we can act beyond a mere sanction and the suspension of a license, and claim Fraud or Scam. Just to mention an example: the Spanish Penal Code, Act 10/1995, dated 23 November, in Chapter VI, on frauds, section 1, on swindles, article 248 (1) states: "Fraud is committed by those who, in order to profit, use gross deceit to lead another into error, inducing them to carry out an act causing personal detriment to them or to others." And then, in Article 251, it sets that "(...) will be subject to imprisonment from one to four years: 1.Whoever, falsely endowing themselves with the faculty to dispose of a good or real estate without having the legal power to do so, either because they have never owned it, or because they owned it before, they sold it, levy it or leased it to another, in detriment of the latter or of a third party". We would then foresee the possibility of introducing the act of selling real estates on the Moon within this legal offense, as they are empowering themselves with the right to sell a property that they do not own.

In the United States, it is worthwhile to ask whether it falls under Investment Fraud crimes the selling of properties one does not own, inducing people to believe that the domain of the property is transferred when the latter is out of commercial status, or it is considered *res extra commercium*, using deceitful means, such as the issuance of property deeds. In addition, there is inducement to deceit by making people believe that there is a legal loophole that allows such appropriation when in fact it does not exist. The FBI itself has included within the crimes of Investment Fraud the real estate investment fraud and the business opportunity fraud [19], which leads us to ask ourselves why this type of criminal behavior has not been investigated in countries where such enterprises have branches.

SPACE POLICIES

It is quite clear that signatory and non-signatory States of space treaties or *Corpus Juris Spatialis* do respect the principle of non-appropriation of outer space in their space activities, but it is also necessary to determine whether or not this norm may be preventing the development of more forward-looking space policies or programs [20].

We might be able to clearly define three macro-level stages in space exploration. The first one refers to the beginning of the space programs by the superpowers during the cold war. At that time, the main booster for conquering space was to show that their national technological capability was far more superior than its opponent's, and to become the first one in achieving some specific goals (place the first man in space; the first one to walk into the space void; the first one to land on the Moon; or the first one to live extensively in such an environs). These goals generated the possibility of counting on a meaningful amount of public funds (budget) as the former became a foremost national interest [21]. This space race did gain meaningful advancements regarding space exploration and use.

The second stage refers to the post-space race period [22], when the States focused their policies on the development of science and their international cooperation programs to develop specific knowledge competences, with ample budgets but lacking unlimited funding for their purposes. This lead to a stage where there is a sense of stagnation regarding the physical space exploration, whose main icon was the human being in action and their *in situ* presence in the exploration. If we contrast it with the first stage, we could observe that the drive to carry out such exploration and to finally reach the goal of a permanent and massive human presence was decelerated, but exploration was increased. It is at this stage that space was explored and investigated in a more detailed manner but focused mainly on the use of autonomous technology such as space satellites, rovers, or space probes.

The third and last stage is the one where private individuals and companies participate in outer space exploration [23]. Even though such participation did also take place in the other two stages which, in fact, combine themselves (that is, the stages have several convergent periods), it is in the latter stage where, we believe, private individuals and companies will be the ones who will pursue space exploration, taking into account the premise under which private enterprise governs itself, that of profitability. Thus, we can see that the main presence of private sector takes place in sectors such a communications or remote sensing, where profit for the provisions is obtained. What happens then with exploration? Without profit, it is unlikely that the private sector gets involved.

It is worthwhile to ask ourselves: How to promote and develop the necessary competences within the private sector for the exploration and conquest of the outer space? It is there where we look into the development of our society, and we see how one of the most important engines for the exploration and conquest of unexplored sites was the desire to acquire private property. Thus, conquerors arrived in America seeking richness or new territories [24], and the colonizers in the American West settled down looking for fortune or private property [25]. Even though this was clearly troublesome, since we do know that the importance of such territories led to nefarious consequences still felt by our society nowadays [26], it is important for us to determine that by allowing for the claim or acquisition of territory in outer space could indeed promote the development of space systems to have access to and remain in space in a more economical, efficient, sensible, and safe way.

PROPERTY AND EXPLORATION

As we proposed in the previous paragraph, it may be possible that the low degree of participation in space exploration by private individuals and companies with their own projects is influenced by the prohibition to appropriate that environs which, according to our analysis, is not subject to discussion, and it is a norm that must be complied with, or *Ius Cogens* [27]. Regarding this issue we can see how two essential arguments are developed. The first one deals with the limited expectations regarding the use of space by private individuals, and how their profits are obtained from mass media or the sale of data gotten mainly from space itself, but not much more; the private sector participates in the development of new exploration technologies only because exist demand by the public sector, which buys this technology for their programs and not because the private sector has a true desire to develop exploration as they are not quite certain of how they will profit at all from it.

The second has to do with the scarce clarity the public sector has as to the legality of exploiting this environs. Is it feasible to exploit outer space? We believe it is not clear; it is quite clear that it can be use and explore, but it cannot be exploited for the individuals own benefit since only in the Moon Treaty mention is made of exploitation which, in any case, must be done for all humankind's benefit, and it is a regulation yet to be determined. This forces the private sector to not have a clear-cut legal view as to where and how to invest.

The capacity to acquire private property would permit to have more clarity regarding the two instances mentioned above because it would provide these owners-to-be with the capacity to dispose of the good as it is done on Earth; that is, it could be exploited by searching for alternatives that would render economic benefits from their properties [28]. We believe that this fact would exponentially increase at a medium- and long-term the participation of private individuals and companies with autonomous programs as well as an extended or permanent permanence of the human being in outer space.

LOOKING FOR AN ALTERNATIVE

The search for incentives to private sector may participate in the exploration and colonization of outer space is crucial for the development of more economical and efficient means to reach it and remain there. And in the search for such alternatives the transformation of norms that facilitate such incentives is imperative. For this reason, we propose:

The authorization of real estate

Since it is quite clear that the appropriation of outer space is not permitted, it is necessary to think of the amendment of article II of the Outer Space Treaty as well as of article XI of the Moon Agreement in order to find a permit for the progressive appropriation of the outer space, together with what – we believe – must be some characteristics that will has the process, as follows:

- Invalidate all the possible claims for sovereignty over celestial bodies currently filed on Earth, establishing precise terms for their invalidity and lack of efficacy.

- To draft an acknowledgement of rights (real estate) by sectors (sectors on the Moon, Mars, etc.) and structuring it by stages (access, permanence, and acknowledgement) that will allow the control of the expansion of property, in relation to the exploration capacity that settlers have. That is, there is not a juridical intervention, nor are there assignations to those who do not have the capacity to arrive and permanently remain there.

- To establish the zones that will be the subject of possible appropriation, and which zones will be under the prohibition to be claimed due to their special conditions.

- To establish as a requirement, with no exception whatsoever, the real fact of accessing and remaining in the site in order to be able to claim those rights (It will be an *aposteriori* rights).

Such an amendment would clearly provide the private sector with a boost by permitting the ownership of parts of outer space since it would greatly reduce the costs involved in the search for economic and effective means to have access to such a possibility as well as the search for alternative ways to exploit them.

In search for a consensus

It can be stated that many of the States participating in space exploration at present have a strong component of incentives for the private sector to participate in space exploration as well as the intent that this sector be the one that carry out the next stage in such a field. This leads us to think that the States could be accept an amendment in international law on the property rights in outer space, should this help to increase the participation of private individuals, as well as to make concrete the permanence of the human being in space [29].

The dilemma between regulating or waiting for a "de facto" event

The issue of private ownership in outer space is polemic and has a number of opponents with solid arguments, such as the protection of the environment, the prevention of an arms race in space, or the rights of future generations to such an environs (all of which we share), but it is also clear that the claim for sovereignty is already present, and it will increase once private entities can access to outer space and remain there.

It is just a matter of time, maybe we have only ten or thirty years, before private individuals develop a technology to live in space and it will be as of this moment when they will start organizing themselves and to disavow the Earth's jurisdiction over them, wanting to own the zones where they remain. So, we believe it is much better to set clear rules for the development of such a principle and to promote the human presence in that environs instead of allowing for the increase of alternatives that only seek the benefit of few and not of the majority [30].

Allowing ownership over outer space may be the best alternative to gain benefits for humankind as a whole since the necessary juridical rules can be enacted to obtain equitable access that, at the same time, continue guaranteeing the protection of the space environment, preventing an arms race, as well as preserving in an adequate manner the right for future generations [31]. The Private Property does not contradict protection and access regulations, and could help in their development and prevent disorganized and pernicious actions, such as the present tries for appropriation.

Finally, we can state that the opportunities provided by Outer Space are limitless; so, it is our right, but also our duty, to take advantage of them, by making use of all the mechanisms within our reach (technical, juridical, social, etc.) in order to definitely access this environs and defeat our own fears regarding our behavior as a society, that will lead us to transcend and reinvent space for ourselves and our future generations.

CONCLUSIONS

- Outer space is presently the object of sovereignty claims by private individuals and companies, and these have no solid foundation under international law, which fully establishes a clear and total prohibition to claim outer space by the States or any other individual.

- It is possible that the low participation of private individuals in space exploration is due to the prohibition to claim by or adjudicate to individuals parts of outer space, since the alternatives for the usufruct of the said outer space by either States or individuals are not clearly defined or are not attractive.

- A clear-cut mechanism that may promote space exploration is to amend the *Corpus Juris Spatialis* to allow Property Rights (real estate) over parts of outer space to promote private programs that lead the human being to explore and remain permanently in outer space.

- There is no contradiction between the protection and preservation of outer space with the permission to claim parts of it. This would even help prevent the *de facto* development and claim to take place due to the collision between regulation and real facts.

REFERENCES

1. United Nations. 1963. *The Declaration of Legal Principles Governing the Activities of States in the Exploration and Uses of Outer Space.* Principle 3. (General Assembly resolution 1962 (XVIII) of 13 December 1963). United Nations, New York.

2. United Nations. 1967.*The Treaty on Principles Governing the Activities of States in the Exploration and Use of Outer Space, including the Moon and Other Celestial Bodies (the "Outer Space Treaty").* Art II. Adopted by the General Assembly in its resolution 2222 (XXI), opened for signature on 27 January 1967, entered into force on 10 October 1967. United Nations, New York.

3. United Nations. 1979. *The Agreement Governing the Activities of States on the Moon and Other Celestial Bodies (the "Moon Agreement").* Art. XI. Adopted by the General Assembly in its resolution 34/68, opened for signature on 18 December 1979, entered into force on 11 July 1984. United Nations, New York.

4. United Nations. 1962. *Document A/5181 and A/C.1/880* United Nations, New York.

5. Lunar Embassy. 2014. See at: http://lunarembassy.com/

6. Velásquez, L.G. 2010. *Bienes*, Duodécima edición, Editorial Temis, Bogotá.

7. Becerra, J. 2014. *El principio de libertad en el derecho especial,* Colección Jus Público, No. 1, Universidad Católica de Colombia, 2014.

8. US State Department. 1998. *"Treaties and other international acts series 12927", Space Station Agreement between the United States of America and other governments*, signed at Washington January 29, 1998. Art. 5, US Government, Washington. Also see, United Nations. 1975. *The Convention on Registration of Objects Launched into Outer Space (the "Registration Convention")*, adopted by the General Assembly in its resolution 3235 (XXIX), opened for signature on 14 January 1975, entered into force on 15 September 1976. Art. II, United Nations, New York, And *Supra cita* 2. Art. VIII.

9. Lee, R.J. and Eylward. October 2005. "Article II of the outer space treaty and human presence on celestial bodies: Prohibition of state sovereignty, exclusive property rights, or both?" *Proceedings of the forty-eight colloquium on the law of the outer space*, Fukuoka, Japan.

10. *Ibid.*

11. *Ibid.*

12. Tronchetti, F. September 2007. "The non-appropriation principle under attack: Using article II of the outer space treaty in its defense" *Proceedings of the 50th colloquium on the law of the outer space*, Hyderabad, India.

13. *Ibid.*

14. Valdeón, J. Agosto 2006. *El hombre que vende la Luna.* Periódico el Mundo, Suplemento crónicas, número 565, España.

15. *Supra note* 4. Título y modo.

16. *Supra note* 5.

17. Byers, M. 2008. "International Law" *The Oxford handbook of international relations,* Reus-Smit, C and Snidal, D.

(ed), p, 625, Oxford university press, New York.

18. Supra note 8. And see at: http://www.semana.com/gente/articulo/se-vende-luna-bajo-precio/336871-3.

19. See at: http://www.fbi.gov/scams-safety/fraud.

20. Subirats, J., Knoepfel, P., Larrue, C. and Varone, F. 2008. *Análisis y gestión de políticas públicas,* Ariel, Barcelona.

21. Launius, R. 2012. "Why go to the moon? The many faces of lunar policy" *Acta Astronautica*, Elsevier, Pages 165–175, Volume 70, January–February 2012.

22. Sagdeev, R. "United States-Soviet Space Cooperation during the Cold War" *Nasa Web page,* See at: http://www.nasa.gov/50th/50th_magazine/coldWarCoOp.html

23. See The European Commission. 2011. *Communication from the commission to the council, the European parliament, the European economic and social committee and the committee of the regions towards a space strategy for the European Union that benefits its citizens.* Com (2011) 152 final. The European Commission, Brussels. And United States Government. 2010. *National space policy of the United States of America,* the White House, Washington.

24. De las Casas, B. 1985. *Brevísima relación de la destrucción de las Indias*, Sarpi, Madrid.

25. Tindall, G.B. and Shi, D.E. *America: A narrative history,* I., W.W. Norton & Company, New York.

26. Ramos Pérez, D. 2008. "La conquista" Historia de Iberoamerica, tomo II. Historia Moderna. Salmora, M. (ed), Ediciones Catedra, Madrid.

27. Gómez Robledo, A. 2003. *El Ius Cogens Internacional. Estudio Histórico-critico.* Serie doctrina jurídica, No. 147, Instituto de investigaciones Jurídicas, Universidad Nacional Autónoma de México, México.

28. *Supra note* 6.

29. *Supra note* 23.

30. Shaw, M.N. *International Law,* sixth edition, Cambridge University Press, 2008.

31. Grinlinton, D. 2011. "Evolution, adaptation, and invention: Property Rights in natural resources in a changing world" *Property Rights and Sustainability. The evolution of property rights to meet ecological challenges.* Page 275, Martinus Nijhoff publishers, Leiden, Boston.

BIS-RS-2014-38

CubeSats to support Mars exploration

Three scenarios for valuable planetary science missions

Sabrina Corpino
Politecnico di Torino
Corso Duca degli Abruzzi, 24; +39 011 090 6867
sabrina.corpino@polito.it

Fabio Nichele
Politecnico di Torino
Corso Duca degli Abruzzi, 24; +39 011 090 6858
fabio.nichele@polito.it

ABSTRACT

Planetary science originally tended to rely on "flagship" missions characterized by large satellites and expensive resources. Interplanetary CubeSat missions represent a radical new approach enabling high quality and impact science to be achieved with ultra-small and low-cost nanosatellites. Launched as second-class payloads, deployed in swarm-like or constellation configurations, they offer spatially distributed measurements and temporal resolution not achievable by single-monolithic-satellite platforms. Reduced development times, standardization of components and deployment systems, platform modularity have been drivers to the growth of CubeSats' launches in Earth orbit in the last decade. Constraints in size, volume and available power on-board, still limit their capabilities for independent planetary exploration. Propulsion, communications, radiation environment protection are top three technological areas to empower for this class of small satellite to support science objectives in the near future. A set of scientific objectives for CubeSats to serve astrobiology goals and support to future human exploration on Mars was selected to the purpose of this work. Missions to accomplish orbital and atmospheric measurement, in situ analyses related to biosignatures detection and environmental characterization have been explored. Three set of mission architectures based on surface penetrators, atmosphere scouts and orbiting fleet, have been assessed in the perspective of the science return value. Mission concepts provided metrics and design options to address the stakeholders' needs and strategic knowledge gaps, as defined by the NASA Mars Exploration Program Analysis Group's definition of top-level required investigations.

KEYWORDS: [CubeSats, Mars exploration, Planetary Science, Mission Architecture, Conceptual Design]

INTRODUCTION

In this paper the authors explore the state of the art in CubeSat missions design and implementation by defining the range of science capabilities for CubeSats beyond LEO, and by enhancing the top technological challenges to support science objectives. The paper highlights the emerging capabilities of distributed small-satellites in the context of a planetary science mission in the Solar System, addressing the high-level objectives defined by formal processes within the scientific community. Planet Mars was chosen as target destination to this purpose, by selecting a set of scientific objectives for CubeSats to serve astrobiology goals in preparation for future human exploration. NASA-MEPAG (Mars Exploration Program Analysis Group) living documents[1] provided the authors the opportunity to explore the key activities necessary to fill the gap of knowledge in a particular area of interest. High-level scientific objectives achievable by distributed platforms have been prioritized, by enhancing measurement and interaction capabilities that are not attainable with single-monolithic structures. The purpose was to generate and explore space mission concepts aimed at gathering unprecedented measurements and data about the planet Mars' ecosystem, enabling in turn the future human exploration. Preference was given to unconventional architectures of distributed space assets, networks of small and replicable satellites, low-cost platforms. Three mission concepts have been generated[2], based on the deployment of a large number of small spacecraft in orbit or on a global distribution of a planet's surface

S. Hatton (ed.), *Proceedings of the 12th Reinventing Space Conference*, DOI 10.1007/978-3-319-34024-1_3

and subsurface landers and scouts. Distributed satellite systems are often overlooked in the preliminary comprehensive science mission proposals, either because their value proposition fails in justifying the risk or expense, or because decision-making is biased by the heritage of traditional monolithic architectures. The needs of alternative solutions are hence not explicitly stated and remain unrevealed throughout the process of concept development and preliminary design. The investigation described in the following sections section consists in searching for evidence of these needs and bringing to light emerging capabilities, issues, risks of distributed small-satellite solutions through the concept exploration of a planetary science mission to Mars. Methodologies for concept exploration and system analysis developed within the team[3] were used to generate different design options and populate a tradespace for the exploration of best alternatives.

THE CUBESAT ERA

The advancement and miniaturization of electronics allowed the shift from mainframes to personal computers up to smartphones, and exponentially increased the number of organizations that could afford this technology. Similarly, in the last decade lower-cost and smaller-size satellites have substituted large monolithic spacecraft architectures, and the number of organizations that gained access to space increased. Small satellites became important in providing cost-affordable access to space to developing countries where space industry was not yet consolidated[4]. NASA's New Frontiers and Discovery programs are two examples of how larger "Flagship" planetary science explorations being complemented by many smaller and more frequent missions using fewer resources and shorter development times. The ultimate example of this diversification is being represented by the proliferation of micro- and nanosatellites, particularly CubeSats. Technology innovation and broaden participation of university and industry are into Space agencies' programs enabled this paradigm shift. In academia, many universities around the world develop nanosatellites as hands-on experience tools to prepare a well-qualified space-engineering workforce in the process of conceiving, implementing and operating a space mission. The support of space agencies over the development and launch of university-driven CubeSats missions manifested in Europe and United States with several initiatives, among which the ESA-Education Office's 2008 "CubeSats on VEGA Maiden Flight" Project[5,6], the 2013 "Fly Your Satellite!" Program[7], and the NASA ELaNa program, which first call for proposals issued in 2010 and counts seven launches and more than 30 CubeSat deployed in orbit to date[8].

CubeSat and nanosatellite missions have been mostly developed for education, technology demonstration in LEO and for Earth observation[9,10]. The possibility to deploy multiple satellites in the same launch, the increased availability of launches (as piggy-back payload) and the advent into the market of private launchers providers, the interest from industries and military organizations in the development of CubeSats as fast-response technology demonstrators, and finally the support of space agencies over the development and launch led to a total of 175 CubeSats launched into Earth orbit in the decade 2003-2013, according to recent surveys[11].

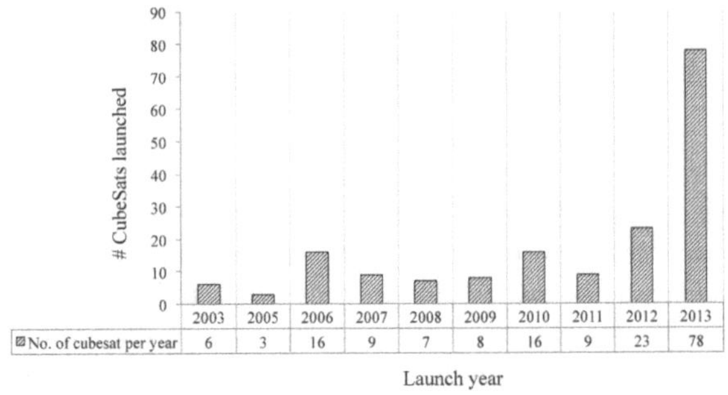

Figure 1: Number of CubeSat launched per year in the decade 2003-2013[11].

Large-scale CubeSat programs established in last years will trigger the trend shown in Figure 1 to grow more rapidly in the next future. QB50, an international project coordinated by Von Karman Institute for Fluid Dynamics, aims at studying temporal and spatial variations of a number of key parameters in lower thermosphere (90-320 km) with a swarm of 40 double CubeSats deployed at the same time[12]. HUMSAT, an international project initiated by the University of Vigo under the patronage of ESA and UNOOSA, aims at monitoring climate changes and supporting humanitarian initiatives. The purpose is to launch a CubeSat constellation to support a communication service based on ground sensors, and to validate a global network of amateur radio ground stations[13,14].

Politecnico di Torino takes part to the latter initiative through the development of a 3U CubeSat (3STAR), designed by the AeroSpace Systems Engineering Team (ASSET) and the CubeSat team of the Department of Mechanical and Aerospace Engineering. The project's challenging scientific objective is to perform on-orbit GNSS-based remote sensing measurements, limb-sounding the atmosphere. The mission will serve weather forecasting systems and eventually warning services[15].

Interplanetary CubeSats

CubeSat community members are starting to propose the use of this class of small satellites for interplanetary missions. A series of yearly workshops and conferences[16] hosted for the first time in 2012 at Massachusetts Institute of Technology (MIT) focusing on interplanetary small satellites and CubeSats missions, draw the attention on the number of challenges that designers will have to overcome: longer lifetimes, propulsion for trajectory changes and potentially entering orbit, power, communications from vastly further distances than Earth orbit, higher radiation outside Earth's protective magnetosphere, and instruments to make meaningful measurements. A study of the NASA Innovative advanced Concepts (NIAC) showed that spacecraft and payloads useful for Solar System exploration, astrophysics, space physics, and heliophysics can utilize a new Interplanetary CubeSat architecture, enabling lower-cost, up-close measurement of distant destinations, including Mars, asteroids, comets and the Moon[17]. MIT EAPS and Draper Lab are building ExoplanetSat[18], a 3U CubeSat that aims to detect superEarth exoplanets by the transit detection method, pushing the limit of potential applications out of the solar system.

CubeSats for interplanetary missions are still seen as secondary payloads: they will keep sharing a launcher, not being able to choose the desired optimal orbit. This translates into the need for a DeltaV available to be spent on operations, transfers, and corrections. This in turn is reflected in the need to have more space on board: 6U is the minimum foreseen format[19] (An example of a 6U configuration is given in Figure 2).

Apart from chemical cold-gas propulsion, electrical propulsion and the solar sails propulsion technology are being considered. In the absence of Earth's magnetic field, the solar radiation pressure represents the highest torque disturbance, but it can also be exploited to obtain thrust. The solar sail technology has established itself in recent years with different missions, such as NanoSail-D, Sunjammer, LightSail-1[20]. Limits in this case are due to the high equivalent specific impulse and low thrust, which reflect on mission duration. Operational lifetime for interplanetary CubeSats shall exceed 5 years. An evolution of the external configurations (e.g. from face-mounted to deployable solar panels) is required to satisfy demanding power requirements.

The approach to communication changes as well. Radio communication system onboard shall be presumably always on, instead of on a cyclic on/off (as can be afforded in certain LEO missions). This translates into a higher average orbit power required (30 W for a 6U, typical). The support of amateur-radio communications would fail: the most common strategies provide for X-band and High Gain Antennas onboard, and the support to the DSN - 34m dish (Beginning of Life) and 70m at the End of Life. The use of the DSN has no reliable alternative to date, and could represent a not-negligible cost percentage. If for a "Discovery-class" mission may be of the order of 1% of total committed costs excluding launch, for a CubeSat mission this could represent the major cost element. Distance limits are also driven by communication. If no data-relay systems are envisaged, the limit for 6U architectures is on the order of 0.5 AU (i.e. Mars). For instance, with X-band, high gain antenna on board and 34m DSN on Earth, the datarate is in the order of 1 kbps @ 0:25 AU, or 250 bps @ 0.5 AU.

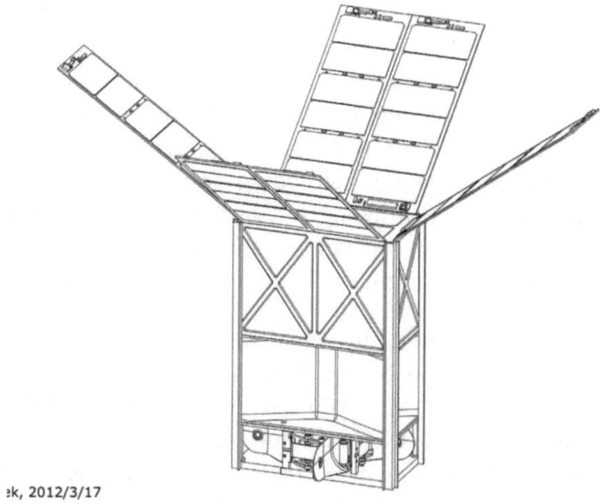

Figure 2: preliminary concept of 6U Interplanetary CubeSat Bus[17] Image copyright: Tomas Svitek, 2012

The common outcome from these concept studies is that a key driver for succeeding resides in the employment of these platforms not as single independent spacecraft, but as part of a large mission deploying them in distributed configurations. Constellations, collaborative networks, fractionated and federated systems are becoming popular between the developers' community, these concepts being able to demonstrate spatially distributed, simultaneous and shared measurements, among other emergent capabilities. Distributed satellites working in concert, used as disposable sensors with reliability and flexibility not achievable by monolithic single-spacecraft platforms, could produce more precise data than a single highly capable large asset, and could open avenues of unprecedented collection of data products[21]. Fleet of nanosatellites or CubeSats are likely to play a role in future planetary missions, but most presumably as daughter craft carried to their destination by larger mothership.

Destination: Mars

Mars has a unique place in solar system exploration: it holds keys to many compelling planetary science questions, and it is accessible enough to allow rapid, systematic exploration to address and answer these questions. The program of Mars exploration over the past 15 years has provided a framework for hypotheses to be formulated and tested and new discoveries to be pursued rapidly and effectively with follow-up observations[22]. According to the Decadal Survey for Planetary Science 2013-2022, the study of Mars as an integrated system will continue well beyond the coming decade, following the approach that produced missions supporting one another both scientifically and through infrastructure, with orbital reconnaissance and site selection, data relay, and critical event coverage. The challenging science objectives will focus on understanding the evolution of the planet as a system, focusing on the interplay between the tectonic and climatic cycles and the implications for habitability and life. Future missions will implement geophysical and atmospheric networks, providing in situ studies of diverse sites, and bringing to Earth additional sample returns, addressing in detail the questions of habitability and the potential origin and evolution of life on Mars.

Over the past decade the Mars science community, as represented by the NASA Mars Exploration Program Analysis Group (MEPAG) has worked to establish consensus priorities for the future scientific exploration of Mars, formulating three major science themes that pertain to understanding Mars as a planetary system: 1) understand the potential for life elsewhere in the universe; 2) characterize the present and past climate and climate processes; and 3) understand the geologic processes affecting Mars's interior, crust, and surface. A fourth theme, the MEPAG Goal IV, identifies the investigations that are still needed to prepare for human exploration. From these themes, MEPAG has derived the key science questions that drive future Mars exploration, providing the science community with updates on the answers found, and shaping future directions[23]. The Goal IV is different in nature from the former three, commonly referred to as Life, Climate, and Geology. Unlike Goals I-III, which focus on answering scientific questions to develop a comprehensive understanding of Mars as a system, Goal IV addresses issues that have relatively specific metrics related to increasing safety, decreasing risk and cost, and increasing the performance of the first crewed mission to the planet[24]. Precursor activities and technology demonstrations in several venues (Earth, LEO, International Space Station, and nearby celestial objects such as the Moon or asteroids) would be involved in the long-term preparation for the human exploration of Mars. Although all represent an important and necessary part of the forward path, the connectivity between these precursor activities and the technology demonstration roadmap are maintained separately and considered complementary to the required science data cited in the MEPAG Goal IV document. For these reasons the precursor activities listed in the document result to be to a lower extent constrained by the necessity of low-term engineering and cost feasibility demonstrations. They are rather explicitly tied to those data products the scientific community requires to fill the gaps of knowledge on critical features of the planet's environment, before planning ahead a manned mission to Mars.

SCIENCE GOAL ANALYSIS

The aim of this work is to generate some space mission concepts where CubeSats play a role in supporting exploration for valuable planetary science beyond LEO. The root problem to be addressed in the formulation of a mission concept was to learn about planet Mars in connection to human exploration. As inferred from MEPAG documents review, in order to prepare the human exploration of Mars it is necessary to fill the gap in the knowledge in, and to address the uncertainties related to specific phenomena in the Mars' environment (orbit, atmosphere, ground). This is especially true on global scale and with coverage of all local times. Science mission concept can be generated by selecting a key observation, measurement, sounding technique that fills the gap of knowledge in a specific area of interest (e.g. ground bio-hazard, atmosphere composition, presence of dust and/or micrometeoroids in orbit) suitable for distributed nanosatellite system architecture. Principal stakeholders of this study have been identified within the scientific community. Space mission planners, strategists, and designers who will be building the future manned missions to Mars would also benefit from mission results. The top-level scenario calls for significant objectives: innovative, unprecedented and visionary concepts have been explored, such as mission architectures based upon constellations, swarms, distributed satellites, single-instrument multiple-units platforms; technology return for Earth-related applications was taken into account, as the prospect to inspire the general public imagination.

The problem statement reads as follows: *to establish a low-cost/fast-delivery space asset at Mars for filling the lack of knowledge on specific phenomena in the Martian orbit, atmosphere and on ground on regional/global scale, that may affect the future human exploration of the planet. To provide the scientific community with unprecedented measurements and data that reduces the level of uncertainty to support the long-term vision of human exploration of Mars.*

Selection of Gap-filling Activities

The science goals analysis opened with the identification of the Strategic Knowledge Gaps (SKG) and with the mapping to the different activities needed. Table 3 lists part of activities and SKG prioritized by the Precursor Strategy Analysis Group (P-SAG). The comprehensive P-SAG science traceability matrix can be found in ref. 25. Authors analysed how activities are mapped to the investigations, or high-level science goals, and how investigations in turn could be driven by single or multiple measurements. Priorities among multiple investigations were determined by the P-SAG first assessing the impact of relevant data within each investigation, and then assessing the value of new precursor data against timing criteria. The result is a classification based on a dual ranking: "timing" and "priority". The first metric indicates which activities are needed earliest, the second is a metric to recognize if a set of activities enables critical need or mitigates high-risk items. The total combined priority indicates that measurements needed earliest were prioritized ahead of measurements of equal priority needed later. Priority and timing levels have been defined as per Table 1 and Table 2. The ranking defined by MEPAG and P-SAG allowed the authors to recognize which activities are being considered critical and what are the needs to be met before others. A further selection has been made on a basis of subject location and type of activity. This allowed to discard those activities planned in Earth orbit or in the vicinity of Phobos/Deimos in preparation for a Mission to Mars, those providing for sample return or demonstration of technologies for rendezvous and docking, entry, descent and landing, and forward contamination. The selection enabled to reduce from 78 to 30 the number of GFAs subject to further analysis. Attention has been given also to the need of spatial and temporal distribution of data products (global coverage, full diurnal cycle, all local time coverage) and to the "class of interaction" between spacecraft and mission subject. The latter refers to a classification made according to two variables, the location of the mission subject (e.g. ground, atmosphere, orbit) and type of sensing (e.g. measurement, observation, in-situ analysis, etc.) Six classes have been identified: A. upward remote sounding of atmosphere; B. downward remote sounding of atmosphere; C. remote sounding of surface; D. in-situ surface measurements; E. in-situ orbit measurements; F. in-situ atmosphere measurements. The resulting distribution of GFAs between classes is shown in Figure 3: remote-sounding classes (i.e. A, B, C) almost equally share the half of the total number of activities, while the three remaining classes include in-situ measurements most needed on the planet surface.

Table 1: Shorthand for human mission goals timing and criteria for setting priorities by P-SAG[24]

Timing	Description
IV	Needed to plan human missions to Mars orbit
IV Early	Needed to plan architecture of the first human missions to the Martian surface
IV Late	Needed to design hardware for first human missions to the Martian surface
IV+	Needed for sustained human presence on the Martian surface
Priority	**Description**
High	Recognized as an enabling critical need or mitigates high-risk items (including crew or performances)
Medium	Less definitive need or mitigates moderate risk items
Low	Need uncertain or mitigates lower risk items

Table 2: Investigation priority levels mapped to Timing and Priority for individual Gap-filling Activities[25].

Priority / Timing	High	Medium	Low
IV-	1	3	4
IV Early	1	3	4
IV Late	2	3	4
IV+	5	5	5

This result tells that on one hand the remote-sounding activities are perceived as important as the direct in-situ measurements, and on the other that being these tasks preliminary and preparatory for a manned mission to the planet, the soil and the subsurface gain most of the interest from scientific community and mission planners. The push towards this interest is also given by the recent success of robotic landers and rovers' missions, which, however, have allowed so far getting a good knowledge of the planet only at the local level in some selected spots. In contrast, the analysis made on the MEPAG and P-SAG documentation already cited, highlights the need of measurements globally distributed in time and space, that robotic missions mentioned above could not offer. The proof is the fact that despite the "class D" necessary activities (in-situ surface measurements) represent 40% of the total, only two of them has been evaluated with a combined score of high priority and timing, while the remaining ones got a medium/low average ranking, their impact on the mitigation of risk being considered moderate and/or the necessity of results in this area not compelling

Table 3: Partial listing of P-SAG Strategic Knowledge Gaps and Gap-filling Activities

SKG	Gap-filling activity	Priority	Timing
A1. Upper Atmosphere	A1-1. Global temperature field	High	IV-
	A1-2. Global aerosol profiles and properties	High	IV-
	A1-3. Global winds and wind profiles	Medium	IV-
A3. Orbital Particulates	A3-1. Orbital particulate environment	Medium	IV-
...			
D1. Water Resources	D1-3. Hydrated mineral compositions	High	IV+
	D1-4. Hydrated mineral occurrences	High	IV+
	D1-5. Shallow water ice composition and properties	Medium	IV+
	D1-6. Shallow water ice occurrences	Medium	IV+

In order to adequately consider the full range of possible designs, and avoid a priori design selections without analysis or consideration of other options, three activities have been selected within the top-ten list illustrated in Table 4: A1-2 Observation of global aerosol composition, B2-1 Detection of biohazards, A3-1 Observation of orbital particulate in high Mars orbit. The choice has been made by selecting activities that were representative of different classes of interaction, different ranking position (combination of priority and timing), and manifested necessity of spatial and temporal data distribution. This approach allowed regarding for the preferences of key decision makers since the early stages of design, still leaving the concept generation open to different options and creative enough to envision in which ways it could be possible to explore planet Mars in the future. For each of the activities identified a mission concept has been generated. The three scenarios are described in the following sections.

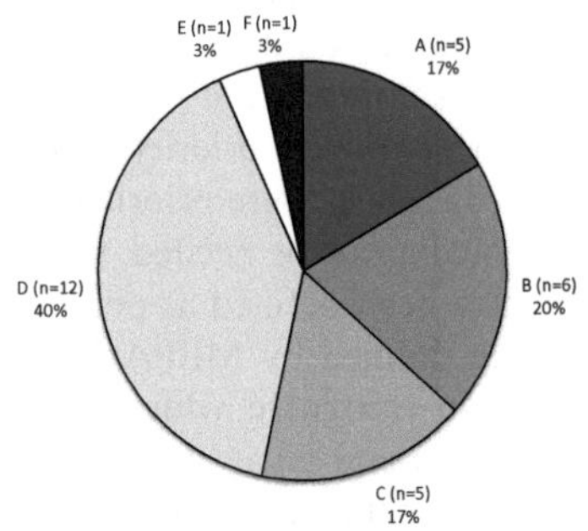

Figure 3: Distribution of Gap-Filling Activities within classes of interaction. Classes A,B,C = remote-sounding, D,E,F = in-situ measurements. See text for in-depth analysis and details.

Table 4: Top ten Gap-Filling Activities as ranked by MEPAG and P-SAG [27-28]. COI = "class of interaction" as defined throughout the analysis. Rank refers to the total priority given by the combination of "timing" and "priority" values. Data distribution needs are deduced from GFAs statements and descriptions as per ref [23]. n/d = not defined. See text for further details.

GFA	Description	Data distribution needs	COI	Rank
A1-1	Observation of global temperature field	Full diurnal coverage	B	1
A1-2	Observation of global aerosol composition	Global coverage, all local times	B	1
B1-2a	Measurement of global surface pressure	Full diurnal cycle, multiple locations	D	2
B1-2b	Observation of local/regional weather	Full diurnal cycle coverage	A	2
B2-1	Detection of biohazards	Multiple environments	D	2
A1-3	Observation of global wind velocity and direction	Global coverage, global distribution	B	3
A3-1	Observation of particulate in high Mars orbit	Equatorial plane, multiple altitudes	E	3
B1-1	Dust and aerosol activity climatology	n/d	B	3
B1-2c	Observation of local weather at multiple sites	Multiple locations, full diurnal cycle	A	3
B5-1	Measurements for presence of ground ice	n/d	C	3

CONCEPT A: ORBITAL PARTICULATE - MARS ORBITAL ENVIRONMENT EXPLORATION (MOEX)

The source of orbital particulate in high Mars orbit is represented by micrometeoroids and dust rings.

Micrometeoroids are fragments of bigger space corps as asteroids or comets, with dimensions ranging millimetres to meter. Because of their high velocity, these corps could jeopardize spacecraft and endanger the success of a mission. Origin, composition and main characteristics of these corps shall be understood performing numerous observations. This could be achieved by gathering enough data to create a consistent statistic model. A possible method consists in to observe the burning trail they left after being entered in Martian atmosphere and gathering information about mass, dimensions, composition and velocity based on a spectrometric analysis.

Regarding dust rings, their existence is still to be proved[26]. Their orbit should be located in the equatorial plane of Mars[27] between Phobos and Deimos. They would induce optical and communication instruments malfunctions, and the particles might have a non-uniform distribution. Thus an adequate number of satellites is necessary to have a good probability to get enough close to discover those particles clouds. A sufficient proximity would be required to surely identify those particles, as their diameter would range under the millimetre. As a result, an impact sensor would be a good solution to detect dust particles.

A mission of CubeSats as distributed systems with the aim of detecting this particulate has both scientific and engineering implications: studying micrometeoroids and dust rings origins and composition will improve the knowledge of the Solar System environment and, at the same time, discovering the position of Mars dust rings and building up a statistical map of the distribution of micrometeoroids could avoid the failure of future mission in Mars environment. Two scientific goals derived from the analysis of needs:

1) To investigate the statistic distribution through the Martian year of Martian micrometeoroids' mass, velocity and composition using meteor trails spectroscopy in order to understand the origins of Martian micrometeoroids and for human exploration hazard mitigation;
2) To search for Martian dust rings and determine the spatial and particle size distributions, composition, origin, density and their time evolution in order to understand Martian system history and evolution and for human exploration hazard mitigation.

Micrometeoroids

A sufficient amount of data is needed to create a statistical distribution and predict the number of events. The optical cameras inside a CubeSat are not likely to have enough resolution to define objects with 1 mm radius. The best option is to look for trails produced by the ablation of micrometeoroids in the atmosphere. Figure 4 shows the concept developed to detect the presence of micrometeoroids. The first analysis combined with a ballistic fall simulation determined the lower limit of the range for the length of micrometeoroids trails. A constellation of nanosatellites in circular orbits around Mars at an altitude that allows optical observations of impact events has been designed.

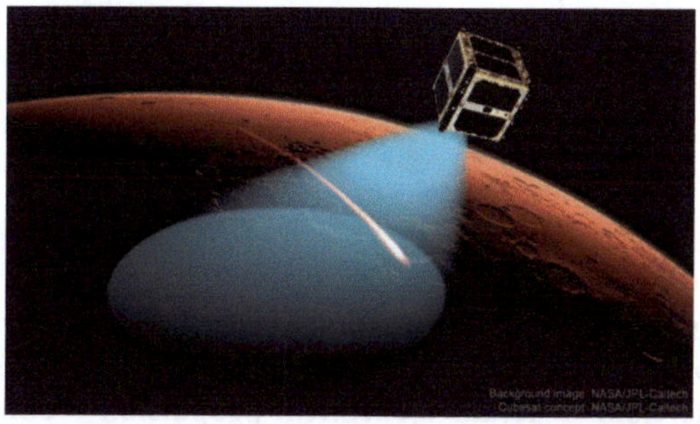

Figure 4: Micrometeoroids mission concept

Two figures of merit have been considered for the trade-off: Resolution and Coverage. A wide coverage pattern allows the highest number of events possible to be detected. Observation payloads for nadir-sounding and limb sounding of the Martian atmosphere have been chosen to address both vertical and horizontal resolution required to see in a satisfactory way what is really happening where they are pointing to, so that the image can be properly analysed and processed in order to obtain the required information. The trade-off result tells that satellites should be positioned in as low as possible orbits to increase the resolution, but this affect negatively the coverage attribute.

Latitude range between 40° north to 40° South has been chosen after trade-off studies on surface spots, future landing sites for human exploration and mission costs. These latitude limits reflect on a total 65% of global surface; increasing the maximum latitude observed will increase this percentage but ΔV limitations for inclination change inclination have to be taken into account. The opportunity to rely on already existing space assets in the proximity of Mars, or on a mothership for orbit insertion in the equatorial plane has been considered in this context.

A second iteration of design allowed to refine instruments models but also fixed the maximum inclination achievable: a constellation of 30 satellites at 2000 km altitude with FOV of 30°, 10 orbit planes has been designed. Two main problems arose with this configuration: the instant coverage was good, but the resolution of each pixel was near to the limit imposed to define a trail. The ΔV needed to increase inclination from equatorial plane to the maximum inclination (Δi=40°) was 2.61 Km/s, too expensive even divided among on-board propulsion system and the mothership. Upper limit for ΔV provided by a CubeSat propulsion system is currently around 1 km/s, that establishes and upper limit of Δi to 15° for plane change (Figure 5).

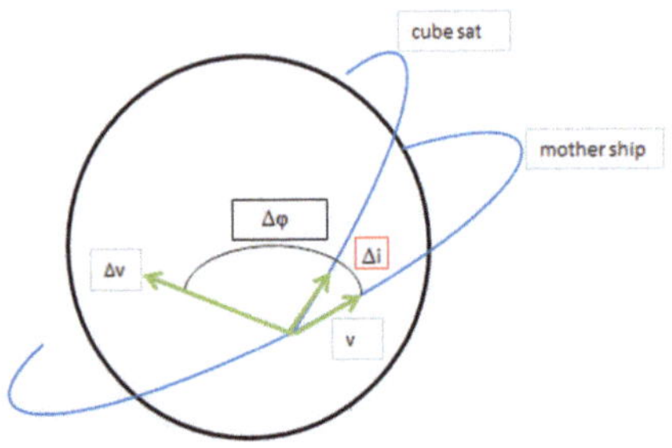

Figure 5: change plane maneuver calculated starting from initial conditions provided by mothership insertion. Actual chemical-propulsion cubesat capabilities limit the DeltaV available and the related inclination change capability.

A 500 km altitude would have led to a very low instant coverage (about 10%). The option of giving to the camera an off-axis pointing angle was taken into account. A good compromise has been found setting altitudes at 1000km, with a starting inclination of 10° provided by a mothership. Combining it with a second manoeuvre provided by the satellites' propulsion system, resulted in a maximum inclination of 25°. Simulation with AGI-STK with a Coverage Definition tool showed an instant mean coverage value of 22.05%, 90% coverage of the planet after 1:30 h, 98% after 3h. Figure 6 shows the instant coverage of the planet.

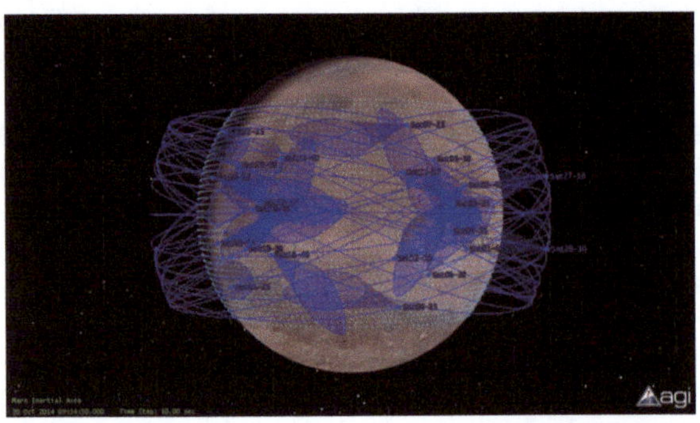

Figure 6: Simulation sample (6 months lifetime).

The mission scenario encompasses three different phases. The first phase will be divided in: 1) arriving in the Martian environment relying on the mothership; 2) to manoeuvre in order to reach the already chosen orbit; 3) to perform commissioning operations for the satellites to be fully functional. The last phase could take months, depending on the kind of propulsion system chosen (e.g. cold gas, electric, solar sail).

The second phase focus mostly on the measurement, i.e. dust images taken of the micrometeoroids trails. This will implicate a precise attitude control, especially during the camera pointing. During the second phase the system will communicate with the chosen network to send data and images to the ground segment, where they will be processed. Image processing could also be achieved on board. In the first case, the satellites would have to store every single image and transmit data very frequently. With an on-board processing system, images without trails and false positives would be discarded and there would be more available memory for data storage and less data to be transmitted. On the other hand this affects the system complexity.

The third and last phase is disposal operation, which would implicate a end-of-life manoeuvre. The disposal could be obtained by crashing on Mars' surface. To fulfil the mission requirements one possible configuration of a single satellite could be a 3U CubeSat, 1U for the camera, 1U with propulsion system and spectrometer analyser, 1U for avionics. Since uncertainties at this stage are pretty high, a 6U configuration has also been assessed, providing more confident results.

Dust rings

This section will focus on the mission concept seeking for dust rings' existence and characteristics. Several studies state that is more likely that dust ring resulted from impact on Phobos and Deimos. The search will be concentrated specifically on the zone from Mars' atmosphere to Deimos, approximately 23500 km far from the planet, and for the most part on a region between the two Martian moons[27] (Figure 7).

A CubeSat mission can accomplish the task by means of impact with the dust or by capturing an image of it. The measurements would require some post processing work; spatial and temporal distribution of dust will be the result of a post processing over the data gathered during the mission, which shall last for one Martian year at minimum in order to gather a number of impacts statistically relevant. Therefore the total space swap by a hypothetical constellation has been considered as figure of merit for the analysis.

Electrically propelled decaying orbits were discarded for time and cost reasons. Polar or equatorial circular orbits were considered at the beginning: circular orbits imply a fixed altitude from the planet, so the regions that would be studied are small, and limit the probability of dust impacts; polar orbits are demanding in terms of expensive manoeuvres for inclination change. Elliptic orbit was the third option evaluated, allowing the constellation to sweep more space with respect to the previous options. A simulation performed with a Simplified Perturbation Propagator (SPG4) revealed that an equatorial orbit with approximately 6000 km perigee and 23000 km apogee would change the ascending node argument of almost 0.5 degrees per sol, being back in on initial position in one Martian year; in this case a multiple-satellite system would swap most of the Martian environment of interest during its mission lifetime.

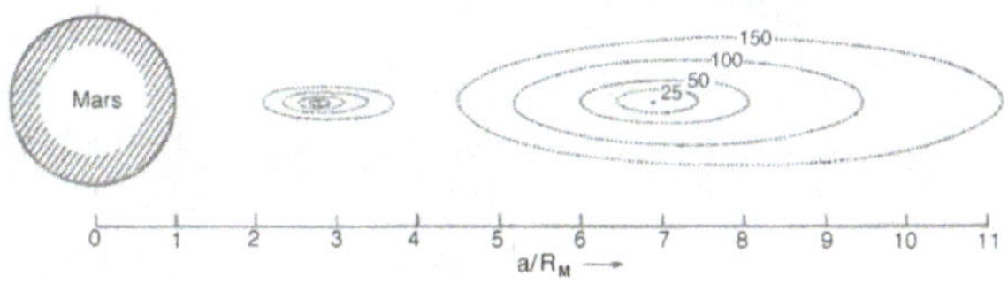

Figure 7: Cross-section of toroidal envelopes containing initial orbits of debris ejected from Deimos dust belt (at right). Smaller Phobos toroids (left) are centered at 2.76 RM. Image copyright by S.Soter [27]

The first part of the mission will involve the Earth-Mars transfer and the orbit insertion by means of a mothership from an equatorial circular orbit of about 30000 km radius. In a second step a Hohmann transfer would place the system inside the zone of interest, where the mothership would operate the deployment of the first satellites. Six satellites deployed with a 60° phasing distance would need to operate themselves an impulse to lower the perigee. The mothership could then perform a manoeuvre to move to a lower circular orbit (with a radius matching the value needed by the micrometeoroids' mission described above) deploying the other set of satellites.

The measurements would probably need post-processing on ground, meaning that one critical aspect of the mission is to communicate back to Earth where those will be analysed. Disposal operations have been considered in preliminary analysis: disposal can be obtained by escaping Mars' influence sphere with an escaping manoeuvre, by crashing on one of the Martian moons or by crashing on Mars. The concept requires the implementation of an impact sensor, of a dust analysing system able to detect charged particles after dust impact, and of an optical sensor for the imaging of orbital dust.

As the impact sensor is concerned, the implementation requires the development of a passive type piezoelectric sensor with a large frontal area to enhance dust impact probability, and the development of an opening system and a structure to support the sensor itself. A quick and simple solution for dust impact sensors' implementation is to use a passive sensor on big surface with electric properties. This type of sensor allows to optimize the available power on board and to increase the impact rate. Piezoelectric polymers materials can be used to detect the deformation of the impact surface.

As the dust analysis system is concerned, the implementation would require the arrangement of a large metal alloy impact surface, a sensors system able to detect charged particles (ions and electrons), and the activation of an electric field between impact surface and sensors system, in order to separate ions and electrons. The dust analyser would measure the electric charge carried by dust particles, the impact direction, the impact speed, mass and chemical composition.

As far as the camera is concerned, this would require the implementation of an optical system that enhances visibility of micrometric dust size in a kilometric range and the development of an attitude control system with an orbital database for optimization of lighting conditions that will help in the visualization.

Given these options, three solutions have been proposed with the aim to be evaluated in a later step of design: dust detector supported by a camera, inflated sail, deployed sail supported by multiple satellites. A drawing concept can be seen in Figure 8.

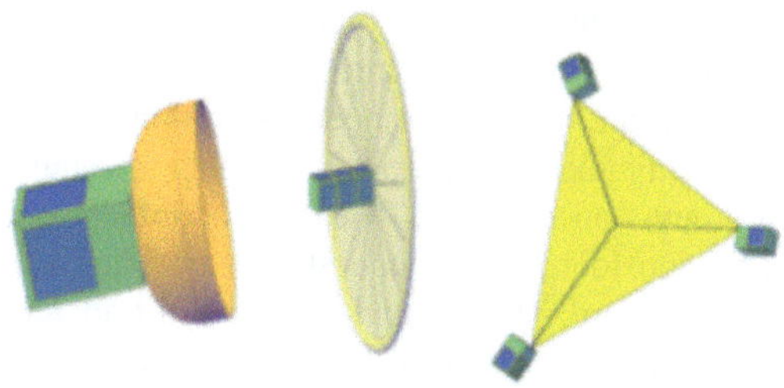

Figure 8: Three system concepts for dust belt detection: dust detector, inflated sail, deployed sail supported by satellites.

For the communication of the mission data and telemetry, a network able to connect the in-orbit systems and the ground stations shall be implemented. Two main parts of this network have been considered in combination: the Deep Space Network (DSN) and a Satellite Relay System (SRS). The DSN must be able to establish a communication link between Mars and the Earth: this means that the nanosatellites must be equipped with a high gain antenna with a high pointing precision and a source unit powerful enough. Recent studies state that the limits for Interplanetary CubeSats are in the order of 250 bps data rate from Mars, with X-band and high beam antennas on-board and relay to DSN 34m antennas (see Interplanetary CubeSats section on this paper). For these reasons the communication segment of the nanosatellite constellation shall be probably supported by a pre-existent orbiter or by the mothership itself.

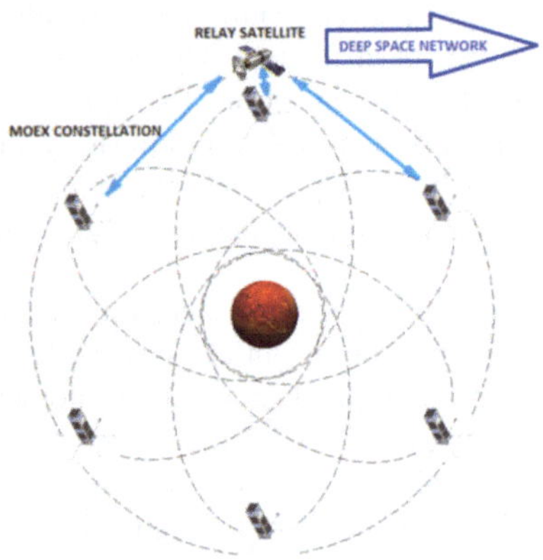

Figure 9: Communication strategy for MOEX mission concept

We considered a relay satellite orbiting on a circular trajectory of 24000 km radius that connects to the satellites when they run on their orbit's apogee (Figure 9). This architecture is expensive because of the implementation of an dedicated relay satellite, but is also appropriate because no need of pointing accuracy, high gain antennas or high power signals to establish a link are envisioned for the CubeSats. Another solution could be considering a ground station on Mars surface as a relay system. This ground station will transfer data to an NASA/ESA orbiter on Low Mars Orbit, which in turn will link to DSN. This strategy however would need an additional link to be implemented. More detailed analyses and trade-offs have to be performed in terms of architecture cost, data volume to transfer, availability of the link, and architecture reliability.

The MOEX mission scenario is illustrated in Figure 10.

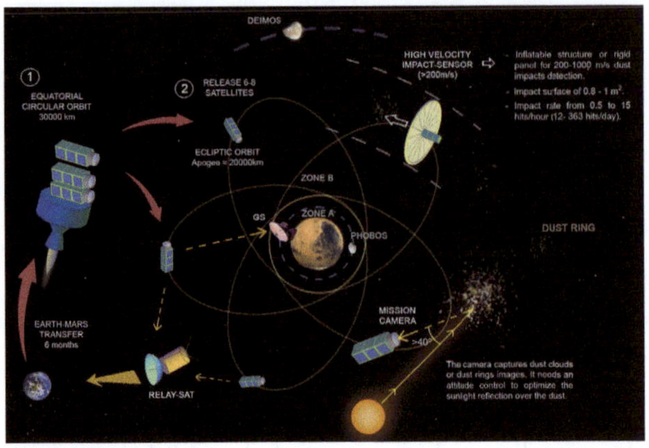

Figure 10: MOEX mission concept

CONCEPT B: AEROSOL COMPOSITION. MARS ATMOSPHERE RESEARCHES WITH IN-SITU OPERATIONS (MARIO)

This mission concept is aimed at the study of the aerosol composition of the Martian upper atmosphere by means of a system of distributed nanosatellites. The mission goal is to provide the scientific community, mission designers and other stakeholders with data capable of improving the confidence on the current Martian atmospheric models, with an eye towards future human habitation, through deployment of low-cost, fast deliverable and multi-purpose platforms.

The scientific goal for this mission concept states: "*To characterize and study atmospheric features and processes of Martian atmosphere and to investigate their interaction with future human in situ missions*".

The broad scope is to understand how the Martian atmosphere affects possible human operations on Mars, with an eye towards future human habitation. This involves the collection of data that help in assessing the feasibility of Martian human exploration and the possibility to support it. The knowledge of atmospheric processes and the interaction between human operations (both crews and equipment) and the atmosphere itself have been considered fundamental, and so included in the scientific goal. The GFA identified requires global measurements of the vertical profile of aerosols (dust and water ice) at all local times between the surface and >60 km. The observations should include optical properties, particle sizes and number densities.

Preliminary mission architectures definition

A preliminary draft of the mission concept was developed through the identification of different types of architectures, including both orbiters and landers, in order to have a more complete and accurate coverage using both points of view. Two main different distributions for those two types of devices have been envisaged: 1) a fleet of few orbiters, and several little landers carrying the instrumentation that cannot be contained on board the former; 2) a fleet of several small orbiters, and a few important landers with almost the full set of payloads.

The best configuration providing low cost, ease of deliverability, use of multipurpose in-situ platforms has been investigated. Another key point established since the very beginning has been the possibility to establish a strong data link between ground and orbit segment. Preliminary trade-offs were performed over the number of satellites within the orbiter fleet, the type of landers within a fleet and the choice of landing sites. Three options have been considered at the beginning for the type of satellites: 1) a ground platform for lower atmosphere measurements with a balloon connected and a retractable wire for measurements in higher altitudes; 2) a lander detecting information about the upper atmosphere during its entry, and then serving as ground station without deployable balloons capability but with the prerogative of a possible re-use by future astronauts; 3) a fleet of landers detecting information about the upper atmosphere during entry, able to extract a balloon during landing to spend more time in atmosphere and working, once landed, as a simple ground station.

A set of 6 Concept of Operations (ConOps) was evaluated.

1. *Tethered Balloons*: several CubeSats, capable of atmospheric entry, performing remote sensing of the upper atmosphere. Some of them used as entry-probe and landers, gathering data during the atmospheric entry and descent. After the landing on Mars, CubeSats are able to gather atmospheric data in a 0–30 m altitude range.
2. *Free-flying Balloons*: During the descent phase, each CubeSat deploys a free-flying balloon, which is inflated with Martian air. After the balloon is deployed atmospheric data is gathered. After a few days of data collection, the balloon lands and the lander gathers atmospheric data from the surface.
3. *Landers shower*: several CubeSats perform remote sensing of the upper atmosphere. After a nominal phase, they perform atmospheric entry and descent, gathering atmospheric data when entry aerodynamics allows it. Once landed, they gather atmospheric data from the surface of Mars (Figure 11).
4. *Ballistic*: Atmospheric data is gathered during the descent, until the touchdown on Martian soil.
5. *Sounding rockets*: CubeSats able to land on the surface of Mars, each one carrying a sounding rocket. During the atmospheric entry and descent, these CubeSats do not gather data regarding the upper atmosphere. After landing on the Martian surface, each lander is able to deploy and operate its sounding rocket that is used to gather atmospheric data from Martian soil to suborbital altitudes.
6. *Rockoons*: CubeSats able to land on the surface of Mars, each one carrying a rockoon (a sounding rocket whose ascent is assisted by a balloon).

Some of these concepts are more realistic than others. Particular speculations have been made on the space segment and on systems able to land on the surface of Mars. Both space segment and ground segment are essential to address the specific knowledge gaps and needed for the mission success. The reliance on a data-relay system, such as a mothership, Mars Reconnaissance Orbiter (MRO), Mars Odyssey, Mars Express or other future data relay satellites, is fundamental for both space and ground assets. Other common features between ConOps evaluated are: 1) all the landers are left by the mothership in an orbit that allows landing in at least two identified landing areas of interest, lately chosen to be Elysium Planitia and Utopia Planitia. 2) Orbiters shall be deployed by a mothership in circular, low-altitude Martian orbits. This is to ensure an effective communications architecture: low altitude orbits allows avoiding powerful transmitters and big receivers, circular patterns allows the signal strength to be uniform along the orbit.

Figure 11: Artist impression of MARIO *Lander Shower* concept

To evaluate the best ConOps, several figures of merit were used, grouped in three families: goals accomplishment, objectives balance and ConOps features. The first family describes how well and completely high priority scientific objectives are met; the second family describes, among the accomplished scientific objectives, if gathered data is more useful to human Martian exploration or Martian atmospheric sciences; the third describes the general features of a given ConOps. Starting from the preliminary configuration chosen and trade-off carried out on the basis of the FoM identified the final mission concept evolved as follows: orbiters fleet: 12 orbiters 3U, 4 orbital planes with 3 orbiters each, to perform remote sensing of the upper atmosphere. The Mars segment includes two kinds of spacecraft: 2 landers, one for each landing site, with a tethered balloon (as described in ConOps 1, and smaller CubeSats as described in ConOps 4). Landers are stored in a CubeSat with a 3U configuration, not including balloons support system (e.g. helium canister), from which one balloon would be inflated and deployed with a retractable wire. The lander swarm is composed by 20 CubeSats in 1U configuration, which dimensions have been imposed downstream of a definition of necessary payloads to make measurements. These CubeSats, smaller and considered more resistant to impact, are released in sets of 5 units from the mothership during the entry phase, passing through the Mars atmosphere. For the final mission concept, two landing sites were chosen: Elysium Planitia and Utopia Planitia, respectively in equatorial and polar planes, enough far in latitudes to have a complete study of lower Mars atmosphere and having considered as landing sites for future human missions or by other Mars missions in development.

Baseline selection

After trade studies and an iterative process the mission architecture has been divided in two main segments: orbit and Mars surface+lower atmosphere. The in-orbit fleet will be constituted by 12 CubeSats arranged in four different orbital planes; all the orbits will be circular and polar, with an inclination of 95° and a 200 km altitude. This architecture will guarantee a complete coverage of the planet's surface during each Martian day. The payloads carried by the orbiters will be cameras, mass spectrometers and radiation sensors. The cameras will allow tracking the dust storms and the formation of clouds, in order to help to characterize their seasonal variations. The mass spectrometers will collect information about the composition and the distribution of the upper Martian atmosphere. Finally, the radiation sensors will gather data about the radiation environment around Mars.

The chosen mission architecture has two other profiles of measurements that require an entry into the Martian atmosphere: these two additional branches are called in this baseline as "CubeSat shower" and "landers". They represent the segment on Mars surface and lower atmosphere. The first one is formed by a series of CubeSats (about 20 units, gathered in groups of 4 or 5 units), which are made to de-orbit in different areas of Mars in order to collect temperature, pressure and wind speed measurements during the descent toward the ground. This will enable the mission to acquire data with an high coverage of the surface of Mars, both from the point of view of latitude and longitude, but also temporally, since it will be possible to make them de-orbit at different times of the Martian year, characterizing its seasons. To collect as much data as possible, landers should be restrained during the descent, so as to extend their measurement life: after reaching the ground, sensors will continue to collect information on atmosphere variations, becoming weather stations on the surface until the exhaustion of their data transmission capacity or the generation of power. This “CubeSat shower” is effective only in the ballistic phase of its components to create a model of the profile variation of the physical properties of the Martian atmosphere, with a vertical resolution otherwise not achievable. The landers form the second part of the ground segment: these are CubeSat sized structures, greater than those presented previously as containing a higher quantity of instrumentation. These components need to reach safely the ground to begin the measurements, for which the descent must be strictly controlled and also the landing zones were chosen in order to avoid areas that are not flat and the various roughness on Mars. After landing, the deployment of instrumentation provides inflation of a balloon in order to make measurements within the chosen altitude of 30 meters. The main feature provided is the repeatability of the measurement: through the balloon and its ability to achieve predetermined heights through the bond with the main lander on the ground given by a special cable, it will be possible to acquire data at different times of the Martian year, but always at the same altitude. The number of lander is reduced to only two units, one for each landing zone (Utopia Planitia and Elysium Planitia) and the type of measurements is different and comprising a large number of aspects of the Martian atmosphere. Only a part of the instrumentation will be embarked on the balloon and flown to the defined heights: these are the probes for pressure, temperature, humidity and wind intensity. The remaining will stay on the ground with the main lander during the entire mission, allowing to collect data about the radiations that reach the Martian surface, as well as those resulting from a mass spectrometer and a sensor for the analysis of dust carried by winds (dust impact sensor). Every CubeSat, both in orbit and on surface, is expected to communicate only with the mothership and with already existing spacecraft orbiting around Mars. No direct Earth communication or crosslink between CubeSats will be implemented, since both these configurations will require too many complex systems. The two segments previously defined will be able to work in synergy in order to provide a more detailed model for the Martian atmosphere that could be used to plan a future human mission on the red planet.

CONCEPT C: SURFACE BIOHAZARDS. PLANETARY PENETRATORS

It is fascinating to imagine to land on Martian surface in order to look for presence of life-evidences. It is also obvious that an extensive in-depth analysis of data from the scientific community should be necessary. The main purpose of this concept generation is to find possible solutions to support future human missions. So it is necessary to look for hazards that could create difficulties for a human crew or even to find evidences of past/present life on Mars so that further missions will investigate more accurately on certain landing sites. In order to improve the knowledge of the potential Martian environment in which a mission is expected to find bio-evidences, the team investigated experiments conducted in laboratories worldwide, focusing on experiments in which astro-biologists tried to repeat Martian conditions on Earth. Then it was defined what kind of biohazards the mission may find on Mars and a way to find bio-evidences was investigated too. All information gathered from this phase were useful to investigate the progress the scientific community made, thus better understanding the probability to find life-related evidences on Mars and how to find them.

The first step consisted of converting the mission statement into a specific objective. In detail, the task were the following: to find life-related molecules on Mars ground using tests, experiments and collecting samples of the Martian soil; to determine if the Martian environments to be contacted by humans are free, to within acceptable risk standards, of biohazards that might have adverse effects on the crew that might be directly exposed. The team focused on identifying in what ways a fleet of CubeSats could have been of support in a search for organic complex molecules, then on the planet soil and subsurface characteristics, on the achievable landing sites and finally on techniques for soil penetrations. The possibility to look for more molecules using the same instrument resulted in extending the purpose and the length of the mission. In this context it was possible to determine at least five achievable landing sites, sorted by different soil types, chemical composition, latitude, temperature, presence of water. Three candidate methods for soil penetration methods were assessed: laser, drill, and impact. Four types of optical analyzers for the search for biomarkers were investigated: Raman Spectroscopy, Infrared Spectroscopy, UV Fluorescence, and Capillary Electrophoresis (Mars Organic Analyzer). Useful information to study the soil was collected: maps of ground ice and sub layered ice, maps of average temperature, humidity and atmosphere composition.

Five landing sites were selected (see Figure 12) after a trade-off based on scientific interest (presence of ice water, atmospheric pressure, etc.) compared with the difficulty of operations (e.g. difficulty of penetration).

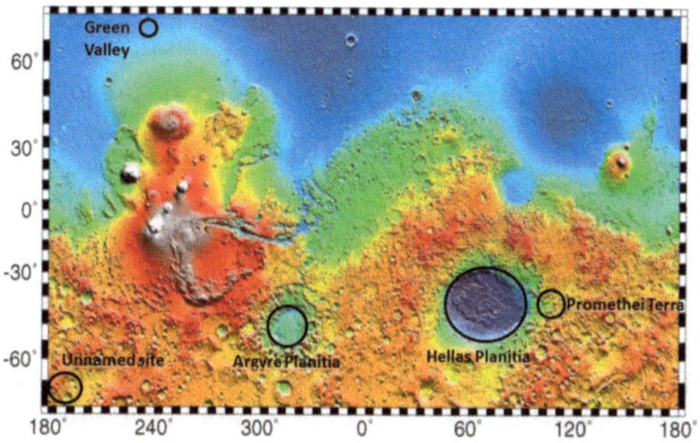

Figure 12: selected landing sites for bio-hazard search

As the penetration methods are concerned, the drilling involves different techniques: force, heat, chemical reactions, and ultrasonic waves. In any case a CubeSat-size system was considered unlikely to provide sufficient force or energy for these solutions. The impact solution with the Space Penetrator System[28] instead seemed very promising at the first glance. Though there have not been successful planetary penetrator missions yet, three systems have been developed and tested on the ground: Deep Space-2 (DS-2), Mars'96 and Lunar-A. Moreover, a lot of work has been done in this field and a great number of new concepts have been developed in the last years. Between the latest mission concepts encountered the idea of penetrator a system providing an alternative way to access the subsurface was interesting: the idea is basically to deliver instrument packages to the subsurface at high speed. This concept have the purpose to take advantage from the high kinetic energy provided by the descent, thus saving weight and power of an heavy decelerator, avoiding as well to carry a complicated drill and all the subsystem needed to sustain its functionality. The truly concern of such a system is to guarantee the survival of the impact. In particular referring to the ESA's Core Technology Programme that has developed the SPS, a 20 kg penetrator 400 mm long and 200 mm wide able to impact the surface at 100 up to 300 m/s. The system is shown in Figure 13.

Figure 13: Space Penetrator System before test[28]

During the test, successfully completed, the penetrator experienced a deceleration of 24 000 g[28]. Such a system can be deployed from carrier, it is designed for hard landing, breaking through ice and regolith (soil) and penetrating to 2-3m depth; instruments might include sample retrieval drill, optical microscope and mass spectrometer.

The necessity of descending on Mars surface in order to fulfill the mission goal was immediately clear and the researches made led to the identification of five landing sites where it is expected to have higher possibilities of finding bio evidences. Landers relatively small, simple and spread across the surface would ensure the capability of exploring all the identified landing sites. The research the group made brought to evidence the necessity of operating during hot seasons both because thicker ice and the highest environmental pressure are expected, which means higher probability for finding water at triple-point conditions (vapor, liquid and ice).

<u>Mission Concepts</u>

The definition of different options led to the creation of three different mission concepts.

The first mission concept is illustrated in Figure 14. It is constituted by five Space Penetrator System SPS that will descend on Mars surface by ballistic fall and penetrate the ground after the impact with the surface. The SPS itself could contain a set of instruments in a CubeSat-form factor (as the ultrasonic drill) needed to sample and make the required analysis. Using five SPS the exploration of the most important landing sites identified by the researches is guaranteed. All the Space Penetrator System involved in Mission Concept 1 will relay on one orbiter that will send all the data to Earth.

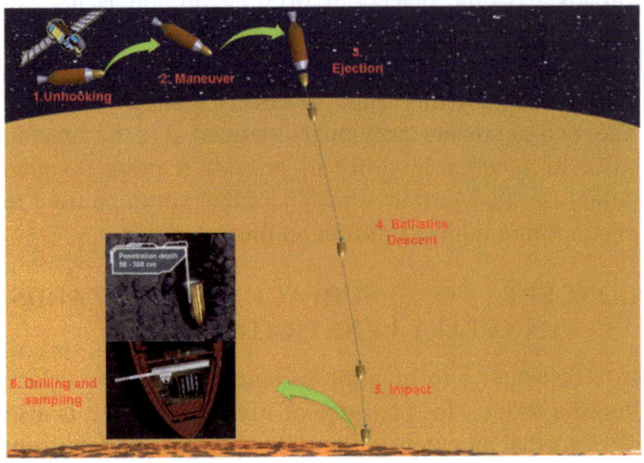

Figure 14: Mission Concept 1 representation

The second mission concept relates to the use of five landers and two orbiters. The five landers will descend on Mars surface through a controlled landing, achieved by using a parachute and airbags. The surface will be penetrated using an ultrasonic drill, then samples will be collected and analyzed: data will be sent to the two orbiters and then to the ground stations on Earth. Having two orbiters may assure a better coverage of the landing sites and scientists on Earth should receive data more frequently.

Mission Concept 3 is a mix of the previous ones, considering both penetrators and landers. SPS will descend on Mars surface by ballistic fall and penetrate the ground after the impact with the surface. On the other hand a controlled landing for the landers is required and will be achieved using a parachute and thrusters. Once on the surface the landers will penetrate the ground using a laser drill and both the landers and SPS will collect and analyze the samples. Data will be sent to an orbiter that will transmit them to the ground stations on Earth. Using landers and SPS the exploration of the most important landing sites identified by the researches is guaranteed. Moreover it will be possible to choose to send the SPS in the most demanding landing sites, for example where a deeper penetration is required.

The critical requirements definition process and the comparison with existing systems, which led to the feasibility assessment and a first sizing estimate, helped the team to define some Figures of Merit such as coverage, resolution, communication, lifetime, payload power, size, weight, and cost. As trade-off result, the mission concept 1 was chosen as baseline, in according with mission and scientific objectives and goals, also considering a total autonomy after the deployment phase. In terms of mission objective the SPS, including the "lab-on-a-chip" and a spectrometer or an ultrasonic drill, should be able to find past or present bio-evidences in the Mars soil with a generically soil analysis, considering a penetration depth from 50 to 300 cm. The mission scenario includes a number of spacecraft equal to the number of landing sites that provide the visit of different portions of Martian surface and taking samples of Martian soil with an autonomous process.

CONCLUSIONS

As part of a research on advanced concepts for future generation of small-satellite missions beyond Earth orbit, the team focused attention on the visionary scenario of networks of CubeSats in support to the human exploration of planet Mars.

The CubeSats-on-Mars-scenarios encompass the perspective of CubeSats as effective tools in support to the envisaged human exploration of Mars' orbit and surface, and contributes to the long-term vision of Mars exploration. An analysis of the potential environmental hazards and of the precursor measurements necessary to support human operations led to the definition of some primary needs prioritized by NASA MEPAG and P-SAG groups.

With the analysis of different levels of conceivable mission scenarios, this work highlighted the necessity for humans on Mars to have a support from a timely responsive and spatially distributed network of highly disposable, replenish-able, and low-cost satellites.

The unique features of CubeSats when used as distributed systems have been evaluated against the need of precursor global measurements: at least three investigations for three different subjects (orbit, atmosphere, surface) seem to fit promisingly. The cost and technical feasibility of the three concepts will be subject to further investigations.

ACKNOWLEDGEMENTS

The study has been carried forward within an exchange program between two research groups at Politecnico and MIT, since January 2014, led respectively by Prof. Sabrina Corpino and Prof. Sara Seager. The MIT team has extensive experience working on the boundary of planetary science and engineering, and contributed to the PoliTo group providing expertise in how to identify science goals for developing scientific missions that can be implemented with CubeSats or small platforms as well as the related science requirements flow down and traceability. The PoliTo research group has expertise in developing technologies for CubeSat and in solving engineering challenges associated with performing high complex mission tasks by using small satellites. As both groups have a good expertise in small satellite and platforms development for Earth orbits, for us the new ambitious paradigm is represented by a research aimed at studying the capabilities of CubeSats beyond LEO for valuable planetary science missions.

The collaboration program between Politecnico and MIT named "Internazionalizzazione della Ricerca 2013/2014" has been sponsored by "Compagnia di San Paolo", Torino.

REFERENCES

1. NASA-MEPAG [online] http://mepag.nasa.gov
2. Nichele F., Corpino S., Seager S. (2014) *Three scenarios fro valuable planetary science missions on Mars: next generation of CubeSats to support space exploration.* In: 65th International Astronautical Congress, Toronto, Canada
3. Ridolfi G., Mooij E., Cardile D., Corpino S., Ferrari G. (2012*) A methodology for system-of-systems design in support of the engineering team.* Acta Astronautica, vol. 73, pp. 88-99. - ISSN 0094-5765
4. D. Wood, A. Weigel, *Architectures of small satellite programs in developing countries,* Acta Astronautica 97 (2014) 109-121
5. ESA Education Office [online] - http://www.esa.int/Education/CubeSats_on_the_Vega_maiden_flight
6. Corpino S., Chiesa S., Stesina F., Viola N. (2010) *Cubesat development at Politecnico di Torino: the e-st@r program.* International Astronautical Congress, Prague (CZ).
7. ESA Edication Office [online] - http://www.esa.int/Education/Call_for_Proposals_Fly_Your_Satellite
8. NASA ElaNa [online] - http://www.nasa.gov/offices/education/centers/kennedy/technology/elana_feature.html
9. Selva D., Krejci D - *A survey and assessment of the capabilities of Cubesats for Earth observation,* Acta Astronautica : Elsevier, 2012, Vol. 74.
10. R., Sandau. *Status and trends of small satellite missions for Earth observation.*1-2, Acta Astronautica : Elsevier, 2010, Vol. 66.
11. S. Corpino, R. Mozzillo, G. Obiols-Rabasa, F. Nichele, *A survey of the first 10 years of CubeSats,* internal report – submitted for peer-review to Acta Astronautica
12. QB50 project. [Online] www.qb50.eu.
13. Aguado Agelet F., Puig-Suari J., Camacho S., Vicente-Vivas E., Castro A. (2010) *HumSat: Humanitarian Satellite constellation. A nano satellite constellation for climate change monitoring and humanitarian initiative.*

Prague, CZ : 61st International Astronautical Congress
14. Ridolfi G., Corpino S., Stesina F., Notarpietro R., Cucca M., Nichele F. (2011) *Constellation of cubesats: 3-star in the humsat/geoid mission*. In: 62nd International Astronautical Congress, Cape Town, South Africa
15. Obiols-Rabasa G., Corpino S., Nichele F., Ridolfi G., Stesina F., Viola N., *3-STAR program at Politecnico di Torino. Small satellites, services and systems* - the 4S symposium. Portoroz, Slovenia, June 2012.
16. ISSC2014 [online] http://www.intersmallsatconference.org
17. R. Staehle et al., *Interplanetary CubeSats: Opening the Solar System to a Broad Community at Lower Cost*, NASA-NIAC Final Report on Phase 1 to NASA Office of the Chief Technologist, 2012
18. Smith, Matthew W., et al. (2012) *ExoplanetSat: Detecting transiting exoplanets using a low-cost CubeSat platform*. SPIE Proceedings, July 2012
19. M.A. Viscio, N. Viola, S. Corpino, F. Stesina, S. Fineschi, F. Fumenti, C. Circi *Interplanetary CubeSats system for space weather evaluations and technology demonstration*. 2014 Acta Astronautica. - ISSN 0094-5765
20. NASA SmallSat [online] http://www.nasa.gov/mission_pages/smallsats.html
21. Nichele F., Corpino S., Ridolfi G. (2012) *System of systems analysis of low-cost small-satellites solutions for end-user oriented space-based services*. 63rd International Astronautical Congress, Naples, Italy
22. Planetary Decadal Survey 2013-2022: The Future of Planetary Science. Visions and Voyages for Planetary Science in the Decade 2013 – 2022. [online] - http://solarsystem.nasa.gov/2013decadal/
23. Mars Science Goals, Objectives, Investigations, and Priorities [online] http://mepag.nasa.gov/reports.cfm
24. Goal IV: Prepare for human exploration. MEPAG 2012 Goal IV update [online] http://mepag.nasa.gov/reports.cfm
25. P-SAG Measurement matrix. MEPAG Topical Analysis Reports. [online] http://mepag.nasa.gov/reports.cfm
26. Mark R. Showalter, Douglas P. Hamilton, Philip D. Nicholson, *A deep search for Martian dust rings and inner moons using the Hubble Space Telescope*
27. S. Soter, *The dust belt of Mars*, 1971 [online] - http://hdl.handle.net/1813/36286
28. SPS [online] http://sci.esa.int/future-missions-office/52782-high-speed-tests-demonstrate-space-penetrator-concept/

Reinventing Space Conference
BIS-RS-2014-31

Technology or Law: Which Will Reach Mars First?

Benjamin G. Davis
Dulles University
Vienna, Virginia, USA
ben.davis@dullesuniversity.org

12th Reinventing Space Conference
18-20 November 2014
London, UK

S. Hatton (ed.), *Proceedings of the 12th Reinventing Space Conference*, DOI 10.1007/978-3-319-34024-1_4

Technology or Law: Which Will Reach Mars First?

Benjamin G. Davis
Dulles University
1934 Old Gallows Road
Vienna, Virginia 22182 USA
+1 703.552.1236
ben.davis@dullesuniversity.org

Abstract

The two major United Nations treaties that relate to property rights in space, the Outer Space treaty (OST) and the Moon Treaty, are of little help. Key ideas and terms are not defined clearly with the exception that they prohibit the establishment of territorial sovereignty on the Moon or other outer space body. There is support in one segment of the legal community for at least some form of property rights in space. The situation will evolve depending on conditions extant at the time a decision is necessary. A laissez-faire approach to property rights in space is the appropriate path.

KEYWORDS: **Property rights in space; Moon Treaty; Space law**

INTRODUCTION AND BACKGROUND TO THE ISSUE

In 1950 Robert Heinlein, acclaimed science fiction master, published the novella *The Man Who Sold the Moon*[1]. The story takes place in the late 20th Century and focuses on the attempt by one man to take control of the Moon. By page 24 of the story, Heinlein's protagonist has separated ownership from use in outer space, something that will be shown below to be of key importance. At the same time, John Cooper wrote the first article specifically about space law.[2] What was science fiction and legal foundations 60 years ago is now about to become science and legal fact, and the issues faced by the protagonist of *The Man Who Sold the Moon* are now being addressed by corporate executives and government bureaucrats around the world.

With the exception of the era of the Apollo Lunar program of the 1960s, the present decade is perhaps the most dynamic in spaceflight. "Russia, ESA, Japan, China, and India all have proposed ambitious missions, including manned missions, to the Moon and planets."[3] The United States is the same. Landing and initial surveying are one thing, but when space activities expand to include the extraction and return of resources from extra-terrestrial bodies, the competition and potential for conflict will only increase.

The *development* of space, particularly by private entities, has yet to occur for a number of reasons. The standard explanation for this lack of development is because the cost of spaceflight is prohibitively high, and there is no question but that the cost is great. One reason why spaceflight costs are high is technical, but as technology develops, cost will fall. A second reason why costs have remained high, some believe, is due to the lack of effective competition for access resulting in a limited incentive to reduce costs. Simberg[4] posits that as long as there remains uncertainty regarding the rights to off-planet resources, private industry will have little incentive to invest competitively. Only when there is clarity about the ability to profit from resource collection activities will costs be driven down.

The major issue to be addressed is one of the legal ability of individuals, firms, and/or countries to access and remove raw materials from the lunar, Martian, or asteroid surfaces for private use and gain. Treaties exist that, some claim, establish rules regarding property rights on lunar property, but this is in significant doubt.

The major concern of this paper is the legal environment surrounding access to and removal of raw materials from the Moon, Mars, and asteroids. The area of specific interest is that of private property rights in space. Property rights related to possession/use of territory and resources, of course, are not the only property rights that need to be defined. "Orbital rights, intellectual property rights, and commercial transactions are other areas in which the law regarding property rights in space need to be developed."[5] Space has long since ceased to be the province of just the United States and Russia. Arising out of scientific research, national pride, or perceived national defense needs, many nations, including those which are considered to be developing, have credible plans for accessing space. In addition, a number of private ventures have demonstrated the ability to operate successfully in space.

Legitimate concern has arisen about the rights of states and of entities within states to access and appropriate resources in space and the extent to which private property rights may be – or may not be – established. Dalton states, "Given the…rise in private ventures into outer space and onto other celestial bodies in our solar system, the set of rights that protect those private ventures should be clearly defined."[6] Clarity about property and the rights surrounding it will do much to both foster rational, non-exploitative development and to reduce the potential for conflict.

Economic development yields pressure for private property. Speaking in an agricultural sense, Feder and Feeny note that "Generally, secure individual property rights over land, or secure and long-term use rights on land induce exertion of higher levels of labor and management effort and higher levels of investment to protect or enhance land fertility."[7] The same may be said of investment in resource development and utilization.

A significant concern with the issue is that private industry may balk at investment in extra-terrestrial landings without legal assurances. "Private enterprises often claim that the ambiguity regarding the status of property rights on celestial bodies is a major barrier to commercial development."[8] Thus, there is the fear that development will not take place without clarification of the issue of property rights.

Therefore, given the above concerns and their importance to the development of outer space, this study addresses the issue of private property rights in space, both title to land and ability to claim natural resources on the Moon, Mars, and the asteroids.

Why Care Now?

A number of factors make this study both timely and relevant. These factors are both qualitative – the overall level of the capability of accessing and living in space has multiplied dramatically – and quantitative – there is an increasing number of national and private space organizations with documented capabilities of space flight.

First, there has been a quantum leap in the technical capabilities of existing and formerly-potential space faring nations, and many other states and non-governmental organizations have demonstrated an interest in space travel. Further, whereas previously interest was limited to the Moon because of its relative proximity, interest now includes the planet Mars where colonization rather than just exploration is likely due to the travel time involved. Each nation or organization has developed its plans essentially in isolation with little evident consideration of the plans and goals of the others. As Schackelford notes, "This lack of multilateral cooperation is now threatening core principles of space law as technology leapfrogs the applicable governance regime."[9] An example of technological advance can be seen in the potential use of 3-D printing to make replacement or new parts and even food in space. NASA has already successfully tested a complex rocket injector[10] and is currently studying how food variety in long duration spaceflight might be improved through literally creation in space.[11] In addition, the creation of hydrogen for fuel while in space is also becoming more likely.[12] Further, as the cost of space operations decreases, additional state and non-state organizations will be able to enter, yielding an even greater potential for commercialization – and conflict.[13] Thus, having an appropriate legal regime in place as far in advance as possible would provide clear parameters within which state and non-state space organizations could function.

A second factor is the potential financial value of access to the Moon, planets, and asteroids. As then-President Bush noted in his call for an expanded program, "The Moon is home to abundant resources. Its soil contains raw materials that might be harvested into rocket fuel and breathable air."[14] In addition, the discovery of significant supplies of helium-3, in extremely limited supply on Earth is in relative abundance in the regolith on the Lunar surface as noted by Slyuta *et al.*[15] The isotope helium-3 is of high value for use as a fuel in nuclear fusion reactors and would generate large amounts of very inexpensive energy. Retrieval of the isotope in the regolith would require the processing of large quantities of surface material to generate the product, so concern over environmental degradation is an issue, but such degradation would have significantly less impact on human lives than if carried out on the surface of the Earth. As Simberg notes, the same resource availability is true of "asteroids [which] are relatively rich in rare earth minerals such as neodymium, scandium, and yttrium,"[16] rare earth minerals used in electronic equipment. Further, the confirmation of water at the poles on the Moon and planets will provide the raw materials for drinking and irrigation water, breathable oxygen, and energy.[17] Finally, "Planetary scientist John S. Lewis has estimated an average sized metallic asteroid would contain more iron nickel, and platinum group metals than have been mined in all of human history."[18] Access to the resources contained in an asteroid could be of significant value to the nation that is able to claim and appropriate these resources.

The current U.S. President has called for landings on asteroids whose orbits bring them within range of U.S. launchers. Studies have been conducted for over two decades that show relatively low-cost flights to asteroids are possible and that high-value resources are available on these bodies *in situ* for use in longer distance space flights as well as for removal and use on Earth itself.[19, 20] More recently, Adamo, *et al.*, have shown that, "No less than 36 potentially accessible human-mission destinations were culled from the SBDB [small-body data base maintained by the Jet Propulsion Laboratory) in July 2009. Over the interval from 2020 to 2050 these destinations give rise to at least 58 potential mission opportunities coinciding with NEO [Near Earth Object] encounters closer than ~0.1 AU from Earth [9.3 million miles]." "As the SBDB is populated with an order of magnitude more NEOs in the coming decades, opportunities using any exploration capability appreciably beyond that achieved during the Apollo program will increase by at least a factor of 10."[21]

An effective, agreed-upon legal system must be in place if Lunar and Martian development is to take place. Collier speaks convincingly of squandered opportunities for economic development in the Two-Thirds World due to the lack of international codes and laws resulting in mistakes and mis-aligned incentives.[22] Opportunities for development could similarly be lost with regarding extra-terrestrial resources if there is not clarity about property rights and resource rights.

Third is the impetus provided by awards such as the Google Lunar X Prize which is designed to encourage the development and launch of a lunar lander. Twenty-six teams are presently entered and competing for the $30 million prize. Modeled after the Ansari X Prize (which itself was modeled after the Orteig Prize won by Charles Lindberg in 1927), the competition will reward the first team that lands a rover on the lunar surface, drives at least 500 meters, and returns images and data.

Fourth, mission planners are exploring alternative mission profiles. When humans are present as was the case in the Apollo program, time in space must be minimized, hence a Hohmann transfer approach to and return from the Moon, for instance, must be taken. When humans are not present, however, other mission profiles may be followed, in particular those which minimize energy and cost while maximizing the size of the launch vehicle and payload. Edwin Belbruno at the Jet Propulsion Laboratory designed such a sample return mission following what is being called the "Interplanetary Superhighway,"(see Lo[23] and Benson[24]) and such an approach would make access to the lunar resources much more economical and, hence, likely. A sample return mission called OSIRIS-Rex and developed by a joint NASA-University of Arizona team to asteroid 101955 Bennu, a potentially Earth-intersecting impactor, is currently planned and on schedule for 2016.

Finally, there is the issue of national prestige. Developing nations, particularly China and India, have a strong desire to become technologically-respected actors on the world stage in much the same manner as the Soviet Union did in the 1950s and 1960s. The Soviet Union used the orbiting of Sputnik to draw ahead of the United States in the perception of the developing world. Now, others wish to follow this model toward international respect.

The Potential Barrier

The reason for this study is the disagreements regarding what a party could or could not do on the lunar or planetary surfaces. Are claims of property rights permissible by states? If not by states, what of non-state entities? Are entities permitted to use but not claim title to property on the lunar surface? Are entities permitted to neither claim nor use lunar property? Finally, is the developing entity permitted to develop-and-*keep* the resources extracted or is it required to develop-and-*share* the resources? The lack of clarity has opened the door for frivolous Lunar land claims by individuals as will be noted below. As Virgiliu Pop states, "While the "extraterrestrial real estate" claims . . . are nothing more than media curiosities, it needs to be agreed that behind their triviality they hide significant legal implications. The advancement of such claims has been only possible because of the lack of a property rights regime in the extraterrestrial realms."[25] The world of trivial claims and forthcoming real ones requires that a coherent statement of property rights in space be understood by all parties.

Types of Property and Property Rights

Feder and Feeny state that "There are four basic categories of property rights in land: none (or open access), communal property, private property, and state (or crown) property. Under open access, rights are left unassigned. The lack of any exclusivity implies the lack of an incentive to conserve, and therefore often results in degradation of scarce resources. Under communal property, exclusive rights are assigned to a group of individuals. Under state property, management of the land is under the authority of the public sector. In private property, rights are held by an individual. These four categories are ideal analytical types. If the group holding exclusive communal rights is large enough, the distinction between communal property and open access becomes moot. If private property rights are not viewed as being legitimate or are not enforced adequately, *de jure* private property becomes *de facto* open access. Nonetheless the simple taxonomy is useful for describing property rights systems."[26] Feder and Feeny go on to note that "Private property rights in land have evolved gradually in response to increases in the scarcity value of land and therefore the benefits to be derived from more precise and secure land rights,"[27] i.e., there is an increasing need for clarification as the value of the resources in question rises. With quite literally trillions of U.S. dollars in untapped resources available on the Moon alone, clarity is essential. Clarity and transparency yield the grounds for trust and the peaceful development of space resources.

For Murray, a key issue is that there is a variety of types of property. Murray cites an opinion rendered in *Moore v. University of California Regents* that "the same bundle of rights does not attach to all forms of property,"[28] and thus land, chattels, and intangible property deserve different treatment under the law. Using the analogy of streets and parks, Murray states that "Communal access to territory...does not preclude all other claims of private property in that territory" or "preclude...all other claims of ownership.[29]

The United Nations treaties refer to property on the Moon and other extraterrestrial bodies as being held in common for all mankind. This being the case, the type of common property must be identified. As Lin states, "we need to understand what it means to own space in common with others. Is our relationship one of 'positive community of ownership,' in that we each own an equal share in space and its contents? On the other hand, if our relationship to space is one of 'negative community of ownership,' then no one has a *prima facie* claim to the property in question, i.e., no one owns anything yet, or we share the common starting point of owning no part of space."[30] As Lin discusses, a "positive" situation brings with it a host of issues including the effective division and accumulation of property today (does the United States own 5% of the Moon because it has 5% of the Earth's population?) and tomorrow (how do we account for the property rights of generations as yet unborn?). Clearly, the definition in use by the United Nations raises more questions than it provides answers.

A further complication arises out of the provision in the United Nations Outer Space Treaty that no nation may claim sovereignty over any body in space. If national sovereignty is required for an individual or firm to hold title to private property or to extract natural resources, clearly these would be precluded by the Outer Space Treaty. However, "Some scholars argue that property rights can exist only under a nation's dominion, but most believe that property rights and sovereignty can be distinct."[31] A determination must be made with regard to this divergence.

The type of legal system at issue can also affect this outcome. Legal systems fall into two broad categories, civil law and common law. In civil law, "the code precedes judgments; the common law follows them."[32] While the treatment in judgment can vary significantly, regardless of whether one functions in a civil law (*a priori* codified) or a common law (precedent derived) jurisdiction, the concept of property is similar. In both systems rights in property include some aspect of the right to possess, the right to use, and the right to enjoy. The concepts of "adverse possession" and "*occupation*" apply, and the occupation and use of real property to which no other person has title results in effective ownership.

"Regarding claiming ownership over asteroidal resources, it appears that the ancient Roman law of *pedis possessio* will apply. *Pedis possessio* is the basis for Western law on ownership, and analogies have long existed in other parts of the world as well."[33] *Pedis possessio* is defined as "The principle that a prospector working on land in the public domain is entitled to freedom from fraudulent or forcible intrusions while actually working on the site."[34] Certainly it must be held that the Moon and asteroids are in the public domain given that it is contrary to the Outer Space Treaty for a nation to claim title to them.

Lin suggests that the starting point for a determination of property rights "would be to consider space development through political thinker John Rawls' Original Position in which we operate under a 'veil of ignorance' or pretend that we don't know any facts about ourselves, including who we are, what economic class we belong to, what nationality we are, and so on. With our biases stripped away, what rules would we set up, knowing that we would have to live by those rules once we find out who we are? Applying the veil of ignorance to rules in space, this helps ensure that the processes we set up are fair and consider the interests of all people, including protecting the worst-off people from an even worse and uncaring fate."[35] If this were followed, it is highly likely that the developed, space-faring nations would approach the issue quite differently from the present situation where their wealth and advantages are known. The concern, however, is obviated if property rights do *not* exist in space.[36] Do property rights exist in on the Moon or elsewhere in space? It is Gilson's position that, "the best prediction given the current state of the law is 'probably not.'"[37] However, although property rights in space may not exist *de jure,* they may well exist *de facto*.

WHERE WE STAND TODAY: TREATIES, CLAIMS, CASE LAW, LEGISLATION, AND THE COMMONS

Background

Space may exist in a vacuum, but space property law does not. There are many environmental factors that affect the landscape and that must be resolved in assessing property rights in space. Some of these environmental factors are potential barriers, others are potential facilitators; all must be examined.

It is informative to review the context within which current space law was developed. Goldman[38] identifies three periods in the evolution of space law, the Classical Period (1957 to 1979), the Transitional Period (1980 to 1991), and the Modern Period (1992 to present). Broadly, space law deals with public, private, and military issues; the first two have evolved whereas military space law has not. The major powers, not wishing to spur a costly arms race in space, yet mindful of the need for an effective defense, have tended to leave military space law extremely flexible. As Kasku-Jackson and Waldrop note, "From a legal perspective. . . it is clear that space law is very permissive for national security space activities. There are also few or no enforcement mechanisms to punish violators."[39] During no period was this more evident than during the Classical Period of space law.

The Classical Period (1957-1979) was dominated by the Cold War. As the Soviet Union and the United States jockeyed for world leadership, space law focused on military and foreign affairs, and the foundation for all space law was laid. "International space law, like all other branches of international law, has two sources: oral or customary law, and written or treaty law."[40] Both forms have affected the evolution and subsequent interpretation of legal claims. An example of the role of customary law put forward by Goldman is that of the altitude at which space begins and aviation space ends given the two very different bodies of law that are applicable. Space law is international in nature whereas aviation law reflects national sovereignty. The specific altitude at which space law begins and aviation law ends has never been established. However, when Sputnik, the first artificial satellite, was placed on orbit, circling the Earth in 1957, no nation filed a protest claiming an invasion of national sovereignty. As a result, "When twenty years later, several equatorial nations raised the issue and claimed sovereignty over the geosynchronous orbits, their claims were not accepted; this rule of law simply had already been established by practice and custom."[41] Given that these orbital positions are a scarce and, hence, valuable resources, the power of customary law and its relevance for private property claims in space may be seen.

During the Classical Period, space flight and exploration were in their infancy. The only major actors on the scene were the national governments of the United States and the Soviet Union. Whereas the space program of the United States included major contributions from private companies, that of the Soviet Union was composed only of government-controlled entities. The issues and concerns of the period were those of the Cold War and national security, and the two nations attempted to provide themselves with the maximum flexibility to pursue national goals. As a result, space law focused on the issues and concerns of governmental actors, and "classical international space law reflects a pro-state, anti-free enterprise orientation."[42]

As the name would imply, the Transitional Period (1980-1992) was one of change in the nature of space law. Additional nations began to assert their presence in space, and commercial enterprises, particularly in the form of providers of telecommunication and remote sensing satellites, became major actors. Compounding this was "that era's political context where third world developing states became a majority in the UN system."[43] While developing nations lacked a space presence, they did hold the majority of votes within the General Assembly of United Nations, and they began to forcefully make their presence felt in New York and Geneva. The result was a stalemate with regard to space law. With veto power in the Security Council, neither the Soviet Union nor the United States would allow space law to be passed that was counter to its broader political and military aims. Therefore, "In this period, international space law entered a hiatus. In its stead, domestic space law and private international law (contractual negotiations among states and corporations) filled the legal void."[44] The result was the creation of private international space law, a variation of the Medieval *lex mercatoria* that oversaw commercial transaction disputes.

Following the legal drought of the Transitional Period, the Modern Period (1992 – to date) has brought a renewed interest in space law. Perhaps, however, the period should be designated as "Transition II" in that, while there has been a "new productivity," the outcome has been one of "only providing declarative and symbolic statements rather than codified treaty."[45] That this is the case should not be surprising. With the fall of the Soviet Union and its focus on state control of space and with the increased importance and dominance of the role of commercial space, space law, particularly United Nations resolutions, have focused on the issues of direct television broadcasting, remote sensing, the sharing of the benefits, and nuclear power in space. (See United Nations[46]) None of the United Nations' resolutions carries with it anything like the force of law, hence passage was significantly easier than if they had been legal statements to which the parties would have been bound to adhere.

It is important to note that all of the major space laws were formulated during the Classical period, an era of domination by two international state powers, both of which were struggling for a position of sole leadership on the world stage. Spaceflight was in its infancy, and private actors were merely supporting cast to the two leading States Parties. Both of the major parties wished to be able to develop space to its advantage and fought to maintain an unfettered openness to its activities. From a legal perspective, then, it is clear that space law is very permissive for national security activities while less open for private and commercial actors. In both cases, there are few or no enforcement mechanisms to punish violators of any sort.

Treaties

Briefly, four issues exist regarding current United Nations treaties. First, according to Article 16 of the Outer Space Treaty and Article 11 of the much-contested Lunar Treaty, signatories to the treaties may leave with one year notice. Second, interpretations of law can vary significantly with multiple parties reading the same text in very different ways. Third, the legal foundations and approaches of the several nations are very different. For example, the Chinese legal perspective, the approach by a major actor in the Lunar drama is quite different from that of the Western democracies.[47] Fourth, the United States is a signatory to the Outer Space Treaty but has refused to sign the Moon Treaty, thus is not bound by the provisions of the latter. However, as Dalton states, "Some of the principles expressed in the Outer Space Treaty have passed from simply binding signatories to the treaty, but into *customary international law* that binds all nations generally. The ideas expressed in the Outer Space Treaty were reaffirmations of a number of UN General Assembly Resolutions." "Therefore, withdrawing from the treaty would not avail a nation from being bound by its principles."[48] Simberg expands this issue by noting that withdrawal from the treaty would not be politically *practicable*. "It [the Outer Space Treaty] is the basis for most current international space law, including follow-on treaties, such as those relating to astronaut rescue and return."[49] It should further be noted that the renegotiation of the treaty is not a viable option because the treaty is now 45 years old and because the developing nations would most certainly block such attempts.

Major Space Related Treaties

There are five major United Nations treaties that directly relate to issues of outer space: the Outer Space Treaty (1967), the Rescue Treaty (1968), the Liability Treaty (1972), the Registration Treaty, an the Moon Treaty (1979).[50]

"The Treaty on Principles Governing the Activities of States in the Exploration and Use of Outer Space, Including the Moon and Other Celestial Bodies" (1967) (known as the "Outer Space Treaty")

This treaty serves as the foundation for all subsequent treaties and, in reality, all of space law. The treaty was very broadly written at the beginning of the space age and attempted to assure that space would be developed peacefully for the benefit of all mankind "with no real profit for any nation or person."[51] The goal was cooperation in international affairs rather than on national actions in space.

The Outer Space Treaty states that nations are free to pursue space exploration which is for the benefit of all mankind (Article I), that bodies in space are not subject to national appropriation (Article II), that there are to be no weapons in space (Article IV), and that states are responsible for damages caused by it or organizations under its jurisdiction (Articles VI and VII). Additional articles include the issue of ownership, i.e., "Ownership of objects launched into space, including objects landed or constructed on a celestial body, and of their component parts, is not affected by their presence in outer space or by their return to Earth" – Article VIII) and non-interference in the space activities of another signatory to the treaty. Articles 9 through 12 either use the word "cooperation" or "reciprocity," furthering the concept of intended mutuality. The treaty has been ratified by 98 states and signed by 27 others. It is "a widely accepted notion that the treaty now constitutes customary international law."[52]

One aspect of the Outer Space Treaty that is of particular relevance for this study is Article II which prohibits the establishment of territorial sovereignty on the Moon or other outer space body. As White notes, there are four reasons why the United States and the Soviet Union and the other signees to the treaty chose to limit sovereignty: "(1) to prevent conflict; (2) to ensure free access to all areas of outer space; (3) because it would be difficult for states to delineate boundaries in outer space; and (4) to enhance national pride, prestige and influence."[53] Item one acknowledges the extremely high cost of the establishment of bases in space and the long history of inter-national conflict on Earth. Item two addresses the issue that nation-state claims on new territory were geographically extensive in the past, essentially preventing other states from access to territory and leading to the long period of armed conflict. Non-state property claims, should any be established, tend to be smaller in scope, leaving territory for newcomers. Item three recognizes the vastness of space, although admittedly the *surface* of outer space bodies is much easier to delineate. Finally, in item four the United States and Russia, who were in intense competition for the allegiance of newly-independent former colonies, recognized that anything that appeared to be a new colonialism in outer space would diminish their influence with the new states. The focus of this treaty was specifically on national states and said nothing about private actors. Therefore, because individuals are specifically not mentioned in the Outer Space Treaty, it is the contention of Stephen Gorove that "at present, an individual acting on his own behalf or on behalf of another individual or a private association or an international organization could lawfully appropriate any part of outer space."[54]

Finally, one key problem with regard to Article II of the Outer Space Treaty is the assumptions that are brought to it. It is clear that the treaty does not mention private entities, only nations. If one follows common law assumptions, this would none-the-less preclude private claims since rights in real property are devolved from the sovereign. Once you move past the assumption that Anglo-Saxon, common law base prevails, that basis is lost.

A present concern with the Outer Space Treaty and other space-related treaties was not envisaged during their drafting. The issue of the definition of what constitutes a "launching state" for liability purposes has become extremely complex. "Take the case of the launch of a Russian-built Soyuz rocket from the spaceport in French Guiana, procured in this hypothetical example from an American company for its satellite. France, Russia, and the United States are all together launching states and accordingly liable to the caused damages if the launched satellite falls back to Earth or collides with another satellite or object in space."[55] While theoretically all three states could be held liable for damages caused as a result of a launch, in reality individual liability would likely have to be established before an award could be granted.

Because the Outer Space Treaty was crafted at the height of both the Cold War and the space race, gaining passage required a herculean effort – and many compromises and ambiguities were a part of the process. As a result, many key ideas and terms are not defined. Whereas the Soviet Union wished to restrict space activities to national governments, the United States wanted to have it open to private companies. The United States got what it wanted, but the Soviet Union did also through the mechanism of making States Parties liable for the activities of their nationals. Space lawyer Ezra Reinstein acknowledges that, "The Outer Space Treaty is riddled with ambiguities. It is silent, outside of affirming freedom of 'exploration and use,' as to what sort of rights parties can claim in celestial bodies. It is silent as to the circumstances under which these unspecified property rights might vest, that is, what a person must do to gain whatever property rights are available."[56] Wasser and Jobes conclude that in dealing with areas of conflict during passage of the Outer Space Treaty, "In many cases, the solution was to insert vague language, that could be interpreted whichever way the reader wanted, but would leave the enactment of any real rules to a future discussion."[57] It is precisely these ambiguities that have opened the door to the confusion of the past 45 years.

Finally, it is worthy to note the parallels between the actors on both sides of the Trade-Related Intellectual Property Rights (TRIPs) Treaty and the outer space resource debates. Developing nations wish to protect their ability to enter the arenas while developed nations wish to protect their investment. These positions were abundantly clear in the development of the Outer Space and Moon treaties.

Despite the prohibition of the establishment of national sovereignty and despite the potential disagreement over the rights of private organizations to establish property claims, the United States ratified the Outer Space Treaty. However because of the prohibition of any portion of the Moon becoming the property of any entity, the United States refused to ratify that treaty.

"The Agreement on the Rescue of Astronauts, the Return of Astronauts and the Return of Objects Launched into Outer Space" (1968) (known as the "Rescue Agreement")

This treaty provides for the safe return of astronauts and space vehicles considered to be the property of the launching state. This treaty has been ratified by 89 states and signed by 24 others.

"The Convention on International Liability for Damage Caused by Space Objects" (1972) (known as the "Liability Convention")

This treaty strengthens the damage provisions of the Outer Space Treaty and holds States Parties responsible for damage caused by property belonging to it and to anyone under its jurisdiction. This treaty has been ratified by 84 states and signed by 24 others.

"The Convention on Registration of Objects Launched Into Outer Space" (1975) (known as the "Registration Convention")

This treaty requires presentation to the United Nations of information regarding the orbital parameters and general function of space craft that have been launched. Much the same as land deeds are registered, spacecraft launched by a state party are to be registered with the United Nations. The treaty has been ratified by 51 states and signed by four others.

"The Agreement Governing the Activities of States on the Moon and Other Celestial Bodies" (1979) (known as the "Moon Treaty")

The goal of the Moon Treaty is to establish a mechanism for a highly controlled development of the Moon. A draft treaty was submitted by Argentina in 1970, one year after the first Apollo landing. In 1971, the Soviet Union submitted its own draft. Interestingly, considering the subsequent actions of most nations, the final version was adopted unanimously. This treaty attempted to settle the question of private property rights on the Moon growing out of the wording of the Outer Space Treaty and to greatly restrict property claims on the Moon and planets. In fact, "The Moon Treaty…very clearly attempts to ban private ownership of land in space. The very existence of the

1979 Moon Treaty is a clear indication that the 1967 Outer Space Treaty does not ban private ownership of land in space and that lawmakers and diplomats recognized that to be the case."[58]

One particularly contentious issue of the Moon Treaty is the term "common heritage of mankind" contained in Article 11.1. Specifically, the treaty states that "The Moon and its natural resources are the common heritage of mankind, which finds its expression in the provisions of this Agreement, in particular in paragraph 5 of this article." (Paragraph 5 relates to the establishment of an international organization along with the necessary procedures to maintain the "common heritage" clause.) Precisely what this term means is not spelled-out and has led to much of the current confusion.

The opening articles of the Moon Treaty parallel the opening articles of the Outer Space Treaty, and many of the later articles address Moon-related, non-controversial issues. It is Article 11, however, that is the major point of contention. The first portion of the Article states that "the Moon and its natural resources are the common heritage of mankind…not subject to national appropriation" and that "neither the surface nor the subsurface of the Moon, nor any part of the natural resources in place, shall become the property" of any entity. The second portion directs States Parties "to establish an international regime, including appropriate procedures, to govern the exploitation of the natural resources of the moon" in order to make certain that there is "an equitable sharing by all States Parties in the benefits derived from those resources, whereby the interests and needs of the developing countries, as well as the efforts of those countries which have contributed either directly or indirectly to the exploration of the moon." Gangale and Dudley-Rowley (p. 5) note, "The American delegation's statement in UNCOPUOS [United Nations Committee on the Peaceful Uses of Outer Space], at the time that the Moon Agreement was negotiated, that the words 'in place' allow private property rights to apply to resources upon extraction, went unchallenged."[59] (see United Nations. 1979. Document No. a/AC.105/PV.203) As of January 1, 2014, the treaty has been ratified by only 15 states (with India being the only truly space-faring signatory) and signed by only four others. The treaty called for a review of its provisions and for the creation of the international regime to oversee lunar property and profits ten years after the treaty was put into force. That date (July 11, 1994) has long since come and gone with no regime established at the review in 1994. Bini speaks of the "Inspirational Principles" of the Moon Treaty, but "inspiration" carries little weight in a court of law. Bini further notes that "As set out by article 18 of the Agreement, ten years after its entry into force COPUOS considered the question of a first review of the Agreement and the prospective of the establishment of an international regime at its thirty-seventh session in 1994. After the discussion, the Committee recommended to the General Assembly at its forty-ninth session that the Assembly should take no further action at the time."[60] Because of all of this, it has been claimed by some that the treaty is, therefore, effectively dead. It is possible, however, that it has been simply resting.

Major Treaty Summary

The concern over private property rights in space arises out of conflicting definitions of property rights along with conflicting interpretations of the Moon Treaty. In the absence of ratification of the Moon treaty by the major actors, the treaties accepted by both nations (the Outer Space, Rescue, Registration, and Liability) provide the legal framework within which States Parties must function. "Space, like the Antarctic and the high seas, is a global commons. However, unlike with the other global commons, the international law that governs space developed rapidly – within months rather than decades or centuries. It was a scant ten months from the end of the Outer Space Treaty negotiations to its entrance into force in 1967. The other four space treaties entered force in under twenty years. The speed with which the international community established this treaty regime demonstrates a clear intent that space was to be governed by international law."[61] Thus, it can be accepted that the four space treaties ratified by the majority of the states and by all space-faring nations will hold sway.

Relevant Non-Space, Other Space Treaties

Beyond the five major space treaties, there are a number of additional treaties and agreements that are relevant. These treaties and agreements provide a background for understanding the thinking that undergirds issues such as property rights in space. Included are the following:

The Antarctic Treaty (1961)

This set of agreements defines relationships between and among the member states that conduct research in the Antarctic. It prohibits territorial claims and requires that the continent be used peacefully for the benefit of mankind. The Antarctic Treaty was an outgrowth of the International Geophysical Year (1957-1958) and was signed by 12 nations on December 1, 1959. The treaty went into force on June 23, 1961. Since then, an additional 16 nations have signed the treaty and 20 more have acceded to it, meaning that they have agreed to abide by the provisions of the treaty.[62] As the Secretariat of the Antarctic Treaty notes, the purpose of the treaty is to ensure "in the interest of all mankind that Antarctica shall continue for ever to be used exclusively for peaceful purposes and shall not become the scene or object of international discord."[63] The continent is be used solely for peaceful purposes with scientific investigation the primary activity. While then-existing territorial claims were continued, the treaty granted open access to the facilities of all nations and provided for an exchange of plans and results. The single treaty has evolved into a treaty system, and "In the 1980s the consultative parties to the Treaty began developing the Regulations of Antarctic Mineral Resource Activities. However, before it was ratified it was agreed to expand their efforts into a more comprehensive system to include the protection of the Antarctic Environment. This developed into the Protocol on Environmental Protection to the Antarctic Treaty,[64] signed on 4 October 1991."[65] While the major focus is on environmental protection, Article 7 of the protocol reads: "Prohibition of Mineral Resource Activities - Any activity relating to mineral resources, other than scientific research, shall be prohibited." Thus, while the treaty itself did not prohibit the removal of resources, the subsequent protocol did. One would anticipate that the Earth-bound snow of the Antarctic would generate a higher level of support for the prohibition of mining activities than would the dusty surface of the Moon.

As a final comment on the impact of the Antarctic Treaty, Gimarc has noted, "Today, Antarctica is an example of what happens when property rights are denied and a government monopoly…is created. Rather than being a job and wealth creator, activities on the continent are net expenditures to the taxpayers of the signatory nations. There is no growing infrastructure in and around the continent. There is no self-sustaining economy."[66] While others may not wish to support economic development in the Antarctic as does Mr. Gimarc, his point is well taken.

"The Agreement Relating to the International Telecommunications Satellite Organization (INTELSAT)" (1971)

This treaty deals with property rights as related to the allocation of the limited number of slots available for telecommunications satellites in geosynchronous orbit. Through the treaty, "It is possible to claim 'property' in geosynchronous orbit, positions which allow a satellite to remain fixed in place over a single small area allowing it to be a part of a global communications network. Both companies and countries can claim a volume of space for their satellite(s), can legally exclude others from this space, and of course can make a private profit from use of this space." These rights are based on "a claim and a fee."[67] These space property rights are extremely valuable, and mechanisms have been put into place to prevent electronic interference with signal transmission.

"The United Nations Convention on the Law of the Sea" (1982)

This agreement defines national property rights as a series of extensions, each with a different set of rights, from the shoreline of a state. This treaty and the Moon Treaty are written, "declaring space and the seabed floor both to be the 'common heritage of mankind,' and thus requiring that any economic activity be conducted by a U.N. agency."[68] This convention has been ratified by 157 states and signed by 9 others, but notably not by the United States. The Law of the Sea has been interpreted to require the sharing of technology and profits, and the United States has acceded only to one small portion of the treaty, that dealing with an enclosed body of water. "U.S. opposition to the Convention focuses on the regulation and use of the deep seabed beyond national jurisdiction. [According to the treaty,] all activities in this area are to be regulated by the International Sea-Bed Authority (ISA), empowered to grant mining rights to both public and private ventures."[69] Written very much in parallel with the Moon Treaty, the Law of the Sea Treaty established a regime to allocate and manage the seabed as does the Moon Treaty. The United States opposed both the relinquishment of power to an international agency and to the forced sharing of the resources developed by a member state.

Rather than accede to the Law of the Sea, the United States has relied on customary maritime salvage law and on the development of its own law. "In maritime salvage law, which also deals with property rights beyond national territory, actually being there is the key: those who reach a wreck first and secure the property are generally entitled to a percentage of what they recover."[70] In adhering to both maritime salvage law and in its own law of the seas, the United States established a precedent that bears weight for property rights in space. "In the 1980 Deep Seabed Hard Mineral Resources Act, the United States recognized deep-sea mining rights outside its own territory without claiming sovereignty over the seabed."[71] As the law states: "By the enactment of this chapter, the United States – (1) exercises its jurisdiction over United States citizens and vessels, and foreign persons and vessels otherwise subject to its jurisdiction, in the exercise of the high seas freedom to engage in exploration for, and recovery of, hard mineral resources of the deep seabed in accordance with generally accepted principles of international law recognized by the United States; but (2) does not thereby assert sovereignty or sovereign or exclusive rights or jurisdiction over, or the ownership of, any areas or resources in the deep seabed." (30 U.S.C.A § 1402)

"International Space Station Intergovernmental Agreement" (1998) (known as the "Space Station Agreement" or the "Intergovernmental Agreement")

This agreement among the partner agencies of the International Space Station (ISS) designates NASA as the lead agency and spells out ownership rights for the component parts supplied by the various team members. Resources developed on a portion of the International Space Station owned by a member state are the property of that state.

"Agreement on Trade-Related Aspects of Intellectual Property Rights" (TRIPs) (1994)

This agreement specifies the rights of intellectual property owners and provides a mechanism for their protection. The agreement has been much debated, and there appear to be close similarities among the types of states which have aligned on each side of the issue of property rights in *trade* and the issue of property rights in *space*, with the developed nations supporting strong intellectual property rights while the less developed nations tending to oppose them.

Other space treaties of less significance to the issue of private property in space include: "Principles Governing the Use by States of Artificial Earth Satellites for International Direct Television Broadcasting," adopted on December 10, 1982; "Principles relating to Remote Sensing of the Earth from Outer Space," adopted on December 3, 1986; "Principles Relevant to the Use of Nuclear Power Sources in Outer Space," adopted on December 14, 1992; and "Declaration on International Cooperation in the Exploration of Outer Space for the Benefit and Interest of All States, Taking into Particular Account the Needs of Developing Countries," adopted on December 13, 1996.

Property Claims

There has been no shortage of claims to lunar and other extra-terrestrial property. Pop has detailed these claims extensively in Chapter 1 of *Who Owns the Moon?*[72] and in the complete book, *Unreal Estate.*[73] The first recorded claim to the Moon dates to 1756 when the Prussian king presented it to a subject out of gratitude. The next claim dates 180 years later and is by A. Dean Lindsay who filed a claim document in Pittsburgh, Pennsylvania. Other claimants have come and gone, but the most serious and influential claim is by Dennis Hope, an American entrepreneur who founded the Lunar Embassy Commission in 1980 and began selling "deeds" to lunar land. Sales moved slowly, reaching only 3,500 by 1998 at which point the ability to market through the Internet increased sales dramatically. The latest figures show that Hope has "sold" over 611 million acres of land on the Moon, 325 million acres of land on Mars, and become a millionaire. Others have followed in Hope's footsteps, offering land on the Moon and Mars. That these deeds have been carefully worded to protect Mr. Hope (and the others) from liability of fraud has not deterred sales. Mr. Hope's claims are invalid for a number of reasons. "The first reason for invalidating the claims…is the lack of *corpus possidendi*. In the acquisition of possession, two concurrent elements – 'the mind' and 'the body' are required. [To own a thing, a person must have the intention to possess it and must actually possess or have a right to the object.] One is insufficient without another; there must be 'both an intention to take the thing and some act of a physical nature giving effect to that intention.'"[74] Mr. Hope is not able to legally document that he owns the land that he purports to sell.

One claim worthy of note because it proceeded into the American courts and became a part of case law is that made by Gregory W. Nemitz for ownership of the asteroid 433, Eros. Mr. Nemitz filed his claim to the asteroid with the Archimedes Institute. Following the 2001 landing by NASA on Eros, Mr. Nemitz billed NASA for parking charges for its spacecraft, claiming special damages of $5,000,000 per day. After an exchange of communication between Mr. Nemitz and both NASA and the U.S. Department of State, Mr. Nemitz filed a complaint in the United States District Court for the District of Nevada. The courts determined that Mr. Nemitz' suit was without merit and dismissed both it and Mr. Nemitz' subsequent appeal to the United States Court of Appeals for the Ninth Circuit. That Mr. Nemitz was willing to pursue his property claim to 433 Eros in the U.S. court system demonstrates that additional such claims may at some future time follow. Some claims are likely to be taken more seriously by the courts, and a client, represented by a trained attorney which Mr. Nemitz who represented himself was not, might have more success given the vagaries of decision making in national and international law.

That there is no official mechanism for recording deeds to extra-terrestrial property has not deterred entrepreneurs. "There are currently three companies selling lunar 'deeds.' The pioneer in the field, Lunar Embassy, founded in 1980; Lunar International, founded in 1996; and Lunar Republic (now called the Luna Society), founded in 1999."[75] These efforts have generated a significant amount of business and revenue for the entrepreneurs. Carefully wording the "deeds" to reflect that they are "novelty items" and of no real value has not deterred sales. In fact, the deeds are invalid. "Their claims are invalid not because the Outer Space Treaty prohibits them per se, but because they have no historically legitimate basis – especially since they do not involve traditional homesteading recognition practices as occupation and improvement."[76]

The sale of these lunar "deeds" with no intrinsic value has shown the level of interest that exists in the public. "Over the past twenty-five years, an entrepreneur named Dennis Hope unwittingly conducted an experiment that indicates the potential market for lunar deeds. In 1980, Hope 'claimed' the Moon and started a business selling lunar land 'deeds.' Thanks to Hope, the average value of lunar land, even on the remote regions of the Moon's surface, is clearly no less than about $200 per acre [the standard sale price] – even with the land undeveloped, completely inaccessible, and barren and airless. Hope has sold over two million of these deeds since 1980."[77] In doing this, Hope may well be the first – and only – person to have directly made a profit from the Moon.

Case Law

To repeat a term used at law, "there is a paucity of case law" regarding property rights in space. In fact, of course, there is a great deal of litigation in the aerospace arena, and societies, including the Forum on Air and Space Law of the American Bar Association of which the author is a member, exist dedicated to the field. The focus of most litigation in aerospace is in the more typical commercial and labor law areas. Gorove notes that "while space law is a distinct discipline among the various branches of the law, the issues which arise in the courts and the procedures used [today] differ little from those encountered in other areas of the law."[78] That similarity may well end when off-Earth cases become the norm in space law.

The first space-related case was filed 11 years before the launch of Sputnik. *United States v. Causby* (328 U.S. 256, 66 S. Ct. 1062 [1946]) dealt with overflight by U.S. military aircraft of an individual's property. As Gorove[79] points out, while the case was decided for the plaintiff, the issue of *cujus est solum, ejus est usque ad coelum* ("he who owns the land owns it to the skies") was part of the case. The court considered and rejected this argument by the plaintiff, stating that the doctrine "had no place in the modern world." The first significant case to deal with space law was *United States v. Safety Steel Services Co.* (S.D. Texas, December 7, 1965) which resulted in enjoining the defendant from operating electronic equipment in a manner that would interfere with communications between a NASA ground station and the Gemini 7 spacecraft.

A case specifically concerning property in space was *COMSAT v. Franchise Tax Board* (156 Cal. App. 3d 726, 203Cal. Rptr. 779 [Cal.App. 1st Dist. 1984]). In this case the court held that a satellite owned by COMSAT, which had offices in California was considered to be tangible personal property owned and used in the state by the appellant. COMSAT maintained ownership and responsibility for a revenue-generating space object operating solely in space, and the State of California taxed it, confirming that property rights extend beyond the surface of the Earth.

What is lacking is litigation specifically regarding property rights in space. With no legitimately recognized property rights claims to challenge, Hertzfeld reports, "Right now there's really very little litigation going on."[80] There has been the occasional suit such as that brought by Gregory William Nemitz (*Nemitz v. US*, Slip Copy, 2004 WL 316704, D. Nev., 2004 (April 26, 2004)) as discussed previously in which Nemitz claimed ownership of the asteroid 433, Eros and that was quickly disposed of in court, but these suits have been few and far between. This does not, however, mean that there is a lack of legal opinion and speculation, informed and otherwise, in the field.

One outcome of the Nemitz suit relates directly to the issue of property rights under the United Nations treaties. Kelly notes that "The failure by the United States to sign the Agreement regarding Activities on the Moon and Other Celestial Bodies or the signing and ratifying of the Principles Governing the Activities of States in the Exploration and Use of Outer Space, including the Moon and Other Celestial Bodies did not provide for a right to private property on asteroids."[81] Kelly further states that Nemitz' claim of denial of his constitutional right under the Tenth Amendment, to whit, "The powers not delegated to the United States by the Constitution, nor prohibited by it to the States, are reserved to the States respectively, or to the people" was not abridged. Kelly reports that Nemitz claimed "that since the power to own or regulate the ownership of lunar and celestial property was not delegated by the Constitution to the Federal Government nor reserved to the States, it is retained by individuals as part of the unenumerated and reserved powers of the Ninth and Tenth Amendments."[82] The courts found this argument to be without merit.

Legislation

Two pieces of legislation regarding property rights in space are currently working their way through committees in the United States Congress. One bill has to do with the establishment of a United States national park at the Apollo landing sites on the Moon and the second is focused on setting property rights for asteroidal resources. The national park bill is H.R. 2617[83] and was introduced by Rep. Donna Edwards and co-sponsored by Rep. Eddie Bernice Johnson. If passed, the bill would establish the landing sites of the Apollo program (as well as crash site of the Saturn IVB stage from Apollo XIII) as a national park under the jurisdiction of the U.S. Park Service. This proposal has received significant criticism particularly in light of the Outer Space Treaty's prohibition of the establishment of national sovereignty on the Moon or other space body.[84] The asteroidal bill is H.R. 5063[85] and was introduced by Rep. Bill Posey and co-sponsored by Rep. Derek Kilmer. This bill would permit resources extracted from asteroids to be the property of the entity that performed the extraction.[86] Legal opinion regarding this bill is split with experts including Joanne Gabrynowicz opposing the bill in its present form[87] while attorney Charles Stotler supports it provided the final bill adheres broadly to the provisions of the Outer Space Treaty.[88] Passage of both bills is uncertain, but it is unlikely that either will be passed in the current Congress.

The Commons

The final concern that must be addressed is a more theoretical one: this concept is that of the commons. The issue was popularized by the publication of Garrett Hardin's 1968 article, "The Tragedy of the Commons."[89] Hardin's contention is that people, following their own self-interest, will over use resources held in common. Hardin applied this in particular to population growth and control, but also discussed applications to environmental issues. Hardin contends that coercion will be required to bring about "cooperation" and that "freedom in a commons brings ruin to all." In economics, land held in common and available to all is considered to be a public good, and with a lack of a legal owner public goods are typically misused. The remedies for goods held in common are to agree to cooperate, to designate a single owner, or to impose government regulation. Private property results in the second option obtaining. Again, economic theory would posit that under a regime of private property, owners will manage the use of the resource such that it will produce the maximum gain over the long run.

There is an ongoing debate among legal scholars about the issue of the commons, however. Do extraterrestrial bodies fall under the definition of *terra nullis* or *terra communis*? Is the concept of *res nullis* (no person's property therefore free to be owned) or *res communis omnium* (belonging to the entire community and not subject to appropriation) more appropriate? With space resources to be utilized for "the benefit of all mankind," there is concern regarding whether or not revenues generated by lunar and other planetary activities would have to be shared among all states. “The crucial determination to be made in interpreting the normative provision of the Outer Space Treaty requiring space activities to be ‘for the benefit and in the interest all countries’ is the determination of the ambit of the norm, namely whether it imposes a positive and specific obligation ‘regarding the sharing the benefits of space exploration and use’ or is merely an expression of desire that the activities should be ‘ beneficial,’ in contrast to being harmful ‘in a general sense.’”[90] Add to this the debate between the “natural law” or idealist school where in laws exist evidently to all rational viewers and the “positivist” or realist school with laws written by a recognized governmental body, and the topic begs further study.

Cherian and Abraham advocate the position of *res communis* as “the most apt concept of space law,”[91] although they go on to state that while “*res communis* prohibits appropriation of property by a person, it does not, however prohibit occupation or use of such property. . . . Possession rights exist, though implicitly.”[92] The benefits from property such as extracted resources would belong to the entity (person or business, not government) that first possesses it.

Hardin’s ideas regarding the commons have been applied widely, but not all have accepted his pessimistic predictions of resource misuse and over-exploitation. For instance, van Vugt in his 2009 article “Triumph of the Commons” states that the decision making of people and organizations is influenced by “four ‘I’s (*eye*-s)” – Information (understanding), Identity (belonging), Institutions (trusting), and Incentives (self-enhancing).[93] Management of these four “I’s” and, particularly making people and by extension states, fully aware of the issues and helping people and states to feel like a part of the “inside” group, has the potential to lead to positive solutions of debated issues.

Treaties, claims, case law, legislation, and the commons – none of these provides a definitive answer as to either the propriety or the legality of private property in space. Each domain has brought light, but not full clarity on the subject. We will now turn to the specific arguments directly put forward in support of and in opposition to the subject.

ARGUMENTS REGARDING PRIVATE PROPERTY RIGHTS IN SPACE

In Support of Some Form of Private Property Rights

There is ample support in the legal community for at least some form of real private property rights in space. Wasser and Jobes[94] provide a summary of the opinions of some of these scholars:

Law professor Glenn Reynolds and columnist Dave Kopel – “It is widely agreed by space-law scholars that the Outer Space Treaty forbids only national sovereignty – not private property rights.”[95]

Joanne Gabrynowicz, Director of the International Institute of Space Law and former Professor of Law and Director of the National Remote Sensing and Space Law Center at the University of Mississippi – “As regards to property rights *per se*, the Outer Space treaty is silent. It contains no prohibition.”[96]

Stephen Gorove, Professor and former Chairman of the Graduate Program of the School of Law at the University of Mississippi – “The Treaty in its present form appears to contain no prohibition regarding individual appropriation or acquisition by a private association or an international organization. Thus, an individual...could lawfully appropriate any part of outer space, including the Moon and other celestial bodies.”[97]

National Space Society Representatives Pat Dasch, Michael Marti-Smith, and Anne Pierce – “Several important principles have been established by customary law and treaty. First, national sovereignty stops where outer space begins. Second, that national appropriation of the Moon, other planets, asteroids, etc., is forbidden. And third, that private property rights are not forbidden.”[98]

Attorney and space law consultant Wayne White – “Some interpret Article II [of the Outer Space Treaty] narrowly to prohibit only national appropriation. Many others interpret the clause broadly to prohibit all forms of appropriation including private and international appropriation. When Article II is compared to similar provisions in other documents, however, it becomes clear that the narrow interpretation is correct.”[99]

Saying so doesn't make it so, as has been demonstrated regarding the lunar land claims made by Dennis Hope, but the collected opinion of legal scholars should carry some level of consequence.

Gorove's interpretation of the Outer Space Treaty[100] (admittedly the OST is of lesser specificity regarding property rights than the unratified Moon Treaty) is that whereas the prohibition of national appropriation of outer space resources applies to these resources *in situ*, it does not once the resources have been extracted and returned to the Earth. No claim can be made while the resources remain on the lunar surface, for instance, but once under the control of an entity and *off* the lunar surface, possession may be claimed. Jakhu and Buzdugan concur, stating that "Natural resources in place can not become the property of anyone [but] once these resources are removed, they may be considered to have become the exclusive property of the entity that caused them to be removed."[101] Further, in the same article Gorove asserts that whereas states are proscribed from asserting a claim to property on the Moon or other extraterrestrial body, private organizations are *not* prohibited from doing so.[102] Jakhu and Buzdugan's position is that governments and private organizations are prohibited from ownership of land under the Outer Space Treaty,[103] but as noted, *not* prohibited to a claim to the resources extracted.

How is a land claim established and who grants the right to claim land? "In countries like France, which follow "civil law" (as opposed to "common law" which the United States inherited from England) property rights have never been based on national sovereignty. Instead they are based on the "Natural Law" theory that individuals mix their labor with the soil and create property rights independent of government. Government merely recognizes those rights." "Throughout history, actual settlement, occupation and use has been the traditional basis for claims of ownership that had no sovereign."[104] As the Space Settlement Institute further notes, "For property rights on the Moon, the U.S. will have to recognize Natural Law's "use and occupation" standard, rather than the common law standard of "gift of the sovereign," because the common law standard cannot be applied on a Moon where sovereignty itself is barred by international treaty."[105]

Brooks, among others, proposes that a new treaty be drawn up that would create a legal framework for managing the occupation and use of extra-terrestrial land, not just the Moon as proposed in the Moon Treaty. His proposal would establish a system or agency, independent of any single government, that would have jurisdiction over the land and which would exercise management of the territory. In speaking of Mars, Brooks states that the agency would "properly regulate the process and provide a legal and reliable regulatory system for purchasing and selling Martian land shares. Since this treaty would not be giving sovereignty of Mars to any nation, but simply granting jurisdiction of the land to an international agency, it would not be in violation of the Outer Space Treaty."[106] While the year 2014 does not have the same competition between the capitalist and socialist powers as when the Outer Space and Lunar treaties were crafted, the era still is beset by a strong developed *versus* developing nation competition that would likely stymie any attempt to create such an agency that would operate on a basis acceptable to the spacefaring nations.

In contrast, Alan Boyle, again among others, calls for the passage in the United States of a Space Settlement Prize Act that would make possible private ownership without national sovereignty. Under this proposal, the United States government would "recognize ownership of extraterrestrial territory if a private venture establishes a permanently inhabited settlement on another world."[107] Boyle quotes Simberg who states, "In some sense, it gives the imprimatur of the U.S. government. But, it doesn't make it a sovereignty question. It's a recognition, not an appropriation."

James Dunstan takes exception with a "recognition" proposal. Dunstan states that, "There is no way that the United States could directly recognize land claims in outer space that were made based on use and occupation, as the legislation Rand proposes would do. The 'loophole,' as Rand calls it, simply doesn't exist."[108] Dunstan, who does support private property rights in space, wishes to effect a system that will internalize the externalities of space activities (although he does not use that terminology). Dunstan points to what he calls "the tragedy of the commons" that exists as a result of space debris left in orbit by defunct satellites. Space, he notes, has become a dumping grounds because nations and firms have not been held responsible for their actions as they should have been under the existing Outer Space Treaty had property rights existed. "Launching states retain jurisdiction and control (as well as liability) over their objects, even after they are abandoned."[109] Hence, the rights of salvage that exist on the seas do not exist in space, and the nation causing the pollution does not bear the cost of it's action yet no other nation may recover and possess the abandoned item.

Pointing to the example of the law of the sea, Tanja Masson-Zwann, President of the International Institute of Space Law, states that it is not necessary to create property rights in space for activities to be profitable for developers. The type of licensing that permits the mining of deep-sea resources solves the issue to her way of thinking. The Law of the Sea approach provides a third way of addressing the extra-terrestrial resource issue. To Masson-Zwann, it is not necessary to grant property rights or to prohibit them. The issue of rights to real property can simply be ignored.[110]

In speaking specifically of the planet Mars, David Collins, Lecturer at Law at City University, London, states his concern for the retention of property rights for land as well as for structures put in place on the surface of the planet. He holds that the principle of first possession as practiced previously on Earth would be appropriate. He believes that, "Rooted in natural law, 'first possession' is compatible with Locke's principle of adding value to an object by investing labor in it."[111] "The most efficient mechanism of real property allocation of an un-owned *res nullis* planet Mars would be a limited form of first possession: the allotment of only a portion of land to the first arriving organization, not the entire planet. This bounded first possession is in keeping with the language of the Outer Space Treaty and Moon Treaties that prohibit sovereign claims to the celestial body, which could be interpreted to mean the planetary sphere itself."[112] He expresses concern, however, that "While relatively straightforward and based on historic precedent, the doctrine of first possession may not be the most efficient...because the first nation to land on Mars is not necessarily the one that will use the planet's land in the most productive way."[113] Collins rejects the idea of competitive bidding in advance for land rights and instead proposes a "use tax on Mars" with those desiring to occupy and use the land paying a tax to be distributed to all nations *a la* the "common benefit of all mankind." This tax would be paid in advance of the first arrival, and the rights to the land could be bought and sold on the open market under the assumption that the land would end up the possession of the nation which placed the greatest value on it and had the capability of exercising that valuation. This must be done, he believes, to assure that those landing on Mars would retain rights to their property.

Ownership of real property transported to the Lunar surface and left in place does not change ownership merely because it is in space – the Outer Space Treaty is quite clear about this. What is less clear is ownership of facilities constructed of local materials in space from a body that by treaty can not be owned by a country or company. "With regard to any structure essentially made from locally available resources, there are no clear rules, and it may be valuable to establish clarity on this subject."[114]

Wayne White believes that the spirit and letter of the law as expressed in the Outer Space Treaty may be observed through a form of limited real property rights in space. White cites the homesteading act in the United States whereby land grants were given to those who would occupy, use, and improve the land for a specified period of time. He believes that this system "would provide greater certainty to investors and entities participating in the development and settlement of outer space. Those entities would be able to look to terrestrial property law for legal precedents. National judicial systems would experience similar benefits, as judges could decide cases on the basis of established legal principles."[115] Precisely whose "established legal principles" would form that basis is not specified by White. As Hertzfeld and von der Dunk, note, "There are many different forms of property rights, each defined by a nation's culture, history, and political priorities. Ownership over immovable property is not a self-evident phenomenon, defined by natural law or divine intervention; it is a concept provided for by national laws that elaborate it in their own fashion as to all relevant details."[116] This being the case, "established principles" should be established in advance for this completely new situation.

It goes without saying that it is a high goal to assure conflict-free space. Desiring this, "If peaceful and efficient coexistence in space is to prevail, then the creation and enforcement of property rights is inevitable."[117] Certainly we wish the rule of law to prevail and wish not to repeat the experience of the United States when it opened its western territories to settlement and use. For a period, the law was "honored more in the breach than in the observance." To make certain that peace prevails and to reduce the possibility of conflict, the law must be in place in advance, waiting for the settlers and prospectors when they arrive and not be developed *a posteriori*.

Opposed to Private Property Rights

It is Keefe's contention that private property rights can not exist on the Moon. Hence, her proposal is that the international regime as called for in the Moon Treaty be established and empowered to implement "a system of leases which will benefit those persons occupying outer space as well as those remaining on Earth."[118] Her concept of an "equitable distribution" is that any profits resulting from activities under the supervision of the regime, after all costs have been recouped and there has been a six-week period in which all profits accrued to the investor, would be split "at a rate of 60% for the investor, 40% for the organization [supervising regime]." Keefe provides no basis for her 60/40 split of the profits.

Thomas Gangale believes that it would be impossible to support private property rights in any form including differentiating between the concepts of "granting" and "recognizing." "Property rights exist only if they are granted or recognized by a government and subject to the protection of law. For Congress to pass 'land claim recognition' legislation legalizing private claims of land in space would be an exercise of state sovereignty, and therefore a violation of international law under provisions of the Outer Space Treaty."[119] Gangale, however, does support the right of a government or private entity to remove resources from an extraterrestrial body and to take title to these goods.[120]

As stated previously, Jakhu and Buzdugan's position is that neither a state nor a private entity may claim private property in space.[121]

Finally, the prestigious International Institute of Space Law (IISL) has taken a position against private claims to property on the Moon. The IISL states that "the activities of non-governmental entities (private parties) are national activities" prohibited under the United Nations treaties and that "the prohibition of national appropriation also precludes the application of any national legislation on a territorial basis to validate a 'private claim.'"[122]

Despite the fact that neither the United States nor Russia has signed the Moon Treaty, one feature of the Moon Treaty that is worth expanding upon is the idea that the Moon specifically and individually may be considered the "common heritage" of mankind. For millennia humans have looked at the Moon and, at least to some extent, formed a bond with it. The same cannot be said of the asteroids. "The Moon may be treated differently from the asteroids. Legal issues aside, the Moon is one large body already in orbit around Earth, is well known by all cultures on Earth since the advent of man (and worshipped by some over the eons), and in some ways is environmentally sensitive, e.g., industrial operations could result in a significant atmosphere around the Moon which would degrade its natural state and could interfere with others' scientific and industrial operations as well."[123] The same cannot be said of the asteroids. The estimates of the number of asteroids in orbit around the Sun is in the neighbourhood of 350,000. The estimate of the number of comets is one trillion. It is not likely that an emotional attachment will be felt toward many of these bodies. Therefore, whereas property rights may be prohibited on the Moon, it may be unnecessary to hold this position with regard to the asteroids.

In Actuality, No Decision May be Required

As a contrast to the preceding, some hold that no *a priori* decision is required: the situation will self-sort depending upon the conditions in existence at the time that a decision is required. Klotz, quoting her interview with Michael Gold, attorney for Bigelow Aerospace, notes that "he doubts that the treaty [U.N. Outer Space Treaty] will be a show-stopper for companies, such as Planetary Resources, that are eying natural resources beyond the earth. . . . Legal justification to mine asteroids likely would follow technical capability, he added. 'It is my belief that in the end, capability will trump law.'"[124] In this, Gold forwards the idea that laws regarding resource extraction off Earth will follow a common law rather than a civil law approach. Instead of establishing laws in advance, Gold posits that applicable laws will follow the need for them.

Furthering the idea that no decision regarding lunar resource extraction is required is Anthony Scott. In "Property Rights and Property Wrongs," Scott notes that the Canadian and, similarly, the American experience with homesteading land grants and sales demonstrate what would likely happen on the surface of the Moon. One of the major goals of the American and Canadian legislatures in establishing criteria for the division of the lands on the vast western prairies of the continent was equity. Equal-sized parcels were established and distributed in a standardized manner. "The new disposal arrangements did not last. Homesteading, for example, did create a new generation of free and equal land owners in freehold tenure. But not for long. The parcels were sold, divided, combined, mortgaged, leased, and abandoned, with scant concern about the original vision. Eighty years later the western ladder of land tenure is little different from that in European regions long under traditional common-law property systems."[125] Scott further denies "that government policy has been an economically important definer of standard real property rights by arguing that there is a process at work outside the government sector." "Government is not needed, because a contractarian process is potentially at work to create and recreate exclusive, enforceable property rights." "Government is not needed because property rights, like all social institutions, simply evolve."[126]

Hertzfeld separates the protection of real property from tangible property. His assessment is that "Ownership of real estate on the Moon is neither a necessary nor a sufficient condition for investing in a lunar business with the ability to earn a fair return on that investment. Even terrestrially, profitable businesses often do not own the land or buildings they occupy. The most important concern for private businesses in space is not property rights. It is the ability for a company to make a rate of return on a new investment that is greater than the return it can get from other investments."[127] Thus, if the right to resources extracted from the Moon can be assured, Hertzfeld believes that need for title to the land from which the resources are extracted is moot.

Finally, Collins, in discussing the likely increase in the number and complexity of space launches as launch costs per kilogram decrease, expresses no concern about the establishment of private property laws in advance. "For practical reasons, the need to be commercially attractive will act as a pressure encouraging the evolution of the most efficient legal regime, as happened in the early days of international trade in Europe."[128] During that period, commercial courts, *lex mercatoria* or law merchants, arose through the actions of those who traded to provide the regulation that best suited their needs. No government action was required for this to come about.

Scott, an academic economist, and Gold, a corporate attorney, would posit that a laissez-faire approach to property rights in space would be the appropriate path to follow. As definitions and laws are required, they will evolve to meet the needs at the time in ways that could not be anticipated a quarter of a million miles away and 20 years in advance. Ultimately, both Scott and Gold agree, "the courts replace the legislature in developing new interpretations of standard property rights."[129]

IN SUMMATION

Ownership of real property by a state, organization, or individual is prohibited under the United Nations Outer Space Treaty. This prohibition was repeated and strengthened through the unratified United Nations Moon Treaty. As a signatory to the Outer Space Treaty, the United States and other like entities are bound by the provisions of that treaty. However, the access, removal, and use of natural resources from the Moon is a permissible action. Precedent has been set for this. Both the United States and Russia have removed soil and minerals from the Moon and returned them to Earth. There has been no legal action taken against either nation for doing so. Given that the bases of the legal systems of the two countries differ, the United States being a common law state while Russia is a civil law state, it has been established that materials may be removed from the Moon and, hence, other extraterrestrial bodies with impunity. "Notably, the Soviet Union sold a very small sample of the dirt that it brought back from the Moon, for $500,000. This set a precedent of a government owning and selling a resource extracted from a celestial body in outer space.130 Functional property rights exist.

Supporting this conclusion are the following arguments. Whereas Article II specifically states that the Moon and other bodies are not subject to national appropriation, other articles point in a different direction. Article VI of the Outer Space Treaty states: "States Parties to the Treaty shall bear international responsibility for national activities in outer space, including the moon and other celestial bodies, whether such activities are carried on by governmental agencies or by non-governmental entities, and for assuring that national activities are carried out in conformity with the provisions set forth in the present Treaty. The activities of non-governmental entities in outer space, including the moon and other celestial bodies, shall require authorization and continuing supervision by the appropriate State Party to the Treaty." Thus, property launched into space and located on the surface of the Moon comes under the legal jurisdiction and control of the states party that launched it. Article VII further establishes that the liability for damages caused by an object launched into space is the responsibility of the State Party of the launching organization, subjecting the object to the laws of that jurisdiction. This liability – and subsequent effective property right – is further established by the United Nations Liability Convention (1972). Finally, the United Nations Registration Convention (1975) states that a States Party from whose territory an object is launched is required to register each item and to file such information with the Secretary-General.

Similar to the Outer Space Treaty, Article 12 of the Moon Treaty declares: "States Parties shall retain jurisdiction and control over their personnel, vehicles, equipment, facilities, stations and installations on the moon. The ownership of space vehicles, equipment, facilities, stations and installations shall not be affected by their presence on the moon." Hence, the laws of the launching state apply to an object or installation placed on the surface of the Moon. While Article 11 states that "Neither the surface nor the subsurface of the Moon, nor any part thereof or natural resources in place, shall become property of any State, international intergovernmental or non- governmental organization, national organization or non-governmental entity or of any natural person," the following article of the treaty undercuts the strength of that argument. To repeat; an object launched from State A and located on the lunar surface remains under the "jurisdiction and control" of that state and the state has liability for any damages. Given that the owned-object rests on the surface and given that both common law and civil law recognize that possession, i.e. occupation and use, constitute effective ownership, a party may work the region surrounding its facility, something permissible under the Moon Treaty. Should a party be aggrieved and claim damages by another party, the suit would be tried in the country from which the object was launched and according to the laws of that state. Given that within the United States commercial activities are considered to be private matters, cases would be tried in U. S. civil courts. Courts in the United States tend not to be guided by international treaties but by local laws, so United States commercial law would apply. The matter would, therefore, be decided in a U.S. court, and, to the extent that the court considered the space treaties, "courts in common law countries will likely interpret treaty language in light of public policy."[131] Appeal to the International Court of Justice is possible, of course, but such a case can only be brought if both nations agree to submit to the decisions of the court. The United States Department of State, which has fundamental differences with the Moon Treaty, is not likely to do this, so there the litigation would end.

There is precedent for the application of functional property rights – the management of the Spitzbergen Islands off the coast of Norway. Ederington provides an insightful review of the issue. Because of its location and presumed lack of natural resources, originally "the states generally recognized the status of the islands as *terra nullis*." When coal deposits were discovered, it became necessary to establish a mechanism to settle conflicting claims in the absence of territorial sovereignty. "The solution was a 1920 treaty in which the nine countries with an interest in Spitzbergen recognized Norway's sovereignty over the archipelago," but in return granted the existing occupiers of the land with "equal freedom of access, commerce, mining, and fishing."[132] Those without sovereignty were granted rights to the resources of the territory. This approach has significant merit for application to outer space territories.

There is another distinct possibility that must be anticipated. In the longer run, individuals will travel to, occupy, and colonize an asteroid or other body. Separated by both time and distance from their home planet, it is not unlikely that the colonizers will declare themselves to be a new nation, independent of the nation and planet from which they originated. When this occurs, the new political entity will no doubt declare itself as a non-signatory of the Earth's Outer Space Treaty to be outside the treaty's jurisdiction. The new nation will proceed to develop its own regulations and laws, including laws regarding the rights to the ownership of private property. Certainly from the many examples of colonies on Earth following this path, there is ample reason to anticipate that this will become the norm rather than the exception. While it has not much been discussed by the world's space agencies, the possibility exists of intentional one-way trips into space – much the same as was the case when people left their home countries in Europe to travel to North America. A return trip was not expected, and settlers relied on a combination of what they brought with them along with what resources might be available at the new location. These space settlers who had severed their relationship with Earth and with no intention to return would quickly establish a self-government and develop a legal system meeting their own unique needs.

While the Outer Space Treaty prohibits the establishment of sovereignty over extraterrestrial lands and could not, therefore, confer title, it could recognize titles to such land claims. It is highly likely that any permanent settlement is likely to contain people of multinational backgrounds. Multinational ventures have become the norm already. "Russia currently launches American astronauts from a base in Kazakhstan, India places satellites in orbit for France and Japan, and the European Space Agency, an international organization with members from 19 states, launches its spacecraft from Kourou in French Guiana."[133] In addition, as with the ESA, the International Space Station is international by design. The same will likely be the case for a settlement which will maintain its own registration of property boundaries regardless of what any state back on Earth might wish to be the case.

CONCLUSION

The issue of private property need not be a barrier to the exploration, development, or use of extraterrestrial bodies. The treaty-based legal framework in existence does not prohibit states or other organizations from accessing and using the resources it needs for its own purposes. The door is open to any and all who would enter.

About the Author –
Benjamin G. Davis, Ph.D., LL.M., serves as Vice President for Legal and Academic Affairs at Dulles University following his retirement from the position of Chancellor to Chancellor Emeritus at the University of North America. Previously he served as a design engineer for General Motors Corporation at the Warren, Michigan, Tech Center and as a research economist in the Office of the President in Washington, DC. He holds ten earned degrees including two doctorates. Among his postgraduate degrees are an M.S. with Distinction in Space Studies, an M.B.A., and an LL.M. with Distinction in Commercial Law with a focus on Space Law. He has been an invited speaker at the 58th International Astronautical Congress in Hyderabad, India; at the 25th International Conference of the International Council for Innovation in Higher Education in Beijing, China; and at the 2005 meeting of the International Social Science Association in Orlando, Florida. He has written publications on property rights in outer space for The SciTech Lawyer journal of the American Bar Association; on manpower adjustments for the United Nations; and on predicting labor mobility for the Operations Research Society of America. His two text books are published by Pearson-PrenticeHall. He resides in Columbia, Maryland, USA.

ENDNOTES

1. Robert A. Heinlein. (2001). *The Man Who Sold the Moon.* Baen Books: Wake Forest, North Carolina.

2. John Cobb Cooper. (1951). " High Altitude Flight and National Sovereignty." *International Law Quarterly* 4 (1951: 411).

3. Ram Ramachandran and Anita Gale. (2010). "Space Colonization." *Aerospace America.* December 2010. p. 64.

4. Rand Simberg. (2012). *Homesteading the Final Frontier; A Practical Proposal for Securing Property Rights in Space.* Issue Analysis 2012 No. 2. Washington: Competitive Enterprise Institute. p. 3.

5. *ibid.* p. 4.

6. *ibid.*

7. Feder, *op. cit.* 146.

8. Dalton, *op. cit.* p. 4.

9. Scott J. Shackelford. (2009). *The Tragedy of the Common Heritage of Mankind.* (May 19, 2009). Stanford Environmental Law Journal, Vol. 27, 2008; Stanford Public Law Working Paper No. 1407332. Available at SSRN: http://ssrn.com/abstract=1407332

10. *Sparks Fly as NASA Pushes the Limits of 3-D Printing Technology.* (2014). Released 14-233 dated August 28, 2014. Accessed from http://www.nasa.gov/press/2014/august/sparks-fly-as-nasa-pushes-the-limits-of-3-d-printing-technology.

11. *3-D Printing: Food in Space.* (2013) Released May 23, 2013; updated July 28, 2013. Accessed from http://www.nasa.gov/directorates/spacetech/home/feature_3d_food.html.

12. Sanderson, Katharine. (2014). *Water-splitter Could Make Hydrogen Fuel on Mars.* From *Science*, DOI: 10.1126/science.1257443 at Astronautical News.

13. Shackelford (2009) *op. cit.*

14. *President Bush Offers New Vision for NASA.* (2004). Retrieved on June 11, 2011 at http://www.nasa.gov/missions/solarsystem/bush_vision.html.

15. E. N. Slyuta, A. M. Abdrakhimov, and E. M. Galimov. (2007). *The Estimation of Helium-3 Probable Reserves in Lunar Regolith.* 38th Lunar and Planetary Science Conference. pp. 2175. http://www.lpi.usra.edu/meetings/lpsc2007/pdf/2175.pdf.

16. Simberg, *op. cit.* p. 7.

17. For a good review of the magnitude of the resources available, see *Mining the Sky* by John S. Lewis, *op. cit.*

18. John S. Lewis. (1997). *Mining the Sky: Untold Riches from Asteroids, Comets, and Planets.* Reading, Massachusetts: Helix Books/Perseus Books. p. 2.

19. M. F. McKay, D. S. McKay, and M. B. Duke. (1992). *Space Resources: Scenarios*. NASA SP-509, Vol. 1. Houston, Texas: NASA – Lyndon B. Johnson Space Center.

20. Lewis, *op. cit.*

21. Daniel R. Adamo, Jon D. Giorgini, Paul A. Abell, and Rob R. Landis. (2010). "Asteroid Destinations Accessible for Human Exploration: A Preliminary Survey in Mid-2009." *Journal of Spacecraft and Rockets*. Volume 47, Number 6. November-December 2010. p. 1002.

22. Paul Collier. (2008). "Law Codes for the Resource Curse." *Yale Human Rights and Development Law Journa*l. 11. Annual 2008. p9ff.

23. Martin W. Lo and Min-Ku Chung. (2002). *Lunar Sample Return via the Interplanetary Superhighway.* Jet Propulsion Laboratory. Accessed on April 29, 2012, from www.gg.caltech.edu/~mwl/publications/papers/LSRvialPS.pdf

24. Benson, Julian. (n.d.). *Interplanetary Superhighway: Finding the Route Mankind Will Take to Colonise the Solar System.* Accessed at https://medium.com/@jbenson/interplanetary-superhighway-8e3e734346ed

25. Virgiliu Pop. (2001). *Lunar Real Estate: Buyer Beware!* Paper presented at the First Convention of Lunar Explorers in Paris in March 2001. Pop quotes Stair, J. Viscount of. (1981*). The Institutions of the Law of Scotland, 1693*: II, I, 18. Edited by Walker, D.M. The University Presses of Edinburgh and Glasgow. p. 7.

26. Gershon Feder and David Feeny. (1991). "Land Tenure and Property Rights: Theory and Implications for Development Policy." *The World Bank Economic Review*, Vol. 5, No. 1. January 1991. p. 137.

27. *ibid.* p. 138

28. Kali N. Murray. (2006). "Of Gardens and Streets: A Differentiated Model of Property in International and National Space Law." *The Journal of Space Law*, Volume 32, Number 2, Winter 2006. p. 361-382. Retrieved on December 2, 2009, from http://www.spacelaw.olemiss.edu/jsl/back_issues/jsl %2032-2.pdf

29. *ibid.*

30. Patrick Lin. (2006). "Space Ethics: Look Before Taking Another Leap for Mankind." *Astropolitics Journal.* Vol. 4. No. 3. Winter 2006. p. 293.

31. Glenn Harland Reynolds. (2008). "Who Owns the Moon? The Case for Lunar Property Rights." Accessed on April 30, 2012, from http://www.popularmechanics.com/science/space/moon-mars/4264325. p. 1. (written in the popular press, but by a professor of law at the University of Tennessee)

32. Bryan A. Garner. (Editor in Chief). (2009). *Black's Law Dictionary, Ninth Edition.* St. Paul, Minnesota: Thompson Reuters.

33. Bryan A. Garner. (Editor in Chief). (2009). *Black's Law Dictionary, Ninth Edition.* St. Paul, Minnesota: Thompson Reuters. p. 1246.

34. Joel D. Scheraga, Joel D. (1997). "Establishing Property Rights in Outer Space." *Cato Journal.* Vol. 6, No. 3. Winter 1987. pp. 890.

35. Patrick Lin. (2006). "Space Ethics: Look Before Taking Another Leap for Mankin." *Astropolitics Journal.* Vol. 4. No. 3. Winter 2006. p. 294.

36. An interesting exercise is to apply the Drake equation, an equation used to estimate the number of technologically advanced civilizations in the Milky Way Galaxy. That number is 10,000 communicative civilizations, some of which must be more advanced than that of the Earth. As Virgiliu Pop states, "Should Drake's equation be true, Earth itself has been claimed, parceled, and sold by Dennis Hop-like aliens. It may be that, someday, one will have to defend one's property against an extraterrestrial armed with a handsomely framed Terrestrial Deed." See Virgiliu Pop. (2001). *Lunar Real Estate: Buyer Beware!* Paper presented at the First Convention of Lunar Explorers in Paris in March 2001. Pop quotes James, Viscount of Stair. (1981). *The Institutions of the Law of Scotland, 1693*: II, I, 18. Edited by Walker, D.M. The University Presses of Edinburgh and Glasgow. p. 7.

37. Blake Gilson. (2011). "Defending Your Client's Property Rights in Space: A Practical Guide for the Lunar Litigator." *Fordham Law Review*. December 2011. Vol. 80, No. 3. p. 1383.

38. Nathan C. Goldman. (2002). "Space Law." In Sadeh, Eligar. *Space Politics and Policy: An Evolutionary Perspective*. Dordrecht, The Netherlands: Kluwer Academic Publishers.

39. Jonty Kasku-Jackson and Elizabeth Waldrop. (2009). "Understanding Space Law: Legal Framework for Space." In Coletta, Damon and Pilch, Frances T., eds. *Space and Defense Policy*. London: Routledge. p. 88.

40. Nathan C. Goldman. *op. cit.* p. 165.

41. Nathan C. Goldman. *op. cit.* p. 164

42. *ibid.*

43. Nathan C. Goldman. *op. cit.* p. 169.

44. Nathan C. Goldman. *op. cit.* p. 164.

45. *ibid.*

46. See: *United Nations treaties and principles on outer space and related General Assembly resolutions: Addendum*. "Status of international agreements relating to activities in outer space as at 1 January 2009." Ref.: Sales No. E.08.1.10 – ST/SPACE/11/Rev/Add 2.

47. For an interesting compare-and-contrast essay regarding international differences, see Laura Wen-yu and Francis S. L. Wang's article "Expatriates Face China's Criminal Law System," in the Spring 2011 issue of the American Bar Association's *International Law News.*

48. Michael Listner. (2012). "The Moon Treaty: It Isn't Dead Yet." *The Space Review*. March 19, 2012. Accessed on March 20, 2012 from http://www.thespacereview.com/article/2047/1. p. 1-2.

49 See National Science Foundation, Office of Polar Programs. (2010). The Antarctic Treaty. Available at http://www.nsf.gov/od/opp/antarct/anttrty.jsp.

50. *ibid.*

51. Heidi Keefe. (1995). "Making the Final Frontier Feasible: A Critical Look at the Current Body of Outer Space Law." *Santa Clara Computer and High Technology Law Journal.* July 1995. Accessed on April 21, 2011, at http://www.lexisnexis.com.exproxy2.apus.edu/lnacui2api/frame.do?reloadEntirePage=true... p. 353

52. *ibid.*, p. 353-354.

53. Wayne N. White, Jr. (1999). *Implications of a Proposal for Real Property Rights in Outer Space.* Retrieved on August 21, 2009, from http://www.spacefuture.com/archive/implications_of_a_proposal_for_real_property_rights...

54. Stephen Gorove. (1968). "Interpreting Article II of the Outer Space Treaty." *Fordham Law Review*. Vol. 37, 1969. Retrieved on August 21, 2009. p. 351.

55. Hassanabadi, Babak Shakouri. (2014). "Complications of the Definition of 'Launching State.'" *The Space Review.* Posted on September 2, 2014, at http://www.thespacereview.com/article/2588/1

56. Ezra J. Reinstein. (1999). "Owning Outer Space." *Northwestern Journal of Law & Business*. Vol. 20. No. 1. p. 71.

57. Allan Wasser and Douglas Jobes. (2008). "Space Settlements, Property Rights, and International Law: Could A Lunar Settlement Claim the Real Estate It Needs to Survive?" *Journal of Air Law and Commerce*. Volume 72, Number 1. Winter 2008. pp. 42.

58. Douglas O. Jobes. (2005). "Lunar Land Claims Recognition: Designing the Ultimate Incentive for Space infrastructure Development." *Space Times*. May/June 2005. p. 4.

59. Thomas Gangale and Marilyn Dudley-Rowley. (2005). *To Build Bifrost: Developing Space Property Rights and Infrastructure.* AIAA-2005-6762. Paper presented at the Space 2005 Conference of the American Institute of Aeronautics and Astronautics, 30 August - 1 September 2005, Long Beach, California. p. 5.

60. Antonella Bini. (2008). "The Moon Agreement: Its Effectiveness in the 21st Century." European Space Policy Institute. *ESPI Perspectives*, No. 15. October 2008. p. 4.

61. Joanne Irene Gabrinowicz. (2010). "One Half Century and Counting: The Evolution of U.S. National Space law and Three Long-Term Emerging Issues." *Harvard Law & Policy Review*. Vol. 4. p. 422.

62. See National Science Foundation, Office of Polar Programs. (2010). *The Antarctic Treaty.* Accessed on 9 November 2011 at http://www.nsf.gov/od/opp/antarct/anttrty.jsp

63. Secretariat of the Antarctic Treaty. (2011). *The Antarctic Treaty*. http://www.ats.aq/index_e.htm.

64. Protocol on Environmental Protection to the Antarctic Treaty.

65. *British Antarctic Survey.* (2007). Accessed on November 9, 2011, from http://www.antarctica.ac.uk/about_antarctica/geopolitical/treaty/faq.php.

66. Alex Gimarc. (2005). "History and Frontiers – What Works, What Doesn't." in Tumlinson, Rick. (ed.). *Return to the Moon*. Burlington, Ontario, Canada: Collector's Guide Publishing/Apogee Books. p. 39.

67. Mark Prado. (2011). *Should the Vagueness of International Law Significantly Discourage Private Ventures Today?* Accessed on May 17, 2011, from www.permanent.com/legal-international-laws.html. p. 3.

68. Berin Szoka and James Dunstan. (2012). "How the U.S. Can Lead the Way to Extraterrestrial Land Deals." *Wired Science*. Accessed on April 26, 2012, from http://www.wired.com/wiredscience/2012/04/opinion-space-property-rights. p. 4.

69. Brian D. Hufford. (1983). "Ideological Rigidity vs. Political Reality: A Critique of Reagan's Policy on the Law of the Sea." From the *Yale Law and Policy Review*, excerpted in Reynolds, Glenn H.

and Merges, Robert P. (1997). *Outer Space: Problems of Law and Policy. Second Edition*. Boulder, Colorado: Westview Press. p. 133.

70. Glenn Harland Reynolds. (2008). "Who Owns the Moon? The Case for Lunar Property Rights." Accessed on April 30, 2012, from http://www.popularmechanics.com/science/space/moon-mars/4264325.

71 *ibid.*

72. Virgiliu Pop. (2009). *Who Owns the Moon? Extraterrestrial Aspects of Land and Mineral Resources Ownership*. Springer Science+Business Media B.V. p. 1-21.

73. Virgiliu Pop. (2006). *Unreal Estate: The Men Who Sold the Moon*. Virgiliu.com Press.

74. Virgiliu Pop. (2001). *Lunar Real Estate: Buyer Beware!* Paper presented at the First Convention of Lunar Explorers in Paris in March 2001. Pop quotes James, Viscount of Stair. (1981*). The Institutions of the Law of Scotland, 1693*: II, I, 18. Edited by Walker, D.M. The University Presses of Edinburgh and Glasgow. p. 4.

75. Rand Simberg, Rand. (2012). *Homesteading the Final Frontier; A Practical Proposal for Securing Property Rights in Space*. Issue Analysis 2012 No. 2. Washington: Competitive Enterprise Institute. p. 13.

76. *ibid.*

77. Douglas O. Jobes. (2005). "Lunar Land Claims Recognition: Designing the Ultimate Incentive for Space infrastructure Development." *Space Times*. May/June 2005. p. 7.

78. Stephen Gorove. (1996). "The Growth of Space Law through the Cases." *Journal of Space Law*. Vol. 24. No. 1. p. 19.

79. ibid., p. 3.

80. Henry R. Hertzfield. (2007). "The Moon is a Land without Sovereignty: Will it be a Business-Friendly Environment?" *High Frontier: The Journal for Space & Missile Professionals*. Vol. 3. No. 2. March 2007. p, 43-45.

81. Robert Kelly. (2004). "Nemitz v. United States, A Case of First Impression: Rights and Space Law before the Federal Courts of the United States." *Journal of Space Law*. Vol. 30. No. 2. Winter 2004. p. 299-300.

82. ibid., p. 301.

83. United States Cong. House. 113th Congress 1st Session. H.R. 2617. Apollo Lunar Landing legacy Act. [introduced in the U.S. House of Representatives; 8 July 2013.

84. David, Leonard. (2013). "Protection of Apollo Moon Landing Sites Sparks Controversy." *Space Insider*, July 26, 2013. Accessed at http://www.space.com/22131-moon-landing-sites-bill-controversy.html

85. United States Cong. House. 113th Congress 2nd Session. H.R. 5063. *American Space technology for Exploring Resource Opportunities in Deep Space Act.* [introduced in the U.S. House of Representatives; 10 July 2014.

86. Smith, Marcia S. (2014). "Posey, Kilmer Introduce ASTEROIDS Act to Grant Property Rights to Asteroid Resources." *Space Policy Online*. Posted July 10, 2014, updates July 17, 2014. Accessed at http://www.spacepolicyonline.com/news/posey-kilmer-introduce-asteroids-act-to-grant-property-rights-to-asteroid-resources

87. Smith, Marcia S. (2014) "Gabrinowicz Warns ASTEROIDS Act Needs More Work." *Space Policy Online.* Posted September 11, 2014 at http://www.spacepolicyonline.com/news/gabrynowicz-warns-asteroids-act-needs-more-work.

88. Stotler, Charles. (2014). "The ASTEROIDS Act and hearing: some observations on international obligations." *The Space Review.* Posted on *September 22, 2014, at* http://www.thespacereview.com/article/2604/1

89. Garrett Hardin. (1968). "The Tragedy of the Commons." *Science.* 162: 1243-1248. Retrieved on December 4, 2009, from http://dieoff.org/page95.htm.

90. Michael Bourbonniere and Ricky J. Lee. (2008). "Legality of the Deployment of Conventional Weapons in Earth Orbit: Balancing Space Law and the Law of Armed Conflict." *The European Journal of International Law.* Vol. 18. No. 5. pp. 873-901. Accessed on March 27, 2012, from *www.ejil.org/pdfs/18/5/251.pdf. p. 884.*

91. Jijo George Cherian and Job Abraham. (2007). "Concept of Private Property in Space – An Analysis." *Journal of International Commercial Law and Technology.* Vol. 2, Issue 4. p. 18.

92. *ibid.* p. 19.

93. Mark Van Vugt. (2009). "Triumph of the Commons." *New Scientist.* August 22, 2009, Vol. 203, Issue 2722, p. 40-43. Retrieved on December 4, 2009, from http://web.ebscohust.com.ezproxy.apollolibrary.com/ehost/delivery?vid=5. p. 21.

94. Allan Wasser and Douglas Jobes. (2008). "Space Settlements, Property Rights, and International Law: Could A Lunar Settlement Claim the Real Estate It Needs to Survive?" *Journal of Air Law and Commerce.* Volume 72, Number 1. Winter 2008. pp. 44-46.

95. Glenn Reynolds and Dave Kopel. (2002). "The New Frontier: Preparing the Law for Settling on Mars. *National Review Online.* June 4, 2002. Accessible at http://nationalreview.com/kopel/kopel060402.asp

96. Joanne Irene Gabrinowicz. (2010). "One Half Century and Counting: The Evolution of U.S. National Space law and Three Long-Term Emerging Issues." *Harvard Law & Policy Review.* Vol. 4. p. 405-426.

97. Stephen Gorove. (1968). "Interpreting Article II of the Outer Space Treaty." Fordham *Law Review.* Vol. 37, 1969. Retrieved on August 21, 2009. p. 349-354.

98. Pat Dasch, Michael Martin-Smith, and Anne Pierce. (1999). Presentation at the 50th International Astronautical Congress: Conference on Space Property Rights: Next Steps. (October 4-8, 1999).

99. Wayne N. White, Jr. (2000). *Implications of a Proposal for Real Property Rights in Outer Space.* Paper presented at the 42nd Colloquium on the Law and Outer Space, IISL.

100. Stephen Gorove. *op. cit.*

101. Ram Jakhu and Maria Buzdugan. (2006). *The Role of Private Actors: Commercial Development of the Outer Space Resources, Including Those of the Moon and other Celestial Bodies: Economic and Legal Implications.* Paper presented at the McGill University Institute of Air and Space Law Conference, June 28-30, 2006, in Montreal. Retrieved August 19, 2009, from http://www.mcgill.ca/iasl/publications/proceedings/

102. Stephen Gorove, *op cit.*

103. Ram Jaku and Maria Buzdugan. *op cit.*

104. Space Settlement Institute. (2012). Frequently Asked Questions – FAQ #3. Accessible at http://www.space-settlement-institute.org/faqs.html.

105. *ibid.*

106. Jeff Brooks. (2006). "The International Agency for the Development of Mars." *The Space Review*. December 11, 2006. Accessed on May 3, 2012, from http://www,thespacereview.com/article/763/1. p. 4-5.

107. Alan Boyle. (2012). *Could Legal "Loophole" Lead to Land Claims on Other Worlds*? Accessed on April 12, 2012, from http://cosmiclog.msnbc.msn.com_news/2012/04/09/11105409-could-legal-loophole-lead-t... p. 1.

108. James E. Dunstan. (2012). Comments at the Competitive Enterprise Institute Hill Briefing, April 5, 2012, on Securing Property Rights in Space. p. 2.

109. *ibid.* p. 3.

110. Tanja Masson-Zwaan quoted in Alan Boyle. (2012) *Private Property in Outer Space: The Other Side of the Argument*. Accessed on April 13, 2012, from http://cosmiclog.msnbc.msn.com/_news/2012/04/13/11189692-private-property-in-outer-space-the-other-side-of-the-argument?lite

111. Patrick Collins. (2008). "Efficient Allocation of Real Property Rights on the Planet Mars." *Boston University Journal of Science and Technology Law*. Vol. 14, No. 201. p. 212.

112. *ibid.* p. 215.

113. *ibid.* p. 212.

114. Henry R. Hertzfeld and Frans von der Dunk. (2005). "Bringing Space Law into the Commercial World: Property Rights without Sovereignty." p. 83. Accessed on August 21, 2009, from *Chicago Journal of International Law*. Summer 2005. 6 Chi. J. Int'l. L 81. Also available at *Space and Telecommunications Law Program Faculty Publications.* Accessed on April 22, 2011, at http://digitalcommons.unl.edu/spacelaw/15.

115. Wayne N. White, Jr. (2000). *Implications of a Proposal for Real Property Rights in Outer Space*. Paper presented at the 42nd Colloquium on the Law and Outer Space, IISL. p. 4.

116. Henry R. Hertzfeld and Frans von der Dunk. Op. cit. 94.

117. Nathan C. Goldman, op. cit. p. 164

118. Heidi Keefe. (1995). "Making the Final Frontier Feasible: A Critical Look at the Current Body of Outer Space Law." *Santa Clara Computer and High Technology Law Journal.* July 1995. p. 367ff. Accessed on April 21, 2011, at http://www.lexisnexis.com.exproxy2.apus.edu/lnacui2api/frame.do?reloadEntirePage=true.....

119. Gangale, Thomas and Dudley-Rowley, Marilyn. (2005). *To Build Bifrost: Developing Space Property Rights and Infrastructure.* AIAA-2005-6762. Paper presented at the Space 2005 Conference of the American Institute of Aeronautics and Astronautics, 30 August - 1 September 2005, Long Beach, California. p. 1.

120. Thomas Gangale. (2009). *The Development of Outer Space: Sovereignty and Property Rights in International Space Law*. Santa Barbara, California: Praeger: an imprint of ABC-CLIO. p. 5.

121. Ram Jaku and Maria Buzdugan. *op cit.*

122. International Institute of Space Law. (2009). *Statement of the Board of Directors of the International Institute of Space Law (IISL)* March 22, 2009. Accessed on July 6, 2011, from http://www.iislweb.org/publications.html

123. Mark Prado. (2011). *Should the Vagueness of International Law Significantly Discourage Private Ventures Today?* Accessed on May 17, 2011, from www.permanent.com/legal-international-laws.html. p. 5.

124. Irene Klotz. (2012). *Who Owns the Asteroids?* Accessed on April 30, 2012, from http://news./discovery.com/space/asteroid-mining-space-treaty-ownership-120427.html#mkcpgn-emnws1. p. 2.

125. Anthony Scott. (1983). "Property Rights and Property Wrongs." *Canadian Journal of Economics*. Vol. 16, No. 4. November 1983. p. 559-560.

126. *ibid.*, p. 560.

127. Henry R. Hertzfield. (2007). "The Moon is a Land without Sovereignty: Will it be a Business-Friendly Environment?" *High Frontier: The Journal for Space & Missile Professionals*. Vol. 3. No. 2. March 2007. p, 43.

128. Patrick Collins. (1992). *Implications of Reduced Launch Costs for Commercial Space Law*. p. 16-17. Retrieved on August 13, 2009 from http://www.spacefuture.com/archive/implications_of_reduced_launch_costs_for_commerci...

129. Scott. *op cit.* p. 563.

130. Prado. *op cit.* p. 2.

131 Blake Gilson. op cit. p. 1383.

132. L. Benjamin Ederington. (1997). "Property as a Natural Institution: The Separation of Property from Sovereignty in International Law." *American University International Law Review*. 13. No. 2. (1997). p. 284-288.

133. Benjamin G. Davis. (2014). "Property Rights in Space: Assuring Economic Development." In the American Bar Association journal *The SciTech Lawyer*, Fall 2014. p. 22.

SELECTED BIBLIOGRAPHY

Adamo, Daniel R., Giorgini, Jon D., Abell, Paul A., and Landis, Rob R. (2010). "Asteroid Destinations Accessible for Human Exploration: A Preliminary Survey in Mid-2009." *Journal of Spacecraft and Rockets*. Volume 47, Number 6. November-December 2010. pp. 994-1002.

Asteroid Resources: The Possibility of Volatile Rich NEOs. (n.d.) Retrieved on June 13, 2011, at http://clowder.net/hop/railroad/asteroidresources.html.

Benson, Julian. (n.d.). *Interplanetary Superhighway: Finding the Route Mankind Will Take to Colonise the Solar System.* Accessed at https://medium.com/@jbenson/interplanetary superhighway-8e3e734346ed

Bini, Antonella. (2008). "The Moon Agreement: Its Effectiveness in the 21st Century." European Space Policy Institute. *ESPI Perspectives*, No. 15. October 2008.

Bourbonniere, Michel, and Lee, Ricky J. Lee. (2008). "Legality of the Deployment of Conventional Weapons in Earth Orbit: Balancing Space Law and the Law of Armed Conflict." *The European Journal of International Law.* Vol. 18. No. 5. pp. 873-901. Accessed on March 27, 2012, from *www.ejil.org/pdfs/18/5/251.pdf.*

Boyle, Alan. (2012). *Could Legal "Loophole" Lead to Land Claims on Other Worlds*? Accessed on April 12, 2012, from http://cosmiclog.msnbc.msn.com_news/2012/04/09/11105409-could-legal-loophole-lead-t...

Boyle, Alan. (2012) *Private Property in Outer Space: The Other Side of the Argument*. Accessed on April 13, 2012, from http://cosmiclog.msnbc.msn.com/_news/2012/04/13/11189692-private-property-in-outer-space-the-other-side-of-the-argument?lite

Brittingham, Byron C. (2010). "Does the World Really Need New Space Law?" *Oregon Review of International Law*. Vol. 12, 31. Pp. 31-54.

Brooks, Jeff. (2006). "The International Agency for the Development of Mars." *The Space Review*. December 11, 2006. Accessed on May 3, 2012, from http://www,thespacereview.com/article/763/1.

Bush, George. W. (2004). *President Bush Offers New Vision for NASA*. Accessed on June 11, 2011, from http://www.nasa.gov/missions/solarsystem/bush_vision.html

Cherian, Jijo George, and Abraham, Job. (2007). "Concept of Private Property in Space – An Analysis." *Journal of International Commercial Law and Technology*. Vol. 2, Issue 4.

Collier, Paul. (2008). "Law Codes for the Resource Curse." *Yale Human Rights and Development Law Journa*l. 11. Annual 2008. p9ff.

Collins, Patrick. (2008). "Efficient Allocation of Real Property Rights on the Planet Mars." B*oston University Journal of Science and Technology Law*. Vol. 14, No. 201. p. 201-219.

Collins, Patrick. (1992). *Implications of Reduced Launch Costs for Commercial Space Law.* Retrieved on August 13, 2009 from http://www.spacefuture.com/archive/implications_of_reduced_launch_costs_for_commerci...

Collins, David. (2008). *Efficient Allocation of Real Property Rights on the Planet Mars.* Retrieved on December 1, 2009, from www.bu.edu/law/central/JD/organizations/journals/volume142/documents/collins.pdf

Cooper, John Cobb. (1951). "High Altitude Flight and National Sovereignty." *International Law Quarterly* 4 (1951: 411).

Dalton, Taylor R. (2010). "Developing the Final Frontier: Defining Private Property Rights on Celestial Bodies for the Benefit of All Mankind." *Cornell Law School Graduate Papers.* Paper 25. Accessed on April 22, 2011, at http://scholarship.law.cornel.edu/lps_papers/25.

Danilenko, Gennady. (n.d.). "Outer Space and the Multilateral Treaty-Making Process." An expanded version of a paper published in 5 *Space Policy*, 321-329 (1989). *Berkeley Technology Law Journal on-line.*

Dasch, Pat, Martin-Smith, Michael, and Pierce, Anne. (1999). Presentation at the 50th International Astronautical Congress: Conference on Space Property Rights: Next Steps. (October 4-8, 1999).

Davis, Benjamin G. (2014). "Property Rights in Space: Assuring Economic Development." In the American Bar Association journal *The SciTech Lawyer*, Fall 2014.

Dinkin, Sam. (2004). *Property Rights and Space Commercialization.* Retrieved on August 21, 2009, from http://www.thespacereview.com/article/141/1

Dunstan, James E. (2012). Comments at the Competitive Enterprise Institute Hill Briefing, April 5, 2012, on Securing Property Rights in Space.

Ederington, L. Benjamin. (1997). "Property as a Natural Institution: The Separation of Property from Sovereignty in International Law." *American University International Law Review.* 13. No. 2. (1997). p. 263-331.

Efrati, Amir. (2012). "A Quixotic Quest to Mine Asteroids." New York *Times.* April 21/22, 2012.

ESA to Push Real-Life Benefits of Space Program Investment. Accessed on May 8, 2012, from http://www.redorbit.com.80/news/space/1112529026/esa-to-push-real-life-benefits-of-space-program-investment/. The ESA website for additional information is http://www.esa.int/esaCP/index.html.

Feder, Gershon and Feeny, David. (1991). "Land Tenure and Property Rights: Theory and Implications for Development Policy." *The World Bank Economic Review*, Vol. 5, No. 1. January 1991. pp. 135-153.

Filiato, Anthony R. (1986). "The Commercial Space launch Act: America's Response to the Moon Treaty?" *Fordham International Law Journal.* Volume 10, Issue 4. pp. 763-781.

Gabrinowicz, Joanne Irene. (2010). "One Half Century and Counting: The Evolution of U.S. National Space law and Three Long-Term Emerging Issues." *Harvard Law & Policy Review.* Vol. 4. p. 405-426.

Gabrinowicz, Joanne Irene. (2005). "The International Space Treaty Regime in the Globalization Era." *Ad Astra.* Fall 2005. p. 30ff.

Gangale, Thomas. (2008). *Common Heritage in Magnificent Desolation.* Paper presented at the January 7-10, 2008, meetings of the American Institute of Aeronautics and Astronautics.

Gangale, Thomas. (2009). *The Development of Outer Space: Sovereignty and Property Rights in International Space Law.* Santa Barbara, California: Praeger: an imprint of ABC-CLIO.

Gangale, Thomas and Dudley-Rowley, Marilyn. (2005). *To Build Bifrost: Developing Space Property Rights and Infrastructure.* AIAA-2005-6762. Paper presented at the Space 2005 Conference of the American Institute of Aeronautics and Astronautics, 30 August - 1 September 2005, Long Beach, California.

Garner, Bryan A. (Editor in Chief). (2009). *Black's Law Dictionary, Ninth Edition.* St. Paul, Minnesota: Thompson Reuters.

Gimarc, Alex. (2005). "History and Frontiers – What Works, What Doesn't." in Tumlinson, Rick. ed. *Return to the Moon.* Burlington, Ontario, Canada: Collector's Guide Publishing/Apogee Books. p. 35-42.

Gilson, Blake. (2011). "Defending Your Client's Property Rights in Space: A Practical Guide for the Lunar Litigator." *Fordham Law Review*. December 2011. Vol. 80, No. 3. p. 1367-1405.

Goldman, Nathan C. (2002). "Space Law." In Sadeh, Eligar. *Space Politics and Policy: An Evolutionary Perspective*. Dordrecht, The Netherlands: Kluwer Academic Publishers.

Gorove, Stephen. (1982). "Implications of International Space Law for Private Enterprise" (1982) 7 *Annals of Air and Space Law*, No 7. p. 321.

Gorove, Stephen. (1968). "Interpreting Article II of the Outer Space Treaty." *Fordham Law Review*. Vol. 37, 1969. Retrieved on August 21, 2009. p. 349-354.

Gorove, Steven. (1996). "The Growth of Space Law through the Cases." *Journal of Space Law*. Vol. 24. No. 1. p. 1-21.

Governments Worldwide Invest a Record $68 Billion in Space Programs. (2010). Document prepared by Euroconsult and reported at www.thespacereview.com/article/1676/1 and retrieved on June 9, 2011.

Hardin, Garrett. (1968). "The Tragedy of the Commons." *Science*. 162: 1243-1248. Retrieved on December 4, 2009, from http://dieoff.org/page95.htm.

Heinlein, Robert A. (2001). *The Man Who Sold the Moon*. Baen Books: Wake Forest, North Carolina.

Hertzfield, Henry R. (2007). "The Moon is a Land without Sovereignty: Will it be a Business-Friendly Environment?" *High Frontier: The Journal for Space & Missile Professionals*. Vol. 3. No. 2. March 2007. p, 43-45.

Hertzfeld, Henry R. and von der Dunk, Frans. (2005). "Bringing Space Law into the Commercial World: Property Rights without Sovereignty." Accessed on August 21, 2009, from *Chicago Journal of International Law*. Summer 2005. 6 Chi. J. Int'l. L 81. Also available at *Space and Telecommunications Law Program Faculty Publications.* Accessed on April 22, 2011, at http://digitalcommons.unl.edu/spacelaw/15.

Hobe, Stephan. (2005). "Current and Future Development of International Space Law." *Proceedings: United Nations/Brazil Workshop on Space Law – Disseminating and Developing International and National Space Law: The Latin America and Caribbean Perspective*. ST/Space/28. No. E.05.I.76. Vienna: United Nations Office for Outer Space Affairs. pp. 1-15.

Hudgins, Edward L (ed.). (2002). *Space: the Free Market Frontier*. Washington, DC: Cato Institute.

Hufford, D. Brian. (1983). "Ideological Rigidity vs. Political Reality: A Critique of Reagan's Policy on the Law of the Sea." From the *Yale Law and Policy Review*, excerpted in Reynolds, Glenn H. and Merges, Robert P. (1997). *Outer Space: Problems of Law and Policy. Second Edition*. Boulder, Colorado: Westview Press. p. 133.

International Institute of Space Law. (2009). *Statement of the Board of Directors of the International Institute of Space Law (IISL)* March 22, 2009. Accessed on July 6, 2011, from http://www.iislweb.org/publications.html.

Jakhu, Ram. (2005). "20 Years of the Moon Agreement: Space Law Challenges for Returning to the Moon." *Proceedings: United Nations/Brazil Workshop on Space Law – Disseminating and Developing International and National Space Law: The Latin America and Caribbean Perspective*. ST/Space/28. No. E.05.I.76. Vienna: United Nations Office for Outer Space Affairs. pp. 339-351.

Jakhu, Ram, and Buzdugan, Maria. (2006). *The Role of Private Actors: Commercial Development of the Outer Space Resources, Including Those of the Moon and other Celestial Bodies: Economic and Legal Implications*. Paper presented at the McGill University Institute of Air and Space Law Conference, June 28-30, 2006, in Montreal. Retrieved August 19, 2009, from http://www.mcgill.ca/iasl/publications/proceedings/

Jobes, Douglas O. (2005). "Lunar Land Claims Recognition: Designing the Ultimate Incentive for Space Infrastructure Development." *Space Times*. May/June 2005.

Kasku-Jackson, Jonty and Waldrop, Elizabeth. (2009). "Understanding Space Law: Legal Framework for Space." In Coletta, Damon and Pilch, Frances T., eds. *Space and Defense Policy*. London: Routledge. pp. 64-103.

Keefe, Heidi. (1995). "Making the Final Frontier Feasible: A Critical Look at the Current Body of Outer Space Law." *Santa Clara Computer and High Technology Law Journal*. July 1995. Accessed on April 21, 2011, at http://www.lexisnexis.com.exproxy2.apus.edu/lnacui2api/frame.do?reloadEntirePage=true...

Kelly, Robert. (2004). "Nemitz v. United States, A Case of First Impression: Rights and Space Law before the Federal Courts of the United States." *Journal of Space Law*. Vol. 30. No. 2. Winter 2004. p. 297-309.

Klotz, Irene. (2012). *Who Owns the Asteroids?* Accessed on April 30, 2012, from http://news./discovery.com/space/asteroid-mining-space-treaty-ownership-120427.html#mkcpgn-emnws1.

Kotler, Steven. (2011). "The Great Galactic Gold Rush." *Playboy* Magazine. April 2011. p. 42ff.

Launius, Roger D. and McCurdy, Howard E., eds. (1997). *Spaceflight and the Myth of Presidential Leadership*. Urbana, Illinois: University of Illinois Press.

Lewis, John S. (1997). *Mining the Sky: Untold Riches from Asteroids, Comets, and Planets.* Reading, Massachusetts: Helix Books/Perseus Books.

Lin, Patrick. (2006). "Space Ethics: Look Before Taking Another Leap for Mankind." *Astropolitics Journal*. Vol. 4. No. 3. Winter 2006. p. 291-294.

Listner, Michael. (2012). "The Moon Treaty: It Isn't Dead Yet." *The Space Review*. March 19, 2012. Accessed on March 20, 2012 from http://www.thespacereview.com/article/2047/1. p. 1-2.

Lo, Martin W. and Chung, Min-Ku. (2002). *Lunar Sample Return via the Interplanetary Superhighway*. Jet Propulsion Laboratory. Accessed on April 29, 2012, from www.gg.caltech.edu/~mwl/publications/papers/LSRvialPS.pdf

McKay, M. F., McKay, D. S., and Duke, M. B. (1992). *Space Resources: Scenarios*. NASA SP-509, Vol. 1. Houston, Texas: NASA – Lyndon B. Johnson Space Center.

Moltz, James Clay. (2009). "Toward Cooperation or Conflict on the Moon? Considering Lunar Governance in Historical Perspective." *Strategic Studies Quarterly*. Fall 2009. p. 82-103.

Moltz, James Clay. (2008). *The Politics of Space Security: Strategic Restraint and the Pursuit of National Interests*. Stanford, California: Stanford University Press.

Murray, Kali N. (2006). "Of Gardens and Streets: A Differentiated Model of Property in International and National Space Law." *The Journal of Space Law*, Volume 32, Number 2, Winter 2006. p. 361-382. Retrieved on December 2, 2009, from http://www.spacelaw.olemiss.edu/jsl/back_issues/jsl%2032-2.pdf

National Aeronautics and Space Administration. *FY 2013 President's Budget Request Summary*. Washington, DC: NASA. Available at http://www.nasa.gov/news/budget/index.html

Office for Outer Space Affairs. (1996). *Space Law: A Bibliography*. New York: United Nations. Retrieved October 10, 2009, from http://www.unoosa.org/pdf/reports/ac105/AC105

Palmer, Brian. (2012). "Prospectors May Soon Go Out of This World for Rare Materials." Washington *Post*. February 7, 2012.

Pop, Virgiliu. (2009). "Is the Moon for Sale?" *Who Owns the Moon? Extraterrestrial Aspects of Land and Mineral Resources Ownership*. Springer: Space Regulations Library.

Pop, Virgiliu. (2001). *Lunar Real Estate: Buyer Beware!* Paper presented at the First Convention of Lunar Explorers in Paris in March 2001. Pop quotes James, Viscount of Stair. (1981). *The Institutions of the Law of Scotland, 1693*: II, I, 18. Edited by Walker, D.M. The University Presses of Edinburgh and Glasgow.

Pop, Virgiliu. (2006). *Unreal Estate: The Men Who Sold the Moon*. Virgiliu.com Press.

Pop, Virgiliu. (2009). *Who Owns the Moon? Extraterrestrial Aspects of Land and Mineral Resources Ownership*. Springer Science+Business Media B.V.

Prado, Mark. (2011). *Should the Vagueness of International Law Significantly Discourage Private Ventures Today?* Accessed on May 17, 2011, from www.permanent.com/legal-international-laws.html.

President Barack Obama on Space Exploration in the 21st Century. Retrieved on June 13, 2011, at http://nasa.gov/news/media/trans/obama_ksc_trans.html.

President Bush Offers New Vision for NASA. (2004). Retrieved on June 11, 2011, at http://www.nasa.gov/missions/solarsystem/bush_vision.html.

Ramachandran, Ram, and Gale, Anita. (2010). "Space Colonization." *Aerospace America*. December 2010. p. 64.

Reinstein, Ezra. J. (1999). "Owning Outer Space." *Northwestern Journal of Law & Business*. Vol. 20. No. 1. p. 59-98.

Research and Innovative Technology Administration. (2012). *December 2011 Airline System Traffic up 0.5 Percent from December 2010*. Accessed on May 6, 2012, from http://www.bts.gov/press_releases/2012/bts014_12/html/bts014_12.html.

Reynolds, Glenn Harland. (2008). "Who Owns the Moon? The Case for Lunar Property Rights." Accessed on April 30, 2012, from http://www.popularmechanics.com/science/space/moon-mars/4264325

Reynolds, Glenn and Kopel, Dave. (2002). "The New Frontier: Preparing the Law for Settling on Mars. *National Review Online*. June 4, 2002. Accessible at http://nationalreview.com/kopel/kopel060402.asp

Reynolds, Glenn H. and Merges, Robert P. (eds.) (1997), *Outer Space: Problems of Law and Policy, Second Edition*. Boulder, Colorado: Westview Press.

Russia Develops Revolutionary Ammonia Rocket Engine. (2012) RIA Novosti. May 5, 2012. Accessed on May 7, 2012 from http://en.rian.ru/russia/20120505/173241862.html.

Sanchez, J. P. and McInnes, C. (2011). "Asteroid Resource Map for Near-Earth Space." *Journal of Spacecraft and Rockets*. Vol. 48. No. 1. January-February 2011. pp. 153-165.

Scheraga, Joel D. (1997). "Establishing Property Rights in Outer Space." *Cato Journal*. Vol. 6, No. 3. Winter 1987. pp. 889-903.

Scott, Anthony. (1983). "Property Rights and Property Wrongs." *Canadian Journal of Economics*. Vol. 16, No. 4. November 1983. p. 555-573.

Shackelford, Scott J. (2009). *The Tragedy of the Common Heritage of Mankind*. (May 19, 2009). Stanford Environmental Law Journal, Vol. 27, 2008; Stanford Public Law Working Paper No. 1407332. Available at SSRN: http://ssrn.com/abstract=1407332

Simberg, Rand. (2012). *Homesteading the Final Frontier; A Practical Proposal for Securing Property Rights in Space.* Issue Analysis 2012 No. 2. Washington: Competitive Enterprise Institute.

Slyuta, E. N.; Abdrakhimov, A. M.; Galimov, E. M. (2007). *The Estimation of Helium-3 Probable Reserves in Lunar Regolith.* 38th Lunar and Planetary Science Conference. pp. 2175. http://www.lpi.usra.edu/meetings/lpsc2007/pdf/2175.pdf.

Space Settlement Institute. (2012). Frequently Asked Questions – FAQ #3. Accessible at http://www.space-settlement-institute.org/faqs.html.

Statement by the Board of Directors of the International Institute of Space Law (IISL) on Claims to Property Rights Regarding the Moon and other Celestial Bodies. (2004). Accessed on August 19, 2009, from http://www.iislweb.org/publications.html.

Swaminathan, Sriram. (2005). *Making Space Law Relevant to Basic Space Science in the Commercial Space Age.* Retrieved on August 20, 2009, from www.sciencedirect.com.

Szoka, Berin and Dunstan, James. (2012). "How the U.S. Can Lead the Way to Extraterrestrial Land Deals." *Wired Science.* Accessed on April 26, 2012, from http://www.wired.com/wiredscience/2012/04/opinion-space-property-rights.

Tatsuzawa, Kunihiko. (1998). *The Regulation of Commercial Space Activities by the Non-Governmental Entities in Space Law.* Retrieved on August 13, 2009, from http://www.spacefuture.com/archive/the_regulation_of_commercial_space_activities_by_t...

United Nations treaties and principles on outer space and related General Assembly resolutions: Addendum. "Status of international agreements relating to activities in outer space as at 1 January 2009." Ref.: Sales No. E.08.1.10 –ST/SPACE/11/Rev/Add 2.

Van Vugt, Mark. (2009). "Triumph of the Commons." *New Scientist.* August 22, 2009. Vol. 203, Issue 2722, p. 40-43. Retrieved on December 4, 2009, from http://web.ebscohust.com.ezproxy.apollolibrary.com/ehost/delivery?vid=5&hid=107&sid=

Van Wyk, J-A. (2008). "Overview of the Implementation Status of the Five United Nations Treaties on Outer Space in African Countries." *African Skies.* No. 12, October 2008. p. 90-98.

Von der Dunk, Frans, Hobe, S., Back-Impallomeni, E., and Ramirez de Arellano, R.M. (2004). "Surreal Estate: Addressing the Issue of 'Immovable Property Rights on the Moon.'" *Space Policy,* (20). p. 149-156.

Wasser, Alan. (1997). "How to Restart a Space Race to the Moon & Mars." Section 6.9.3.2.103 of the *Artemis Data Book.* Accessed on may 3, 2012, from www.asi.org/adb/06/09/03/02/103/space-race.html

Wasser, Allan and Jobes, Douglas. (2008). "Space Settlements, Property Rights, and International Law: Could A Lunar Settlement Claim the Real Estate It Needs to Survive?" *Journal of Air Law and Commerce.* Volume 72, Number 1. Winter 2008. pp. 37-78.

Wen-yu, Laura and Wang, Francis S. L. (2011). "Expatriates Face China's Criminal Law System," *International Law News.* Spring 2011. pp. 10-13.

White, Wayne N., Jr. (1999). *Implications of a Proposal for Real Property Rights in Outer Space.* Retrieved on August 21, 2009, from http://www.spacefuture.com/archive/implications_of_a_proposal_for_real_property_rights...

White, Wayne N., Jr. (1999). *Real Property Rights in Outer Space.* Retrieved on August 21, 2009, from http://www.spacefuture.com/archive/real_property_rights_in_outer_space.shtml.

White, Wayne N., Jr. (2000). *Implications of a Proposal for Real Property Rights in Outer Space*. Paper presented at the 42nd Colloquium on the Law and Outer Space, IISL.

Wingo, Dennis. (2004). *Moonrush: Improving Life on Earth with the Moon's Resources*. Burlington, Ontario, Canada: Apogee Books.
.

BIS-RS-2014-62

Paradigm Change in Earth Observation - Skybox Imaging and SkySat-1

Jonathan M. Dyer and Jim McClelland

Skybox Imaging, Inc., Mountain View, CA 94043

Skybox Imaging is building a constellation of high-resolution Earth imaging micro-satellites in order to revolutionize access to information about the changes happening across Earth. In 2013, Skybox launched its first satellite, SkySat-1 as the pathfinder to a constellation of small spacecraft.

We begin with a high-level overview of Skybox Imaging and its mission. Then we discuss the unique design approach at Skybox and the critical engineering ingredients that have enabled such a powerful spacecraft in a small package and at low cost.

Finally we discuss our plans for a constellation of small imaging satellites.

I. Introduction

Skybox Imaging, Inc. was founded in 2009 with a vision to revolutionize access to information generated from timely, very high resolution satellite image data with cutting edge data extraction and analysis. Our first two satellites, SkySat-1 and SkySat-2, were launched November 2013 and July 2014 respectively and have provided Skybox with an initial source of timely, sub-meter multispectral imagery and panchromatic high definition video.

We will discuss where Skybox Imaging fits in the greater continuum of Earth Remote Sensing, and what we see in the future for Skybox and the industry at large.

A. A Brief History of Satellite Earth Observation

For more than 50 years [10] humans have taken high quality images of the Earth from space, starting with the United States' Corona program of the early 1960's. The Corona ("KH"

S. Hatton (ed.), *Proceedings of the 12th Reinventing Space Conference*, DOI 10.1007/978-3-319-34024-1_5

for Keyhole) missions were the first satellites specifically designed to take high resolution pictures of things on Earth. They were film-based and had an effective resolving capability that started around 12-14m in early missions and rapidly improved to better than 1m over subsequent Corona satellite missions and those of its successors[9], Lanyard and Argon.

The Gambit and Hexagon missions followed with additional film-based satellites capable of resolutions better than 15cm[2] (Gambit-3) and, at lower resolution, covering huge areas of Earth (Hexagon). All of these systems were, however, fundamentally limited by a finite supply of photographic film that had to be returned to Earth for "read out" - a difficult and slow proposition.

(a) Corona image of the Pentagon, 1967

(b) Gambit (KH-7) image of the US Capitol, 1966

Figure 1. Corona and Gambit images of Washington, DC

The charge coupled device (CCD) was invented in 1969 and ushered in a revolution in imaging. Applications formerly requiring film to be returned from space could now utilize these sensitive, compact and, most importantly, purely electronic devices to capture images for transmission via radio. This technology was quickly recognized as revolutionary for remote sensing and the use of CCD-based, electro-optical imaging in the next generation of Keyhole systems (KH-11+, still classified) allowed them to collect more data, with lower latency and over much longer time spans than the previous film-based systems.

Under the Clinton Administration in the late 1990's and early 2000's a revolution in the availability of high resolution remote sensing data was kicked off through the following actions:

- Congress passed the 1992 Land Remote Sensing Act, making it legal for commercial entities to capture high resolution satellite imagery and sell it commercially

- Declassification of details and photos from the Corona, Lanyard and Argon programs due to Executive Order 12951 (1995)
- Declassification of details and photos from the Gambit and Hexagon programs due to Executive Order 12951 (2002)

The change in regulatory environment and demonstrated value of the declassified imagery catalyzed three American companies to build high resolution Earth imaging spacecraft for commercial use - Space Imaging, Inc., EarthWatch, Inc. (later to become DigitalGlobe, Inc.) and Orbital Imaging Corporation (ORBIMAGE - later to become GeoEye). Over the next two decades, these companies, as well as other European and Indian entities, deployed progressively more capable imaging systems culminating in systems capable of collecting extremely high quality 0.5m (and soon 0.25m) imagery.

B. State of the Market and Path Forward

As remarkable as the rapid progress in the capabilities of commercial remote sensing satellites has been since the 1990's, the trend has been towards more and more costly systems, with the newest generation of spacecraft costing more than US $600M[5] to-orbit. This large and rising capital requirement has limited the number of such systems on orbit, and therefore the availability of fresh data to a wide variety of customers. It has also precluded private investment in such systems, the majority of the capital coming from state entities such as the National Geospatial Agency (NGA) within the US.

Several trends are setting the stage for what we believe will be a second revolution in the commercial remote sensing industry:

- Increased awareness of and demand for data in the commercial, civil and consumer sectors (thanks in part to popular mapping services such as Google Maps and Earth)
- Demonstration of high performance, low-cost "SmallSat" missions such as ChipSat
- Commoditization of scalable cloud storage and processing for image and GIS data
- High value data extraction through crowd-sourcing and machine learning algorithms
- Exponential performance improvement of COTS processing, solid state storage and image sensors

- Faster engineering development cycles through rapid prototyping tools, flexible manufacturing options

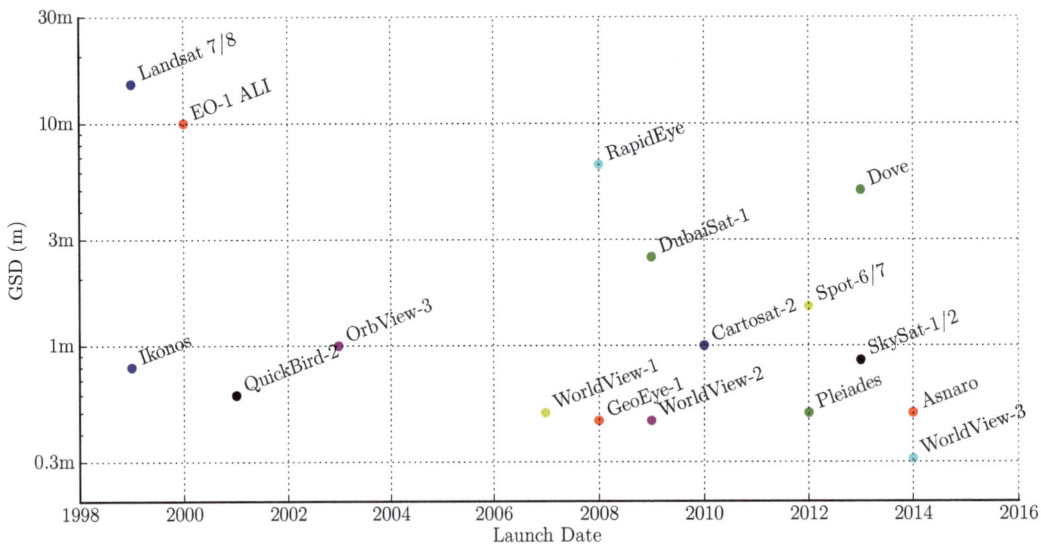

Figure 2. Resolution of Commercial Remote Sensing Satellites vs. Time

Together these factors have enabled Skybox Imaging to deploy two sub-meter resolution small imaging spacecraft, SkySat-1 and SkySat-2, at a cost point that enables constellations of 10's of satellites. And, as importantly, this was achieved with private capital brought to bear by commercial market demand rather than government investment.

The excitement created by the success of SkySat-1 and SkySat-2 in the remote sensing and private capital industries is apparent and fascinating. In this paper, however, we will focus primarily on the spacecraft themselves.

We assert that trends in many areas are poised to enable fleets of small, inexpensive satellites to collect Earth remote sensing data of high business value. We will discuss the trends and critical enablers driving the paradigm shift to more, smaller satellites after describing SkySat-1 and SkySat-2 in a bit more detail. Then we will define a Figure of Merit useful in quantitatively assessing capability or value in a space-based Remote Sensing system. And finally we look at the capability a constellation of remote sensing satellites like SkySat-1 can provide.

II. SkySat-1 and SkySat-2

SkySat-1 was launched on November 21, 2013 and SkySat-2 on July 8, 2014. They are nearly identical high resolution imaging spacecraft consisting of precise 3-axis attitude control, a 35cm primary optical instrument, high performance X-band downlink system and all necessary C&DH support avionics. First-light images for both spacecraft are shown in Fig. 3.

(a) Nice, France - SkySat-1

(b) Port-au-Prince, Haiti - SkySat-2

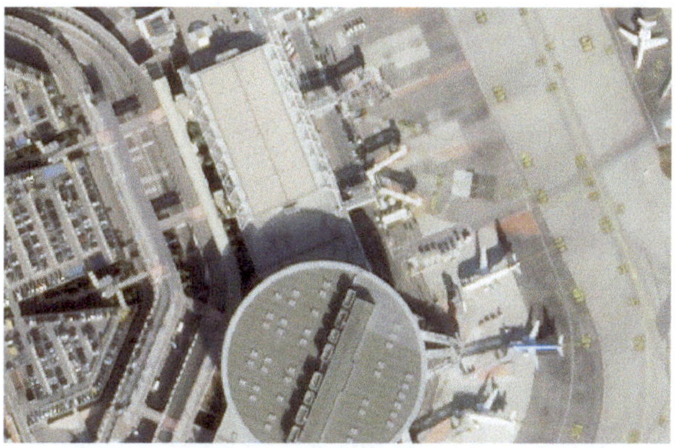

(c) Nice, France Zoomed

(d) Port-au-Prince, Haiti Zoomed

Figure 3. First-light images captured by SkySat-1 and SkySat-2

A. Approach and Philosophy

In building SkySat-1, the team kept developed a philosophy for design and build with several tenants (see Dyer [3, 4] for more details):

- Keep it simple
- Modularity is key
- Leverage other industries

- Take time to do the right thing
- Design to fail gracefully - redundancy is **NOT** the root of all evil
- Build, test and fail early and often
- Balanced engineering decisions - don't let the tail wag the dog
- Perform "right-sized" analysis
- Vertical integration is key

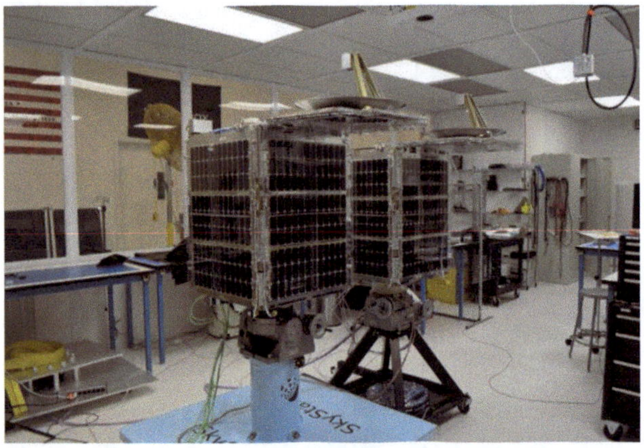

Figure 4. SkySat-1 and SkySat-2 in the clean-room awaiting launch

Having the luxury of good in-house facilities for prototyping and testing allowed us to build hardware early and learn from our mistakes. SkySats consist of almost exclusively Commercial-off-the-shelf (COTS) electronic components, allowing us to build avionics quickly and inexpensively. We **do the right thing** by screening these electronics for the space thermal, vacuum and radiation environment extensively[11].

Skybox' satellite engineering work was performed with team numbering less than 30. Additionally we leverage the knowledge of a senior advisory group we call the Technical Advisory Group (TAG) which has injected several hundred years of combined space experience into our team knowledge base. The small team size coupled with a flexible, iterative design build and qualification approach allowed us to go from a clean-sheet design to a flight-ready spacecraft in less than 3 years.

We believe this approach is incredibly powerful and applicable far outside the realm of just Remote Sensing satellites. And the results obtained from SkySat-1 and SkySat-2 speak for themselves.

B. Results

Fig. 5 illustrates the diversity and quality of imagery SkySat-1 and SkySat-2 have collected to-date.

(a) Football field lines

(b) SkySat-2's Soyuz launch pad

(c) Runway markers and paint scheme on aircraft

(d) An oblique shot

Figure 5. Example images from SkySat-1 and SkySat-2

While the pictures in Fig. 5 are all visible color reproductions, the spacecraft actually capture imagery in 5 bands - panchromatic, red, green, blue and near-infrared.

Fig. 6 shows an example of what can be seen by fusing NIR with visible imagery.

(a) Comparison of RGB and NRG renderings of the same scene

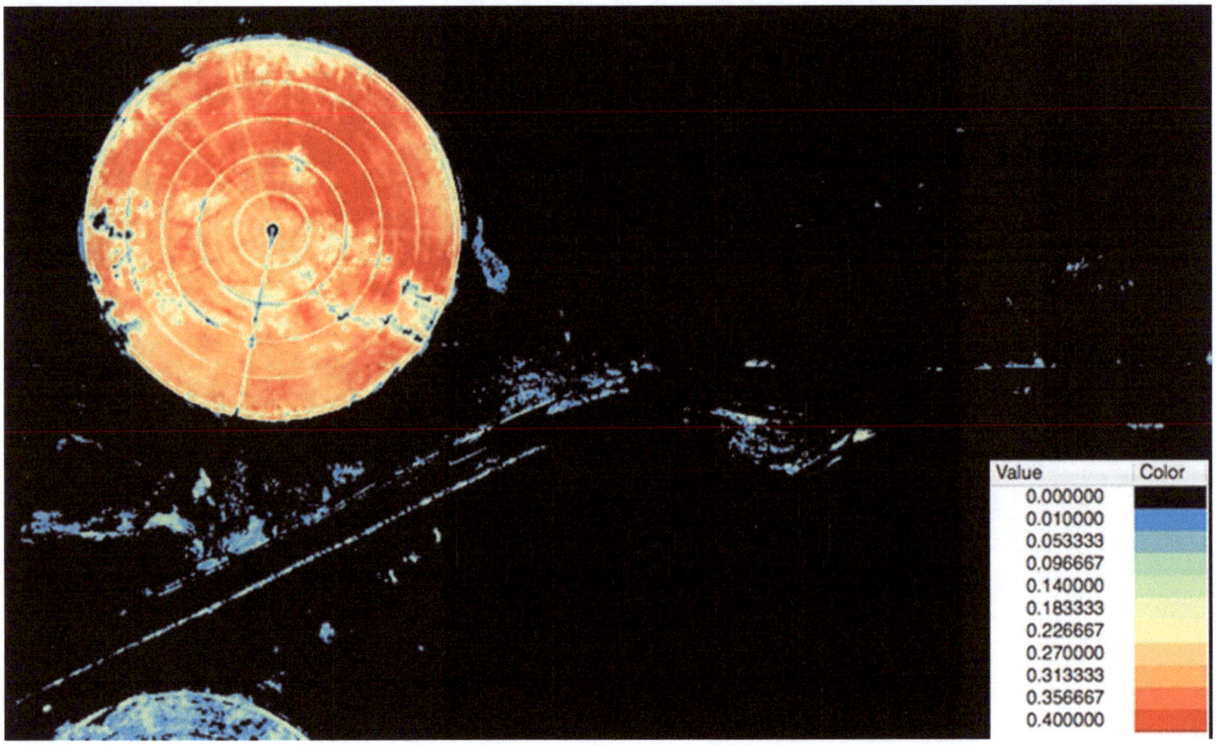

(b) MSAVI vegetation index computed form SkySat-1 imagery

Figure 6. The value of NIR

Critical assessment of the data being produced by SkySat-1 and SkySat-2 shows their ability to clearly resolve sub-meter sized features. For more details on SkySat-1 and SkySat-2 imaging performance, see Murthy et al. [7, 8].

III. Trends and Critical Enablers

There is a growing recognition that smaller systems generally fair much better in **performance-per-cost** than large systems. This is a critical component of return-on-investment and, therefore, one of the most important factors in determining commercial viability and enabling scalability. We believe strongly that smaller systems do provide greater value and will elaborate on the forces at play in this section.

Figure 7. Small is the new big![a]

A. Small Satellites

"SmallSats", roughly defined as <400kg mass, are very trendy right now. There are a number of commercial endeavors (e.g. Planet-Labs Inc., Planetary Resources Inc., PlanetIQ Inc., Dauria Labs) as well as University (e.g. Can-Xn), science (e.g. QuakeSat) and Government-funded (e.g. TechDemoSat-1) systems in development or on orbit today. The popularity of the smaller systems is well-deserved and the result of a great number of factors, not least of which is huge success of the Cubesat form factor. However SmallSats are not a new thing; most of the earliest satellites fit within the definition of SmallSat's and tended to benefit from many of the same advantages that modern SmallSat's leverage.

Examples of these advantages are simplicity, shorter design cycles, higher programmatic tolerance to risk and physical scaling laws that makes thermal, structural and electrical design easier.

Small satellites also require significantly less capital to deploy. The ability to launch satellites as secondary payloads or cluster several smaller satellites on a single launch results in much less expensive launch.

Recognition of the recent success small teams in the University and small science community have had in building very successful and capable SmallSats in very short time periods

[a]GeoEye-2 image Credit DigitalGlobe, Inc.

and at very low cost has catalyzed an explosion of interest in small systems. We firmly believe that fleets of smaller, less expensive remote sensing satellites will prove revolutionary to the industry.

IV. Remote Sensing Figure of Merit

We have extolled the virtues of SmallSats and the associated philosophy and will now propose a Figure of Merit (FOM) by which to quantitatively demonstrate the advantage provided by systems like SkySat-1/2.

In optical engineering and image processing, the "Space-Bandwidth Product" (SW) is a common metric used to assess performance[6]. It is fundamentally related to the information content in an image in that you can trade image extent (or system field of view) against resolution (resolving power). Lohmann et al. [6] have shown that the space-bandwidth product for an image, SWI, is not necessarily equivalent to that of the system that produced it, SWY. Because we are looking for qualitative FOM by which to compare relative performance of systems, we will consider SWI and SWY equivalent and compared the SW of various systems by looking at the SW of the image data they produce.

The SW for a remote sensing system should be representative of the **number of dimensionally independent samples produced per time**. A dimensionally independent sample in a multispectral image is one spatial sample (pixel) of one channel. We want to assess this per time, because time becomes one more dimension of information content; or thought of differently, inverse time is a form of bandwidth in the time/frequency duality sense.

Because "channels" are not necessarily dimensionally independent if they overlap, we will define the following spectral independence fraction:

$$\phi_i = \frac{\int_\lambda f_i(\lambda)d\lambda}{\int_\lambda f_i(\lambda)d\lambda + \sum_{j\neq i}\int_\lambda f_i(\lambda)f_j(\lambda)d\lambda} \tag{1}$$

where i is the channel being evaluated, and $f(\lambda)$ is the non-dimensional spectral sensitivity function for that channel and $\{\phi_i \in \mathbb{R}, 0 \leq \phi_i \leq 1\}$.

The number of samples per time that a system generates can be conveniently defined as:

$$SW_i = \frac{A_i\phi_i}{\Delta t \times \mathrm{GSD}_i^2} \tag{2}$$

where A_i is the collected area for ith channel over Δt, the time collection time and GSD_i is the Ground Sample Distance for the ith channel. For convenience we will use $\Delta t = 1$ day because most systems report capacity in square kilometers per day.

The system space-bandwidth product thus becomes:

$$SW = \sum_{i=1 \to N} \frac{A_i/\Delta t}{\text{GSD}_i^2} \times \frac{\int\limits_{\lambda} f_i(\lambda) d\lambda}{\int\limits_{\lambda} f_i(\lambda) d\lambda + \sum\limits_{j \neq i} \int\limits_{\lambda} f_i(\lambda) f_j(\lambda) d\lambda} \tag{3}$$

Note that if we use consistent units for GSD_i and A_i, the units of SW become samples per day.

SW is shown for several remote sensing system launched in the last 20 years in Fig. 8.

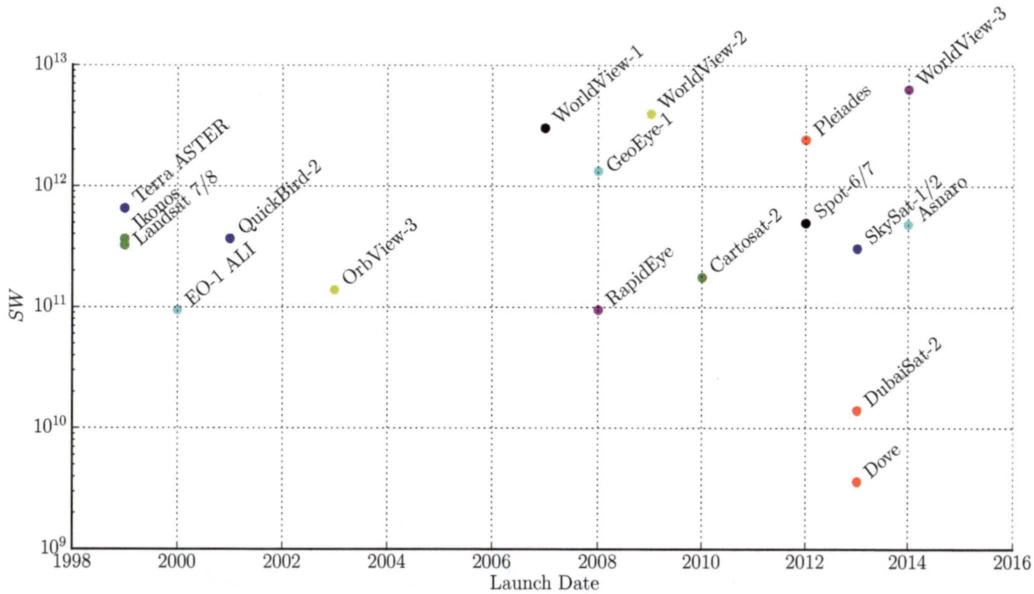

Figure 8. ***SW*** **vs. time for several systems**

As can be seen there is a general trend upwards in SW over time although there are strong outliers. This is due to the fact that the missions evaluated differed by 2-3 orders of magnitude in cost and many system metrics such as mass, pointing accuracy, and resolution. This highlights a few inadequacies of using just SW as a figure of merit:

- Spectral and spatial resolution are treated identically - in reality, many applications weigh one as more valuable than the other
- Performance is not normalized by cost or complexity

The first inadequacy we will not address as it is very application dependent. There are good arguments that state that for many commercial applications spatial resolution should be weighed more heavily and for many Earth Science applications, spectral resolution.

To address the second inadequacy, we seek to generate a "cost-normalized" figure of merit by dividing SW by a cost metric.

Bearden et al. [1] provides a variety of statistical cost models for small spacecraft and, while imperfect, mass is generally the most popular variable correlated with cost. Bearden et al. [1]'s mass / cost correlation nominally is only applicable up to a 400kg spacecraft mass, but the authors have observed good predictive performance in systems of known cost with masses > 3000 kg as well.

$$\text{cost} \propto m^{1.261} \tag{4}$$

Finally we will define our primary figure of merit (FOM) for remote sensing systems as

$$\text{FOM} = \frac{SW}{m^{1.261}} \tag{5}$$

Fig. 9 shows SW normalized by mass-predicted cost.

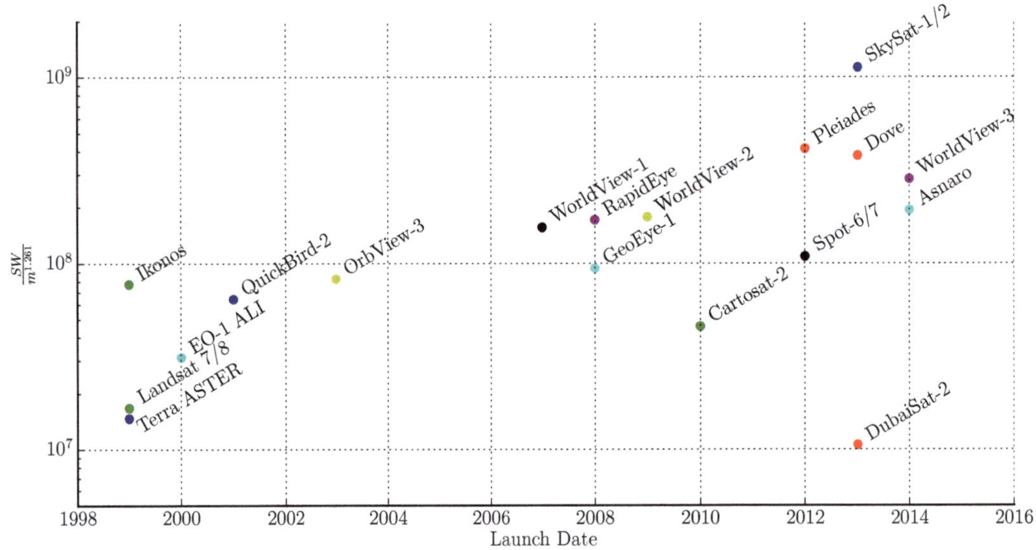

Figure 9. FOM vs. time for several systems

V. Constellation

A cornerstone of the Skybox vision is a belief that by leveraging Silicon Valleys big data technology and innovation engine we can fuel a revolution in the use of remote sensed imagery to fundamentally transform our understanding of the ever-changing world around us. We realized early on, however, that to bring about this transformation we would need a set of imagery sources that met minimum thresholds for **both utility and availability**. These thresholds for utility and availability are driven by the need to distinguish activities occurring on a human scale in both space and time.

The existing market has already fairly clearly established the utility threshold at around 1m GSD with multiple suppliers offering products at or below this level. However, while these suppliers do have reasonably stable business bases, nearly all are fairly heavily subsidized by their respective governments.

The slow growth of these commercial markets has been primarily driven by the very limited availability that these suppliers are able to offer. By way of comparison, the current availability for imagery is a bit like having a GPS system that can give you a position fix once every couple hours. While there are some applications that may be able to tolerate such a sparse availability, the real potential of the commercial markets cannot be unlocked until the refresh rates are much closer to the timescales over which most things are changing.

So, to tip the commercial market scales, the set of imagery sources need to be not only capable of refreshing the changing parts of the world on a timescale of hours to days versus the current timescale of weeks to months but also provide relatively uniform access over a significant portion of the day. Having developed and demonstrated a satellite design capable of delivering the threshold sub-meter imaging performance capability, the solution to the availability challenge is basically a numbers game where technical success is measured by access uniformity and overall area capacity of the system but business success is measured by the cost per unit of GSD-normalized area per unit time.

A. SkySat-3 - The Constellation Pathfinder

One of the most important factors in the systems ability to ensure timely refresh of the worlds most interesting places is the ability to establish and maintain the required orbital configurations. And while there are numerous options for precisely how the satellites get deployed, one requirement common to every option is the capability to modify the characteristics of a given satellites orbit.

Following the successful launch and demonstration of SkySats 1 and 2, Skybox immediately kicked off the design process to update the design in a few areas that are essential for operations within a much larger constellation context. The most critical of these design updates was the addition of basic propulsion capabilities to facilitate the proper initial establishment and long-term maintenance of each satellites orbit.

Along with a few other design improvements, SkySat 3s design now includes a propulsion system using a high-performance green propellant capable of delivering over 170 m/s of total ΔV. This level of performance facilitates not only the correction of the initial launch vehicle dispersions but also affords the capability to establish and later correct modest relative sun-sync drift rates.

B. The Constellation

As discussed previously, the fuel needed to revolutionize the remote sensing industry is a set of imagery sources that meet both the minimum utility and availability thresholds. To address the availability challenge, Skybox has selected a minimum viable configuration "MVC" for the Constellation that is sized to achieve an average of daily refresh for the "interesting" areas of the world. While all of these areas may not require daily refresh, by setting the minimum threshold at this level we baseline a minimum system capacity that is capable of a refresh mix that includes both intra-day and multi-day refresh rates.

In addition, we have structured our development cadence and selected our partners to facilitate scaling flexibility, allowing us to rapidly adjust our plans to better respond to the evolving remote sensing landscape.

In parallel with the SkySat 3 pathfinder development, Skybox is ramping production of additional spacecraft. Figure 10 provides a notional view of revisit (access opportunities per day) performance for an evolving constellation of satellites.

C. The future

As is the case with all thriving companies, continuous exploration and innovation is essential to remain current, competitive, and connected with the ever-changing world. Skybox is no different in this regard and to ensure that we remain at the forefront of the aerospace development frontier we have structured the program to inject new and/or updated technology into each Block revision. Leveraging this steady and rapid deployment cadence allows us to remain in lock step with the latest developments in technology and constantly push the boundaries of aerospace innovation.

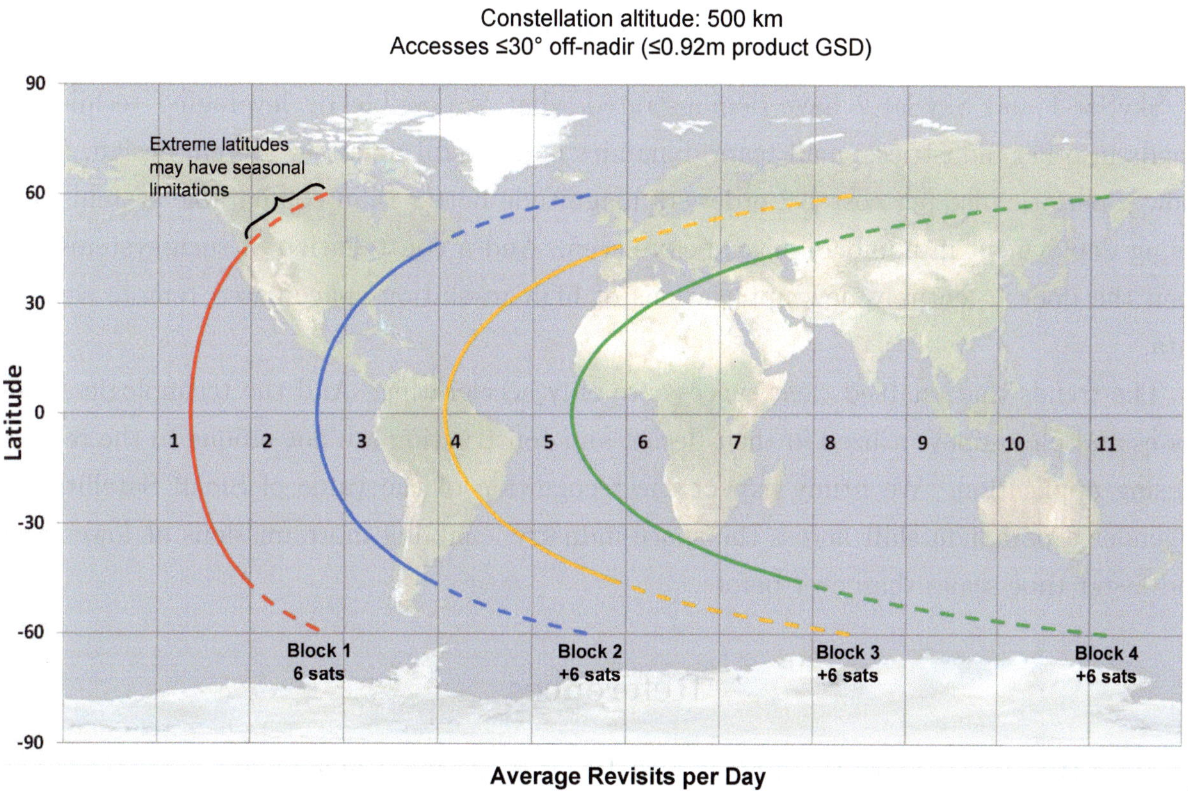

Figure 10. Constellation average revisit performance

VI. Conclusion

SkySat-1 and SkySat-2 have demonstrated what is possible by leveraging technology trends in other industries, small team dynamics and careful SmallSat system design. They show a performance per cost 1-2 orders of magnitude better than comparable systems, enabling deployment in numbers never before seen. And a constellation of such systems will open the door to entirely new applications of high resolution space-based remote sensing data.

The trends that enabled their success are only accelerating. And the technologies, processes and philosophy utilized in their design and construction are not unique to the remote sensing application. We firmly expect the recognition of the value of Small Satellites to engender a paradigm shift across the space industry, enabling more missions at lower cost and faster time scales than ever before.

References

[1] David Bearden, Richard Boudreault, and James R. Wertz. *Reducing Space Mission Cost*, chapter Cost Modeling, pages 253–284. Microcosm Press, 401 Coral Circle, El Segundo, CA 90245-4622 USA, 1996.

[2] Dwayne A. Day. Ike's gambit: The kh-8 reconnaissance satellite. *The Space Review*, January 2009.

[3] Jonathan Dyer. Building the world's most capable microsatellites, part i, November 2014. URL `http://www.skyboximaging.com/news/the-worlds-most-capable-microsats-part1`.

[4] Jonathan Dyer. Building the world's most capable microsatellites, part ii, November 2014. URL `http://skyboximaging.com/blog/microsatellites-pt-2`.

[5] Warren Ferster. Digital globe adding infrared capability to worldview-3 satellite. *SpaceNews*, 2012.

[6] Adolf W. Lohmann, Rainer G. Dorsch, David Mendlovic, Zeev Zalevsky, and Carlos Ferreira. Space-bandwidth product of optical signals and systems. *Jouran of Optical Society of America*, 13(3):470–473, March 1996.

[7] Kiran Murthy, Alexandra H. Chau, Minesh B. Amin, and M. Dirk Robinson. An earth imaging camera simulation using wide-scale construction of reflectance surfaces, 2013. URL `http://dx.doi.org/10.1117/12.2028432`.

[8] Kiran Murthy, Michael Shearn, Byron D. Smiley, Alexandra H. Chau, Josh Levine, and Dirk Robinson. Skysat-1: very high-resolution imagery from a small satellite, 2014. URL `http://dx.doi.org/10.1117/12.2074163`.

[9] National Reconnaissance Office. The corona story, June 2010.

[10] National Reconnaissance Office. The national reconnaissance office at 50 years: A brief history. Government Printing Office, 732 North Capitol Street, NW Washington, DC 20401, September 2011.

[11] Doug Sinclair and Jonathan Dyer. Radiation effects and cots parts in smallsats. *Conference on Small Sateilltes*, (SSC13-IV-3), August 2013.

VII. Appendix A - Tabulated system performance parameters

Table 1: Remote Sensing System Data

System	Mass	Bands	GSD	km^2 / day
Terra ASTER	4850 kg	10250-10950 nm	90.0 m	31,949k
		10950-11650 nm	90.0 m	31,949k
		8475-8825 nm	90.0 m	31,949k
		8925-9275 nm	90.0 m	31,949k
		2360-2430 nm	30.0 m	31,949k
		8124-8475 nm	90.0 m	31,949k
		1600-1700 nm	30.0 m	31,949k
		2145-2185 nm	30.0 m	31,949k
		2185-2225 nm	30.0 m	31,949k
		520-600 nm	15.0 m	31,949k
		630-690 nm	15.0 m	31,949k
		760-860 nm	15.0 m	31,949k

		2235-2285 nm	30.0 m	31,949k
		2295-2365 nm	30.0 m	31,949k
Cartosat-2	695 kg	479-830 nm	0.8 m	175k
DubaiSat-1	200 kg	420-510 nm	5.0 m	12k
		760-890 nm	5.0 m	12k
		510-580 nm	5.0 m	12k
		600-720 nm	5.0 m	12k
		420-720 nm	2.5 m	12k
DubaiSat-2	300 kg	450-520 nm	4.0 m	18k
		770-890 nm	4.0 m	18k
		520-590 nm	4.0 m	18k
		630-690 nm	4.0 m	18k
		550-900 nm	1.0 m	18k
EO-1 ALI	573 kg	433-453 nm	30.0 m	6,815k
		450-515 nm	30.0 m	6,815k
		2080-2350 nm	30.0 m	6,815k
		845-890 nm	30.0 m	6,815k
		1550-1750 nm	30.0 m	6,815k
		775-805 nm	30.0 m	6,815k
		630-690 nm	30.0 m	6,815k
		525-605 nm	30.0 m	6,815k
		479-690 nm	10.0 m	6,815k
		1200-1300 nm	30.0 m	6,815k
WorldView-3	2800 kg	450-510 nm	1.2 m	630k
		1710-1750 nm	3.7 m	630k
		2185-2225 nm	3.7 m	630k
		2235-2285 nm	3.7 m	630k
		770-895 nm	1.2 m	630k
		1195-1225 nm	3.7 m	630k
		1550-1590 nm	3.7 m	630k
		860-900 nm	1.2 m	630k

		2295-2365 nm	3.7 m	630k
		1640-1680 nm	3.7 m	630k
		705-745 nm	1.2 m	630k
		585-625 nm	1.2 m	630k
		2145-2185 nm	3.7 m	630k
		510-580 nm	1.2 m	630k
		630-690 nm	1.2 m	630k
		400-450 nm	1.2 m	630k
		450-800 nm	0.3 m	630k
		450-510 nm	1.8 m	785k
		630-690 nm	1.8 m	785k
		510-580 nm	1.8 m	785k
		585-625 nm	1.8 m	785k
WorldView-2	2800 kg	400-450 nm	1.8 m	785k
		860-900 nm	1.8 m	785k
		770-895 nm	1.8 m	785k
		705-745 nm	1.8 m	785k
		450-800 nm	0.5 m	785k
WorldView-1	2500 kg	450-900 nm	0.5 m	750k
		450-515 nm	2.4 m	200k
		740-900 nm	2.4 m	200k
QuickBird-2	951 kg	515-595 nm	2.4 m	200k
		605-695 nm	2.4 m	200k
		450-900 nm	0.6 m	200k
		445-516 nm	4.0 m	350k
		757-853 nm	4.0 m	350k
Ikonos	817 kg	506-595 nm	4.0 m	350k
		632-698 nm	4.0 m	350k
		450-900 nm	0.8 m	350k
		400-550 nm	5.0 m	43k
Dove	6 kg	475-600 nm	5.0 m	43k

		575-750 nm	5.0 m	43k
Asnaro	495 kg	430-550 nm	2.0 m	200k
		750-950 nm	2.0 m	200k
		489-610 nm	2.0 m	200k
		600-720 nm	2.0 m	200k
		479-830 nm	0.5 m	200k
RapidEye	150 kg	440-510 nm	6.5 m	800k
		690-730 nm	6.5 m	800k
		760-850 nm	6.5 m	800k
		520-590 nm	6.5 m	800k
		630-685 nm	6.5 m	800k
OrbView-3	360 kg	450-515 nm	4.0 m	210k
		740-900 nm	4.0 m	210k
		515-595 nm	4.0 m	210k
		605-695 nm	4.0 m	210k
		450-900 nm	1.0 m	210k
Spot-6/7	800 kg	450-525 nm	6.0 m	1,500k
		760-890 nm	6.0 m	1,500k
		530-590 nm	6.0 m	1,500k
		625-695 nm	6.0 m	1,500k
		450-745 nm	1.5 m	1,500k
SkySat-1/2	85 kg	450-515 nm	1.1 m	125k
		740-900 nm	1.1 m	125k
		515-595 nm	1.1 m	125k
		605-695 nm	1.1 m	125k
		450-900 nm	0.9 m	125k
Landsat 7/8	2500 kg	11500-12500 nm	100.0 m	31,949k
		10300-11300 nm	100.0 m	31,949k
		433-453 nm	30.0 m	31,949k
		525-600 nm	30.0 m	31,949k
		450-515 nm	30.0 m	31,949k

		845-885 nm	30.0 m	31,949k
		630-680 nm	30.0 m	31,949k
		2100-2300 nm	30.0 m	31,949k
		1560-1660 nm	30.0 m	31,949k
		1360-1390 nm	30.0 m	31,949k
		500-680 nm	15.0 m	31,949k
		430-550 nm	2.0 m	1,000k
		750-950 nm	2.0 m	1,000k
Pleiades	970 kg	489-610 nm	2.0 m	1,000k
		600-720 nm	2.0 m	1,000k
		479-830 nm	0.5 m	1,000k
		450-510 nm	1.8 m	500k
		780-920 nm	1.8 m	500k
GeoEye-1	1955 kg	510-580 nm	1.8 m	500k
		655-690 nm	1.8 m	500k
		450-800 nm	0.5 m	500k

BIS-RS-2014-59

NEXT GENERATION NOVASAR DEVELOPMENT

Owen HAWKINS[1], Phil WHITTAKER[1], Kevin HALL[1], Martin COHEN[2], Andrew CAWTHORNE[1], Caroline SLIM[1]

[1]Surrey Satellite Technology Ltd, Guildford, UK, [2]Airbus Defence and Space, Portsmouth, UK

E-mail: O.Hawkins@sstl.co.uk

ABSTRACT

The NovaSAR program builds on the synergies between SSTL and Airbus Defence and Space in the UK translating their expertise to small satellite based synthetic aperture radar. The UK government has supported the development of the first spacecraft in the series, and further satellites are planned. The current system works in S-Band and focuses on applications including maritime monitoring and forestry.

The roadmap for the development of this program may take a number of different avenues. One possibility is the enlargement of the antenna array to realise improvements in the payload performance. Another is the development of the platform to reflect mass reductions made in other SSTL platforms, through SSTL's ongoing development activities. Alternatively, the system could be modified to allow operation at X-Band, the shorter wavelength offering enhanced access and different physical observables. Airborne testing has been performed at both S and X band to investigate and demonstrate the differing potential data products. Each of these alterations has varying advantages, disadvantages and impacts on the concept of operations, the possible applications of the data, the payload imaging modes and other aspects of the mission and system. The system changes are presented and discussed.

KEYWORDS:

THE NOVASAR PROGRAMME

NovaSAR is an initiative brought about by SSTL and Airbus Defence and Space under the support of the UK government to realize a small satellite design capable of supporting the requirements of a Synthetic Aperture Radar (SAR) payload. Previously, small satellites (i.e. those below 500kg) have supported a wide range of payloads, but the size and high power requirements of an active SAR payload makes accommodation on a small platform difficult. While many small satellite SAR mission designs exist they are rarely implemented due to performance issues, or planned small satellites during detailed design result in satellites weighing tonnes.[i]

Figure 1 – NovaSAR-S

Recent years have seen a number of advances that have enabled small satellite platforms to support more resource hungry payloads. Some examples are higher efficiency solar cells, higher power density and more reliable batteries and more compact and lighter data handling and on board computing technologies.

On the SAR payload side, advances have also been made. Gallium Nitride amplifiers offer higher peak RF power and higher efficiency of power conversion into RF, enabling a power efficient solution with far fewer phase centres than traditional systems. An antenna solution has been devised as a non-deployable design, and the thermal handling is simplified, thus significantly reducing mass and risk.

The first NovaSAR mission is in manufacture in the clean rooms in Guildford and Portsmouth for flight readiness review early 2016.

The specifications of NovaSAR meet a wide range of application needs, and it is particularly suitable for maritime and forestry monitoring. The antenna and payload design offer modes with wide observation angles, including a novel maritime mode operating with high along-track ambiguity specifically to spot bright targets on the surface of the ocean.[ii]

S. Hatton (ed.), *Proceedings of the 12th Reinventing Space Conference*, DOI 10.1007/978-3-319-34024-1_6

Figure 2 – NovaSAR-S in SSTL cleanroom

The selection of S-Band for imaging is less common than the other more widely used X- C- and L- bands. However, at the time of mission design the high efficiencies of new GaN based amplifiers (40-45%) at S-Band made this highly desirable. Also, the band offers different performances across various applications. Airborne tests have demonstrated performance improvements for forestry application compared to shorter wavelength observations, expected to also provide benefits to agricultural applications.[iii] The Almaz-1[iv,v] and HJ-1C[vi] satellites have demonstrated various applications in S-Band, and the planned Kondor-E satellite will build on this experience.

FUTURE DEVELOPMENTS

The NovaSAR system is intended to work both alone or as part of a constellation, but each individual owner has their own interests and application requirements to meet and may require specific modifications. Whereas in some cases, these may be classed as a new generation of the mission design, others require relatively simple alterations. As a reference, below is a table of key specifications for the first satellite.

Imaging band	S-band (3.1-3.3GHz)
Lifetime	7 years
Mass	450kg
Lead time	24 months KO to FRR
Antenna array	Microstrip patch phased (3x1m)
Imaging polarisations	Single, dual or tri-polar (HH,HV,VH,VV) incoherent
Optimum orbit	580 km (SSO or low inclination Equatorial orbit)
Payload duty cycle	Average at least 2 minutes per orbit (single image strip >800km long)
Typical area coverage	>1 million km2 per day (mode dependent)

Table 1 – First NovaSAR-S specifications

Some possible modifications are now described and discussed.

Increased Antenna Area

The NovaSAR-S SAR payload antenna is made up of an array of 6x3 phase centres. This configuration allows for a 3m by 1m total antenna area with the required element spacing, allowing launcher accommodation of 3 in a Dnepr, Vega or PSLV faring or similar. Adding an extra two columns would allow for enhanced performance, but could limit the launcher options, particularly in the case of Dnepr. This is one of the simpler possible antenna modifications.

Having an 8x3 array would require a slightly higher standby power due to the increased phase centres, but will increase access to higher incidence angles and reduce the power needed during imaging.

The increased antenna area would enable a reduced pulse duty from 25% for the existing configuration to 10.55% for the alternative configuration.

This will reduce the peak power requirement of the mission, and providing the imaging time dominates will reduce the total average power consumption or increase the imaging time capability of the mission.

Other impacts however would need to be considered with respect to their impact on other payload aspects, for example data storage and downlink power.

Power (W)	18 p/centres	24 p/centres
Standby Back End	126	126
Standby Front End	222	296
Standby Total	**348**	**442**
Imaging Back End	126	126
Imaging Front End	1993	1176
Imaging Total	**2119**	**1302**

Table 2 – Payload power versus phase centres

Deployable antennas can be considered, but may add to the mission complexity and mass. Most SAR satellites use deployable phased array antennas but there are design challenges including: ensuring the alignment of the panels is good; that the panels are stable and that each array element has an equal path length from the RF source/s. These challenges are not insurmountable, but do add complexity.

Increasing Orbit Control

Many satellites operate on a fixed orbit, using on-board propulsion systems to correct the orbit as it decays. One advantage of this is that it allows for the smallest baseline variation at regular time intervals for the purposes of interferometric measurements. A disadvantage is that for a particular imaging mode, the revisit is fixed, whereas allowing the orbit to change can offer varying revisit rates over different sites, and less predictable imaging times.

The NovaSAR mission under manufacture does not allow for such a high level of orbital control, with mission Δv of 29m/s. Interferometric measurements will be possible, but will be opportunistic, using image pairs that have similar observation conditions. Temporal decorrelation therefore will have to be taken into account, which will reduce the available image pairs. The particular type of interferometry needed for measuring ground deformation, particularly of interest for mining, relies on maintaining the imaging baseline, which will also vary. Improving the propellant volume and/or reducing the spacecraft mass will allow for better orbital control.

Alternatively, the extra Δv could allow for the spacecraft to fly at a lower orbit, which requires more propellant to maintain said orbit. This would allow for modification and potential improvement of the imaging modes to account for the shorter time to target and increased received RF power at the sensor array.

X-Band Variant

A recent evaluation by the Earth Observation Mission Advisory Group lead by the Centre for Earth Observation Implementation saw good prospects for an X-Band variant of NovaSAR. X-Band is a popular SAR mission waveband due to its potential to achieve higher imaging resolutions with its smaller wavelength and broader available imaging bandwidth. However, unlike the first generation NovaSAR, aimed primarily at maritime and forestry services, an X-Band variant would alter the proposed application areas due to multiple factors, but particularly the typically reduced imaging swath and reduced RF penetration of the target. This may make it more suitable for urban and mapping applications, but limit its utility when needing to image very large areas or gain more information about the structure below the surface observed. An X-Band variant will importantly benefit from significantly greater access to higher incidence angles, providing shorter revisit intervals.

Those Gallium Nitride Solid State Power Amplifiers (SSPAs) that afforded NovaSAR an improvement in RF conversion are becoming available in X-Band with good levels of efficiency (35-40%). Though these may need more space qualification, the GaN devices are well suited to space with their inherent radiation hardness and are expected to perform well.

Increasing Payload Duty Cycle

While the GaN High Electron Mobility Transistor based SSPAs to be launched in the first mission are running at efficiencies of 40-45%, the latest S-Band high power products just to reach the market have efficiencies of 50-55%, and development is planned for the higher power variants needed for SAR imaging from space. These products need space qualification, but will improve the power requirement.[vii]

The power availability to the platform may also be improved by adding additional deployable solar panels. Unlike extending the phased array antenna, adding solar panels is a less challenging engineering task, with simpler connections required to the platform and less stringent alignment requirements.

Other Improvements

SSTL is continually improving the specifications and capabilities of its various satellite platforms and subsystems. Through a variety of mission programs and technology development activities, various subsystems are being redesigned and improved, and space proven in a piecewise fashion – the risk managed through redundancy on operational missions and technology demonstration missions.

The SSTL avionics suite has been improved, along with data storage and on board computing. Within the timescale of the next NovaSAR mission, these improvements may help realise mass reductions in the order of 50kg on this platform type with corresponding volume and power consumption improvements. The extra platform resource may also be used to offer increased mass storage.

Parallel activities include the modification of the X-Band downlink chain currently on board NovaSAR-S to a Ka band variant.[viii] While using Ka band results in increased platform stability and control requirements, the narrower beam allows for higher power and a much higher bit rate for data downlink. Though this band may experience a higher level of signal drop out due to atmospheric moisture, a variable bit rate can be implemented to mitigate this.

Figure 3 – 'Selfie' of SSTL high gain antenna pointing mechanism on board TechDemoSat-1

Software and Ground Segment Modifications

The SSTL mission planning software solution is designed to function in a distributed manner, allowing multiple users to interface with the central mission planning system. However, the various requirements of the satellite, for example the need to reorient the satellite for different imaging modes, the thermal control and power regulation requirements mean that on occasion imaging requests must be prioritised to maximise the mission utility. This prioritisation is done automatically, but occasionally requires human intervention. As the number of users increases the complexity of scheduling the imaging tasks increases. Systems can be developed to improve the automated sorting of the imaging tasks, making a many user interface feasible.

One particularly interesting area of application development will come following the commissioning of the first mission, the data fusion of near-synchronously acquired SAR imagery and AIS signals transmitted by ships. NovaSAR is likely to be the only satellite with this capability at time of launch, and if not first will be the only one with very wide swath imaging modes. Previously, where images are acquired at a different time to the AIS data, there is a need to extrapolate and predict ships past or future positions based on old data. It is likely that simultaneously (or near simultaneously as ships do not transmit AIS constantly) acquired data will allow for automated processing of fused data, with AIS unique identifiers bringing together tertiary AIS data to provide a fuller maritime picture.

Within the context of image processing, there are also opportunities to automate and productise output data on the fly to provide the best application performance. The Satellite Applications Catapult promises to kick off a broad range of activities in this area.

Improvements Offered through Constellation

NovaSAR was always intended to function in a similar manner to the Disaster Monitoring Constellation, where individual satellite owners come together to benefit from each other's resources[ix]. Low earth orbiting satellites spend only a proportion of their time over the owner's country. During other times the satellite capacity may be traded to offer improved imaging capacity and revisit rates over other areas of the interest.

The first NovaSAR satellite will be placed in a polar orbit, which has global reach but revisit in the equatorial regions is not as good at the poles. Future NovaSAR owners may opt for further polar orbiting satellites, or may go for lower inclination orbits. The revisit rate will then be improved.

Figure 4 – NovaSAR average revisit time in a polar orbit – 1 satellite, imaging mode 4, yellow is 3 days, green <1 day

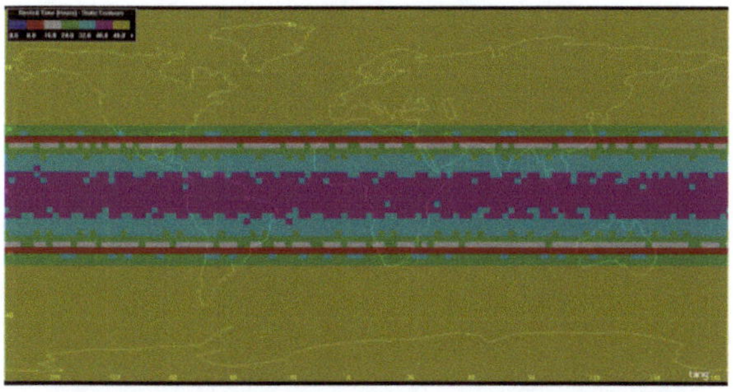

Figure 5 – NovaSAR average revisit time in 30 degree inclined orbit – 1 satellite, imaging mode 4, purple is 40 hours, green is 24 hours

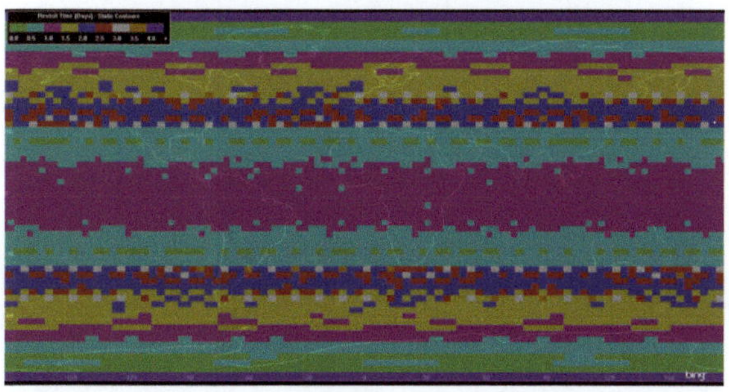

Figure 6 – NovaSAR average revisit time in combined 30 degree inclined and polar orbits – 2 satellites, imaging mode 4, light blue is 12 hours, purple 24 hours

DISCUSSION

The NovaSAR program is intended to remain on a small platform, to offer an affordable and accessible solution for those requiring SAR data. The lower cost of the system that makes constellations more affordable is key to its success. It is possible to envisage a NovaSAR system that conforms to the historic 'all things to all men' archetype. However, this would not offer anything substantively new and already exists elsewhere. The benefits to be found in the NovaSAR program are those where the changes can be done without breaking the bank. Of the above changes, increasing the phase centres, adding deployable solar arrays and implementing the improved SSTL subsystems may be achieved at low system resource 'cost', in some cases reducing the overall mission mass and volume. These improvements may well allow the imaging payload to be run for longer each orbit, a key user requirement. This offers the opportunity to propose a wider range of SAR satellite variants to the NovaSAR constellation and meet a more user needs.

REFERENCES

1 Davies, P (2012) NovaSAR - Bringing Radar Capability to the Disaster Monitoring Constellation, 26th Annual AIAA/USU Conference on Small Satellites

2 Iervolino, P, Guida, R and Whittaker, P (2013) NovaSAR-S and Maritime Surveillance In: IEEE Geoscience and Remote Sensing Symposium, 2013-07-21 - 2013-07-26, Melbourne, Australia.

3 Guida, R. Validation of S-band data performance for future spaceborne SAR missions.

4 □ Fransson, J. E. S., Walter, F., & Olsson, H. (1999). Identification of clear felled areas using SPOT P and Almaz-1 SAR data. *International Journal of Remote Sensing*, *20*(18), 3583-3593.

5 Litovchenko, K., Ivanov, A., & Ermakov, S. (1999). Detection of oil slicks parameters from ALMAZ-1 and ERS-1 SAR imagery. In *Geoscience and Remote Sensing Symposium, 1999. IGARSS'99 Proceedings. IEEE 1999 International* (Vol. 3, pp. 1484-1486). IEEE.

6 CHEN, Q., LI, Z., & WANG, L. (2005). AN ANALYSIS OF APPLYING S-BAND SAR OF HJ-1C TO SOIL MOISTURE CHANGE RETRIEVAL [J]. *Remote Sensing For Land & Resources*, *2*, 003.

7 http://www.macom.com/about/news-and-events/news-archive/row-col1/news-archive/macom-news-list-archive/macom-extends-industry-leading-1

8 Ferris, M., & Phillips, N. (2011). The use and advancement of an affordable, adaptable antenna pointing mechanism. In *Proc. 14th Eur. Space Mech. Tribol. Symp* (pp. 227-234).

9 Stephens, P., Cooksley, J., da Silva Curiel, A., Boland, L., Jason, S., Northham, J., ... & Sweeting, M. (2003, November). Launch of the International Disaster Monitoring Constellation; the development of a novel international partnership in space. In *Recent Advances in Space Technologies, 2003. RAST'03. International Conference on. Proceedings of* (pp. 525-535). IEEE.

BIS-RS-2014-12

Designing for Cost Effectiveness Results in Responsiveness: Demonstrating The SSTL X-Series

Shaun Kenyon
Surrey Satellite Technology Ltd
Tycho House, Guildford, GU2 7YE, UK
s.kenyon@sstl.co.uk

Dai Stanton
Surrey Satellite Technology Ltd
Tycho House, Guildford, GU2 7YE, UK
d.stanton@sstl.co.uk

ABSTRACT

Demand for small low-cost spacecraft – either as individual missions or in groups such as clusters, swarms or constellations – has been increasing significantly over recent years. In parallel, the microsatellite market has seen an increase in the rate of innovation, as more players create new micro (and nano) satellite offerings. In order to address this increased demand, SSTL has investigated, qualified and implemented a new satellite platform production process and associated new set of avionics as part of a larger innovation framework. The first fruits of this development approach are the new X-series of spacecraft. The process makes significant use of modern automated manufacture and test techniques, and the avionics are designed taking this into consideration. The consequence of this is that significant savings in production costs and schedule are achieved, which are quantified in this paper. A secondary and serendipitous effect of this development is the X-series family of spacecraft (and an expandable space technology framework) which is tremendously well matched to operationally responsive space applications. The first mission to use an X-series platform has been contracted for delivery. This paper outlines how the X-series was designed primarily for cost effectiveness, and how these cost savings manifest themselves as time savings. Additionally, the use of an X-series spacecraft as the launch vehicle avionics is discussed, highlighting the cost, schedule and mass benefits to the launch segment of the mission. This paper then summarises how all the design features combine and compound on each other resulting in a game changing approach to cost effective responsive space, ultimately leading to a demonstrable ability to design, build and test a cost effective microsatellite platform in less than 6 months.

KEYWORDS: Batch Manufacture, Reflow, Schedule, Mission Flexibility, Standard Interfacing, Build-from-stock

INTRODUCTION

Over the last 33 years, SSTL has delivered 43 small satellites for a range of applications including Earth Observation, Communications, Navigation, Science and Technology Demonstration. Last year, SSTL released the first of a new range of mass efficient, high performance low cost satellite platforms – the SSTL X-Series.

Initially implemented on a high performance 50kg class platform, the X-series core architecture and avionics will be implemented in all SSTL's small platforms over the next three to five years, thereby lowering the cost and improving delivery times for all X-series platforms.

SSTLs current range of platforms has been extremely successful in the market place, offering high performance at low cost. A special SSTL internal R&D programme was started in 2011, whose main objective has been to improve performance to cost ratios even further, by exploiting new commercially developed technologies, protocols and processes in order to offer the equivalent performance of current SSTL small satellite platforms at a significantly improved mass, schedule and price point. This internal development programme has therefore realised the X-Series.

The X-Series platform developments span the range of current and future SSTL platform classes.

S. Hatton (ed.), *Proceedings of the 12th Reinventing Space Conference*, DOI 10.1007/978-3-319-34024-1_7

Figure 1 illustrates the envelope of X-series platforms in the context of current SSTL mission and satellite platform offerings

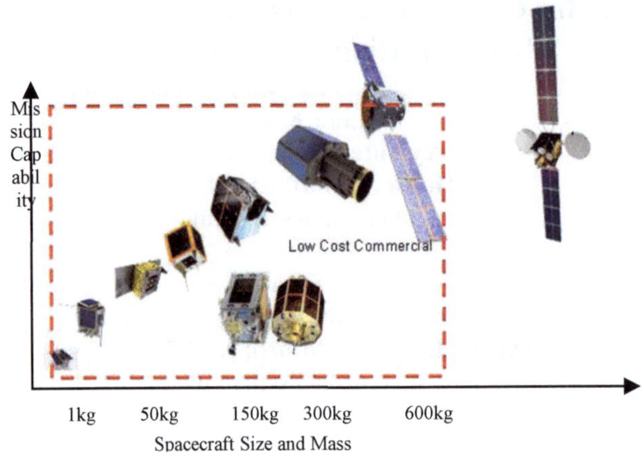

Figure 1: Capability-mass analysis of SSTL missions

THE X-SERIES

At the time of writing, there are two X-series platforms under development; the X-15 and the X-50. The X-series platforms are designed primarily for users who want to develop their own payload, or want to pilot an idea in a cost-effective manner quickly. The low recurrent cost and high performance to mass ratio offered by the X-Series satellites make them ideal for low cost science and in orbit demonstrations missions, where an affordable and rapid solution can, for example:

- Provide a platform for on orbit demonstration of new technologies, instruments, services or techniques,
- Provide risk reduction, pre-cursor or gap filler type capabilities by accommodating scientific and/or imaging payloads.

The available payload volume for the X-50 variant is a 530x430x400mm contiguous volume, with access to 5 sides of the cuboid shaped spacecraft. The 50kg bus can accommodate 45kg of payload, and provide 35W of Orbit Average power and 85W peak power in a standard spacecraft configuration.

The spacecraft structure can be modified as demanded for different missions, and in its simplest form a payload is carried on top of the core avionics providing an instrument field of view over more than an entire hemisphere.

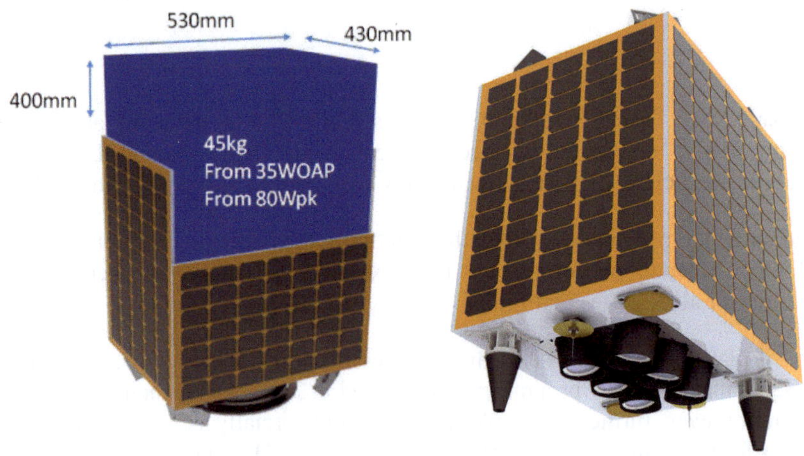

The X-15 variant of the X-series of platforms is targeted at missions that require a very low cost-per-spacecraft metric; primarily entry-level missions, extremely rapid demonstrations or large constellation deployments with modest payloads. The platform can accommodate up to 15kg of payload mass within its ~30cm cuboid structure, and can generate an orbit average power of approximately 20W. Development of this platform offering continues.

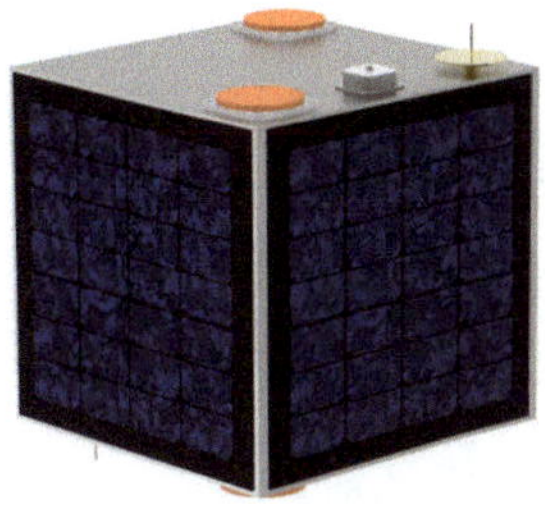

COST AND TIME SAVINGS

Schedule, price and reliability are very important factors for all customers, and increasingly commercial spacecraft operators are considering groups of satellites working as part of a constellation. SSTL has been involved with a number of such constellations in various capacities, including DMC, RapidEye, KANOPUS, FORMOSAT-7 at spacecraft level, and at payload / subsystem level in ORBCOMM-SG, CYGNUS and the 22 Galileo FOC payloads. The production philosophy typically adopted for such batches of satellites to consider the closest existing product, modify this to the specific mission needs, build a proto flight unit, and then carry out a limited batch production run. As a result, the cost savings that can be achieved in batch production are often limited, as the base design was never optimised for such a batch production.

For individual spacecraft, the cost of the spacecraft production can also be prohibitive for science and technology demonstration missions, or restrict the feasibility of new business ideas in space. Increasingly, the labour costs involved in spacecraft production are dominating.

SSTL has developed a concept for spacecraft manufacture which has so far not been used elsewhere in the space industry, which removes some of these limitations and cost drivers. This is targeted at improving production cost, speeding up mission delivery, and maintaining and improving reliability.

The low cost, batch build philosophy allows for a module level rather than component level stores inventory, drastically reducing the period from customer order to spacecraft delivery. Coupled with the low cost of the recurrent parts, this results in extremely low recurrent costs for the core X-Series elements.

Key to this approach is moving to card frame based hardware architecture at the heart of all of the X-Series platforms, developed with maximum flexibility and modularity in mind. A number of standard building block "cards" deliver the main platform functions such as power conditioning, TT&C, on-board processing and storage. The card-frame can be configured in an expandable arrangement to deliver the platform performance required by different payloads and associated mission classes.

Savings in Mission Design Phase

The rewards of this approach are first reaped in the mission design stage. A more traditional approach to new mission design is to take an existing spacecraft design as a baseline, identify any changes necessary to the design and implement the non-recurring engineering (NRE) to realise the differences. Often, the NRE occurs at module level, where electronics may have to go through a design, review, prototype, test and manufacture cycle.

In the new SSTL X-series approach, module-level NRE is limited to payload interfacing only, and changes to mission requirements such as power generation or data storage and handling are dealt with at system design level, by including more or less modules of the same design. This creates a truly modular framework that simplifies initial spacecraft system design.

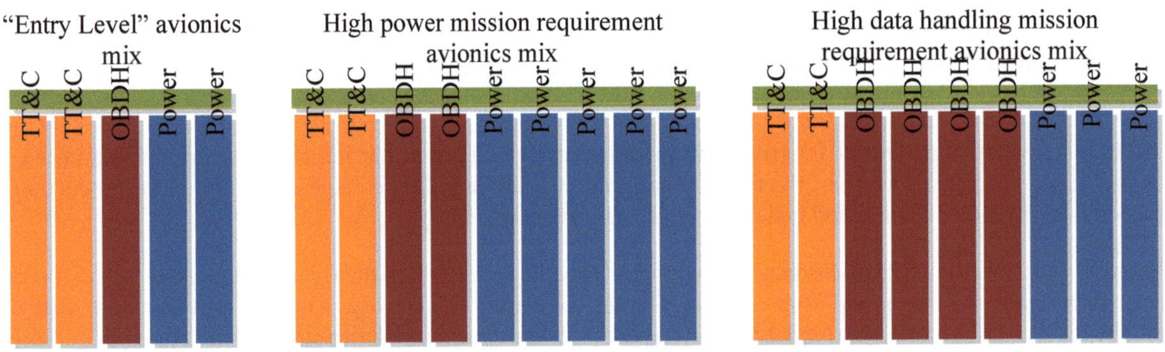

Savings in Manufacturing Phase

The X-Series avionics and core platform elements are designed with maximum production efficiency as a driver. Hence the design takes advantage of automated batch manufacture and test processes. This entails management of a controlled preferred parts list, use of pick and place machines, reflow soldering process, press fit connectors and comprehensive test coverage via automatic test equipment and built-in self-test.

In recent years SSTL have invested heavily in mass production facilities and processes. These in-house capabilities are now installed and qualified, and are being levered to drastically reduce the production costs of the next generation X-Series elements. The key capabilities are:

- Pick & Place of Components
- Automated soldering (reflow)
- Automatic Inspection, including X-ray inspection of re-flowed electronics
- Automated Test

These factors result in a step change in production approach: Typically when producing space hardware, significant costs are expended both on the raw materials and parts ('Space Qualified' parts being extremely expensive with associated long lead times), and the skilled labour associated with largely manual assembly and test activities. Even if lower cost (e.g. COTs) parts are used, the significant labour elements results in a high cost investment in any space equipment produced in this way. A secondary effect is that the resulting hardware is typically extensively analysed, tested and repaired if it develops a fault – due to the high value, discarding the equipment at the point where the fault is discovered is not a realistic option. This pattern is broken by the production approach for the X-Series; low cost parts are procured, automated process are used extensively to manufacture, inspect and test the hardware. Therefore if a fault is detected, during test, for example, the option to simply discard the element in question becomes an easier and more attractive option in the cases where faults can be shown not to be systematic and/or design related.

The following table illustrates in broad labour cost terms (man hours/weeks/days/months) the differences between three production approaches:

- A traditional space industry approach, where Space Qualified Parts are employed and the manufacture and test phases are largely bespoke and manual
- SSTL's historical approach, where COTs low cost parts are utilised, but the remaining production phases remain labour intensive,
- The X-Series approach, where COTs parts are used and the remaining production activities are all automated and therefore shortened in duration.

	Approach Scenario	*Traditional/ Bespoke*	*COTs Manual*	*COTS Automated*
Production Phase Durations	Procurement	Weeks	Days	Days
	Assembly	Weeks	Weeks	Hours
	Inspection	Days	Days	Hours
	Test	Weeks	Weeks	Hours
	Overall	Months	Weeks/ Months	Days

Table 1: Comparison of production of equipment between traditional labour intensive methods and SSTL X-Series automated techniques

It can clearly be seen that a dramatic saving in production time and therefore cost is attained with the X-Series approach. This is further illustrated in the graph below which illustrates broadly how the proportion of costs associated with each production phase and activity changes with the different approaches adopted.

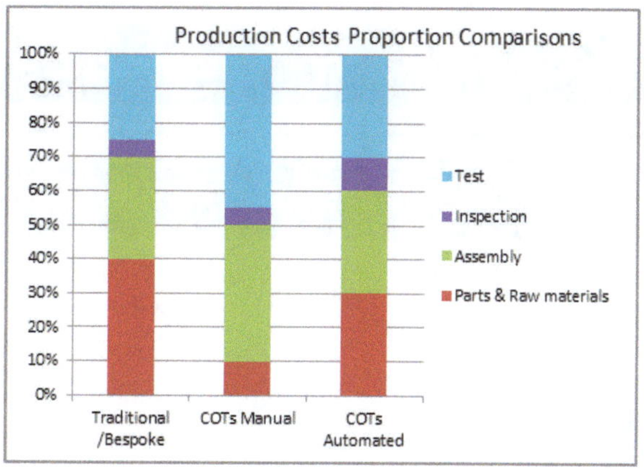

Figure 4: Breakdown of production of equipment between traditional labour intensive methods and SSTL X-Series automated techniques

The manufacturing approach has recently been put to the test, the first batch of X-series cards has been manufactured and the effort savings have been shown to be approximately 55%.

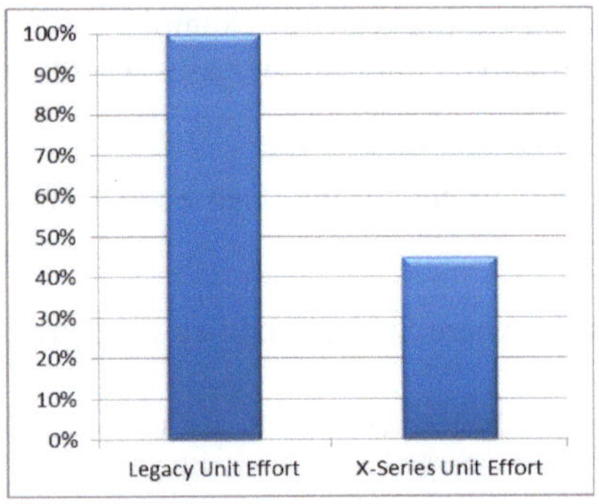

The effort savings are largely from moving to an almost completely pick & place, reflowed design. The process is fully automated and so the effort savings increase with manufacture batch size. This reduces the time to build the units, and if still using a build-to-order approach, would still yield some time savings on the whole mission timeline. However, since the build effort is reduced, this dramatically reduces the value of the built units, allowing the units to become a low enough value item to be stocked at risk without specific missions associated to each of the units. If known good (i.e. fully tested) stock is held, then further time savings on the top level mission schedule can be made.

Savings in AIT Phase

The knock-on effects of lower cost avionics being ready in stock are significant for the AIT phase of a mission. In a legacy AIT phase, the manufacture of all the various modules (to order) has to be carefully managed such that they arrive for integration in to the spacecraft in the correct order. Managing lead times for various units becomes a major part of AIT schedule management, and defines the critical path.

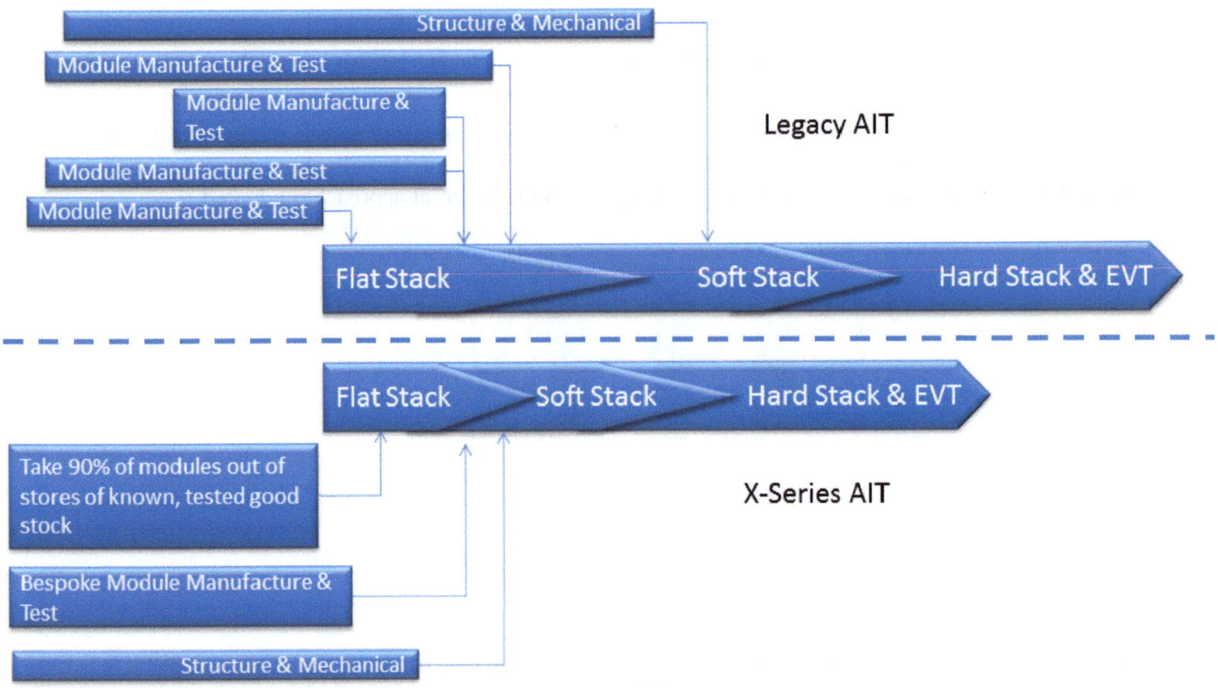

With all the core avionics in stock ready for integration, however, integration of the spacecraft can begin immediately after the mission critical design review is complete. This allows for a more streamlined schedule to manage, with long-lead items being limited to mission-specific bespoke items.

LAUNCH AVIONICS

The primary focus of the development of the X-Series avionics was a disruptive reduction in the build costs of an SSTL spacecraft. For launch safety and simplicity, it is normal for SSTL spacecraft to be launched unpowered, with the power to the spacecraft only being switched on at the point of separation from the launch vehicle. Therefore vibration tests have tended to only validate the functionality of SSTL spacecraft *post*-launch, not *during* launch.

The X-series avionics however, have been designed to not rule out a "live" launch, or indeed be used as launch vehicle avionics in and of themselves.

Additionally, the OBDH module and equipment bus have been sized to allow for guidance and control loops in the order of tens of hertz, partly because of the potential for very agile mission requirements (such as rendezvous and docking missions or rapid off-pointing of payloads), but also in part to enable the avionics to be used as launch vehicle control avionics.

APPLICABILITY TO RESPONSIVE SPACE

SSTL has a rich history of building spacecraft faster than the industry standard. Mission design and build lifecycles tend to be less than 4 years. However SSTL has also on occasion built spacecraft in less than 12 months. The X-series takes a further step towards a sub-12 month design and build schedule becoming the norm. It does this by building on the robust and proven design philosophies of all SSTL spacecraft (for example keeping the same FDIR and ConOps strategies, using stable automotive industry standards where applicable), whilst capitalising on new electronic manufacturing techniques developed for terrestrial industries. In terms of responsive space, being able design, build, test and deliver a spacecraft platform ready for launch in less than 12 months opens many different options for rapid deployment of space assets to react and respond to various scenarios.

The existing SSTL policy of maximising the use of industry standard interfaces to reduce development costs is maintained and extended in the X-series, with extensive use of CAN, LIN, LVDS, RS422/485, SpaceWire and internet protocol (IP). Further capability for USB and CameraLink is also possible. In terms of applicability to Responsive Space, the use of industry standard interfaces allows for abstraction of payload development, or the rapid integration of payloads based on modern terrestrial COTS technologies, with interfaces that most engineers are familiar with regardless of field.

The development of the X-series capability to launch live or indeed even act as launch vehicle avionics also opens up possibilities for extremely rapid LEOP procedures and in-orbit operations within hours of launch.

CONCLUSIONS

SSTL continues to develop its product range become ever more cost effective. The most recent development to that end – the X-series range of spacecraft platforms – aim to reduce total mission cost by reducing the time required to design the mission and manufacture the units, and by allowing unit costs to be low enough such that stock can be held at risk so that spacecraft integration reaps the benefits of having all core avionics ready for integration at the beginning of AIT.

By reducing the time aspect of mission development for cost purposes, SSTL has also enabled a framework for developing missions extremely rapidly, which may then also have uses in its own right, in the responsive space community.

REFERENCES

1. Bradford, A.E. and D. Stanton. October 2013. "Lowering the Cost of Space Access – A New generation of Low Cost SSTL platforms" *International Astronautical Congress*, Beijing, China.

2. Cawthorne, A. and A. Da Silva Curiel. October 2014 "The SSTL-X50 TrueColour Mission" *International Astronautical Congress*, Toronto, Canada.

3. Bridges, C.P. and S.R. Kenyon. August 2013. "A Baptism of Fire: The STRaND-1 Nanosatellite" *Small Satellite Conference,* Logan, Utah.

BIS-RS-2014-18

Thinking Differently about Standard Smallsat Interfaces- Let Adapters Take the Brunt

daniel Lim
TriSept Corporation
14425 Penrose Place, Suite 270, Chantilly, VA 20151; (703) 297-4622, ext. 7170
dlim@trisept.com

ABSTRACT

With the increase in interest of small satellite technologies and solutions comes a strong desire by the industry to simplify integration of the spacecraft with the launch vehicle (LV) interfaces. Those that have been developing, integrating and launching smallsats for some time are seeking to reduce the need for "one-off" mechanical, electrical, environmental, and operational interfaces between the satellites and LV. The prominent solution to this desire for better-defined interfaces is a call to develop standardized interfaces. Smallsat providers are interested in a standard interface to help bound their spacecraft design trade spaces (it is important to be efficient this this fiscally constrained environment) and to simplify the integration process when their systems are ready for LV integration and launch. LV providers are keen on standardized interfaces with smallsats to reduce complexity and analyses costs in the integration process so that they can focus on their main objective: The successful integration and launch of the primary spacecraft provider for a rideshare mission.

This paper contends that the creation of standardized interfaces by a governmental body is not the most optimal answer to meet the overall needs of smallsat and LV providers: The desire to have a clear expectation of what the interfaces are, to simplify interfaces to reduce one-off analyses, and enough bounds to guide smallsat developers to ensure the maximum probability that their spacecraft is successfully integrated and launched. Instead, the paper asserts that the integration hardware (e.g., LV adapters and dispensers) providers and mission integrators should be given the responsibility of providing a consistent, well-defined interface to each LV provider and for the smallsats that are integrated with the hardware. Defining consistent interfaces in this nuanced fashion allows the interface hardware providers and integrators to absorb the complexity of defining and managing interfaces between the smallsats and the LV providers. Further, the paper presents how interface hardware and integration service providers are best suited to lead efforts to setting clear and constant interfaces and expectations to each stakeholder because of their typical role as the intermediary across the interfaces.

Moreover, the paper presents a clear argument for allowing the United States (US) and international space market to define "industry standards" based on viable and successful interface systems versus creating an interface standard by a governmental entity and requiring the industry to adhere to the defined standard. This paper describes how the latter practice is much less likely to be adopted by the industry as a whole and why this method of defining the interfaces creates definitions that are much less responsive to new technological and methodological advances and lessons learned. Even more, the paper describes how the latter is more likely to over-constrain smallsat developers. The paper cites several examples that the former means to convey clear and consistent expectations to smallsat and LV providers increases the probability of more widespread adoption, while at the same time, allows for responsive evolution of interface requirements and maximizes the design space and flexibility for smallsat creators and developers.

KEYWORDS:

Standard interfaces, rideshare, small satellites, smallsats, piggy-backing, adapters, dispensers, industry standards, ESPA, CubeSat, P-POD, FANTM-RiDE

THE NEED FOR CHANGE

On the Cusp of Progress…

The United States (US) and international space industry is at the precipice of major transformation from "business as usual" in the manner in which programs acquire, develop, launch and field space assets to orbit. The global economic

This document does not contain technical data as defined by the International Traffic in Arms Regulations, 22 CFR 120.10.

S. Hatton (ed.), *Proceedings of the 12th Reinventing Space Conference*, DOI 10.1007/978-3-319-34024-1_8

downturn that began in 2008 has driven the international community to re-evaluate how to more efficiently and affordably deploy and operate space systems and architectures while not appreciably diminishing the effectiveness of the systems currently on orbit. Leadership worldwide has been evaluating how to reach this vision through architecture studies and analyses, and both the government and commercial space industries have been strongly considering small satellite (smallsat) solutions to reducing costs, shortening the time of technology infusion to orbital assets, and bring resilience to space architectures. This evolution follows the phenomena of Moore's Law and the increase in technical and mission capability of smallsats, as depicted in Figure 1. This figure, adapted from this author's presentation "Shaping Our Future: Practical Actions to Bring Real Industry Change" given at the 2014 Small Payload Rideshare Association Rideshare Conference, depicts how the intersection between Moore's Law and smallsat innovation brings resilient, effective and lower cost space architectures, consisting of more agile large satellites, constellations of smallsats, as well as hybrid constellations of interconnected networks of small and large satellite systems[1]. Although this phenomena should occur naturally in the advancement of space development, the process has been stymied by institutional and programmatic barriers that hinder opportunities for smallsat programs to demonstrate true mission utility to decision makers. Therefore, the entire space industry stands at the cusp of advancing towards more affordable, effective and resilient space architectures, yet wholly dependent on changes in the mindsets of leadership of government and commercial space programs to progress forward.

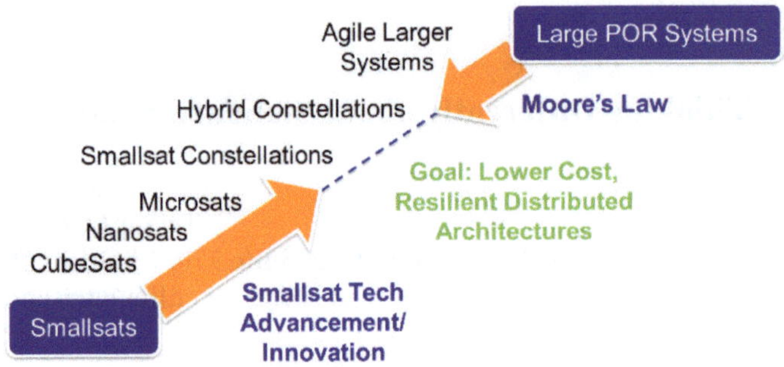

Figure 1. Intersection of Moore's Law and Smallsat Innovation Brings Robust Space Architectures. [Adapted from 1]

Smallsats Are Critical for Progress

In order to progress towards the goal of robust, advanced, lower cost, and resilient space architectures, smallsat technologies must be allowed to demonstrate their worth in terms of real operational and scientific capabilities. Furthermore, as commonly applied today, smallsat systems are necessary in the ecosystem to demonstrate emerging technologies and applications that will be incorporated into larger space systems. In his Master's thesis entitled "EELV Secondary Payload Adapter (ESPA) Ring: Overcoming Challenges to Enable Responsive Space," Robert Atkins argued that smallsat technologies were a necessary step in reducing risk to larger space systems by increasing the technical readiness level (TRL) of technologies that need first to be proven out prior to integration into the larger architectures. Atkins states, "By taking the new technology and testing it in the operational environment, the feasibility and concept is proven, which greatly reduces the unknowns in development [of larger spacecraft systems]."[2]

But Something's Holding Back Forward Movement…

As alluded to earlier, several barriers restrict the number of smallsat technologies that are put on orbit, and thus, hold back the entire space industry. As detailed in this author's paper entitled "Defining a Roadmap to Bringing the US Space Industry Back to Health," two phenomena stunt progress of US (and international) space programs: The "launch cost dichotomy" and "institutional inertia".[3] Without going into much detail on these phenomena, the launch cost dichotomy is summarized as follows: Decision makers desire distributed (disaggregated) architectures of smaller spacecraft, in theory, to reduce the cost of larger systems and overall architectures. However, to prove out these theoretical disaggregated architectures, smallsat technologies must be put on orbit to demonstrate system performance. Yet, high launch costs preclude smallsat access to space for many small payload providers, thus leaving these distributed architecture as theoretical. In short, systems must be distributed to reduce overall architecture costs, but the high price point for launching the smallsats that will constitute the disaggregated architectures dissuade decision makers to embrace distributed architectures in their trade studies.

Institutional inertia also constrains access to space for the smallsat technologies that would bring more robust and advanced systems on orbit. Five major mindsets held by key decision makers in the space industry are outlined in this author's aforementioned paper that hinder adoption of smallsat technologies in operational and scientific programs. In the physical realm, objects at rest tend to stay at rest and objects in motion tend to stay in motion; accordingly, the same phenomenon occurs for space programs, methodologies, and ideologies. The driving factor that hampers smallsat access to space is *not* technical complexity; institutional mindsets are the main reason for reduced frequency of smallsat launch opportunities.

Dismantling the Barriers

To break through the launch cost dichotomy, the international space industry is increasingly turning to "ridesharing" (also known as "piggy-backing") of smallsats with other spacecraft to divide up launch costs for each spacecraft on the manifest. Instead of a smallsat provider having to obtain funding for an entire launch vehicle (LV) to gain access to space, by ridesharing spacecraft, the smallsat provider would only be responsible for paying for a smaller portion of the total launch costs. Two types of rideshare missions have emerged to facilitate smallsat access to space: Standard rideshare (sRS) and dedicated rideshare (dRS) missions, as outlined in this author's paper entitled "Practical Knowledge on Opening Up Low-Cost US Launch Opportunities for International Smallsats."[4]

For the sake of brevity, sRS missions are those that are comprised of a primary spacecraft provider that wholly dictates the mission requirements (e.g., launch date, orbital parameters, spacecraft separation timing, etc.), with one or more smaller spacecraft flying as secondary (also called auxiliary) payloads (APLs). With sRS missions, the primary spacecraft provider must be willing to take on additional APLs on their manifest. Currently, many of them decline the potential of sharing the overall launch costs with rideshares because of high perceived and actual risks/complexity added by including ridesharers on their missions. In contrast, dRS missions do not have a primary spacecraft and is composed of a number of smallsats ridesharing on a single mission. These missions provide greater control for the ridesharers over the mission requirements, such as launch date and final orbits. However, dRS missions are more difficult to build because it requires a complex effort in aggregating enough smallsats to fully fund the overall launch costs for that mission.

Both types of rideshare missions provide the potential of significant cost savings for the smallsat providers. As an example, Atkins provides the following example scenario in his thesis: The US Air Force Space Test Program (STP)-1 mission in 2007 consisted of five distinct spacecraft with nine total experiments launched on a single $90 million (M) US dollars (USD) mission. He stated that that single mission was "equivalent to seven launches spanning a year's worth of launches on the [Evolved Expendable Launch Vehicle (EELV)] manifest and costing approximately $700M (USD) in launch vehicle cost."[5] Obviously, there was a substantial cost benefit to using rideshare to divide up the launch costs and provide a greater utilization of launch capabilities for these spacecraft customers.

Despite the financial advantages brought by ridesharing spacecraft, widespread adoption of these modes of smallsat access to space have yet to materialize except for pockets in the international space industry. This is due to the institutional inertia that continues to throttle forward progress. Primary spacecraft and LV providers continue to view rideshare as more hassle than benefit, and government and commercial decision makers continue to hesitate in funding enablers and opportunities for smallsats to exploit launch opportunities.

In order to shift the paradigm towards frequent, lower cost access to space for smallsats, and subsequently, improvement of all international space programs, the smallsat industry must work together to remove all excuses for not allow ridesharing to occur on the majority of launch missions around the world. Specifically, the smallsat industry must provide technologies that will show clear applicability to future integration into the next generation of space architectures, either with payloads with direct operational or scientific applicability, or indirectly by launching technologies that raise TRLs of enablers for future systems. Simultaneously, the smallsat industry must work to make their spacecraft least intrusive and non-interfering to primary spacecraft providers and low risk and complexity to LV providers. They must be "transparent" to these two entities in order to encourage them to allow for more rideshare missions.

Standardized Interfaces for Smallsats Facilitate Change

Aside from specific technological enablers, one (of several) facilitators to reducing complexity and risk in adding rideshare spacecraft to missions is a well-defined standardized interface between the APL rideshares and the LV*. To establish a baseline lexicon for this paper, a standard is defined as Jeff Ganley from the Air Force Research Laboratory

* The purpose of this paper is not to outline other enablers to addressing the barriers to smallsat rideshare, such as technological solutions, innovative methodologies, and changes to national and international space policies and regulations; the paper seeks only to address how standardized interfaces contribute to the advancement of the international space industry.

(AFRL) stated in his paper "Small Satellite Standards Development": "A standard is a rule or requirement that is determined by a consensus opinion of users and that prescribes the accepted criteria for a product, process or procedure."[6]

By defining and enforcing a standard APL to LV interface, primary spacecraft and LV providers will have a predetermined set of expectations and requirements for the integration of rideshares on their missions, significantly reducing risk and complexity on these launches. Atkins comments "a standardized launch service aboard the majority of all missions manifested… will allow many programs to deploy new, faster, smaller satellites into orbit…"[7] He continues by conveying that the "standardization process" will allow for "safe deployment of the secondary payloads on a noninterference basis with the primary payload, which will benefit US space programs."[8] This truth certainly applies to space programs throughout the international community as well.

Additionally, Reese, Martin, and Acton from STP stated in their presentation at the Small Satellite Conference in Logan, Utah entitled "STPSat-3: The Benefits of a Multiple- Build, Standard Payload Interface Spacecraft Bus" that standardized interfaces maximize potential launch opportunities for their smallsats by allowing for integration on a wider selection of LVs, as well as by reducing risk and schedule during integration.[9] Analogously, the Futron Corporation speaks of the benefits of standard interfaces of payloads onto host spacecraft buses in their "Hosted Payloads Guidebook" prepared for NASA Langley Research Center. The guidebook states that standard interfaces create greater "flexibility" in choosing hosts because it introduces a level of interchangeability of payloads because the interface requirements would remain uniform across all candidate hosts.[10] The guidebook also asserts that utilization of standard interfaces also proffers a "price reduction" in launch costs for the host spacecraft as they contract with the rideshare hosted payload for the slot on their bus. These statements are made with respect to hosted payloads, i.e., small payloads without smallsat buses that rideshare on host spacecraft buses versus on the LV itself. However, the benefits of standardized interfaces directly applies to the discussion of this paper because the integration challenges between hosted payloads and host buses are very similar to the issues in integrating whole smallsats onto LVs.

Historically, the international community has agreed with the necessity to standardize interfaces. In his article in the International Standards Organization (ISO) Focus magazine entitled "The launch business: Standard formats for launch vehicle - spacecraft interface documents", Phillipe Boland strongly advocates for the establishment of standard interfaces between launch vehicles and spacecraft, citing several ISO standards development efforts to clearly define these interface requirements.[11] Boland cites in an earlier article "Developments and Initiatives: Interfaces between launch vehicles and spacecraft" that delegates from Europe, Japan, Russia, the US, China and Brazil have been involved in the working group to develop LV to spacecraft standards.[12] Specific to smallsat integration, ISO has published a standard interface control document (ICD) for the "small-auxiliary-spacecraft (SASC)-to- launch- vehicle interface" (ISO 26869:2012) to help define the types of requirements necessary for integration between the LV to the smallsat APL.[13]

Why Standardized Interfaces Help Increase Smallsat Launch Rates

The potential benefits of defining standardized interfaces for smallsat rideshare are four-fold, summarized in Figure 2. First of all, they help alleviate concerns of the primary space vehicle (SV) and/or LV provider by providing a clear understanding of the requirements levied on the rideshare spacecraft to ensure that they will "do no harm" to the primary mission (for sRS launches), or the LV and other ridesharers (on dRS missions). This clear definition of requirements will enable the primary SV and LV providers to baseline the risks and expectations of having one or more rideshare spacecraft on each mission, helping to alleviate the institutional psychological stigma against rideshare spacecraft on their missions. For instance, standardizing the mechanical attachment interface of the rideshare APLs to the LV or its adapter provides the LV provider confidence that mission unique analyses for the mechanical mating surfaces will not be necessary for the predominance of rideshare missions. An operational standard interface requirement, like stipulating that rideshare

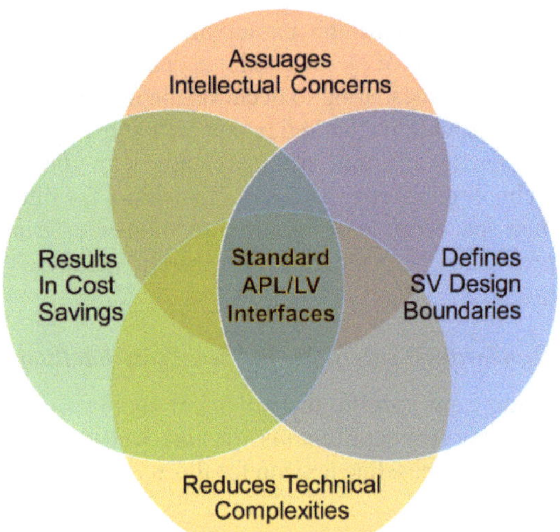

Figure 2. Potential Benefits of Clearly Defined Standard Rideshare to LV Interfaces

spacecraft will not power on nor transmit data until after the primary SV has separated, negates the concern of electromagnetic interference (EMI) of rideshare spacecraft that may damage sensitive primary SV systems. Standardized interfaces define these requirements and expectations up front, reducing the concern of possible degradation, destruction, or delay of the primary mission because of complex rideshare APLs.

Secondly, a standard smallsat to LV interface provides the added benefit of delineating design bounds for spacecraft development. Although this could potentially serve as a double edge sword to over-constrain spacecraft design, defining a standard interface can provide helpful technological development direction for the smallsat providers in terms of parameters such as overall spacecraft mass, the physical bus dimensions and mechanical interface to the LV, electrical harnessing and signal protocols, or allowable materials to build spacecraft components. The standard can also provide smallsat providers a clearer understanding of what to expect during the overall mission integration process or orbital deployment operations. By building to the requirements set in the standards, the smallsat APL increases its probability of obtaining an opportunity to be integrated and launched to orbit because the standard requirements should ease the concerns of technological and programmatic risk and complexity of rideshare.

Additionally, standardized interfaces reduce the technical complexities associated with integrating rideshare spacecraft by reducing the amount of recurring engineering analyses for each mission. Sparing the details of the technical nuances of these analyses, defining standard interfaces allows for APL, LV and primary SV providers to design a APL to LV interface only once as non-recurring engineering (NRE) versus having to re-design the interface for a unique mating of numerous differing spacecraft interfaces. This cuts down on the design effort and analyses required to demonstrate compatibility of the rideshare spacecraft with the LV interface. For example, by defining the mechanical attachment interface between the APL and LV once as a standard interface, the LV provider would only have to conduct a single NRE analysis of the mechanical interface, and every subsequent spacecraft that would mate to this interface would not require re-design given that the standard requirements are upheld, such as overall spacecraft mass, number and sizes of attachment bolts, etc.

Subsequently, appreciable cost savings are possible with the reduction of interface re-design and/or interface requirements re-definition through the use of a standardized interface. For instance, a standardized electrical interface harness and pin-out would provide design guidance for APL smallsat providers to electrically mate the spacecraft to the LV interface. NASA Goddard Space Flight Center's (GSFC's) *Design and Manufacturing Standard for Electrical Harnesses* specifically warns of the "cost and schedule" risks associated with progressing with spacecraft design with an "inadequate" model of the electrical interface.[14] Standardization of interface requirements on the APL rideshares reduces costs by allowing for a smaller number of recurring engineering tasks because the upfront design and analyses are completed as NRE only one time during the first implementation of the standard. Ganley states that standards move "a significant amount of the technical work to the front of the process," which "can greatly reduce the recurring engineering costs of a product."[15]

CHALLENGES IN DEVELOPING A STANDARDIZED INTERFACE

Despite the benefits of defining a standard APL to LV interface and the great interest in the international industry to creating these requirements sets, the overall national and international communities have not yet been able to establish a unifying single standard interface for any size/class of spacecraft. As stated earlier, several ISO standards have been published to bring a consensus on a standard interface, in addition to other attempts to define a single standard for SV to LV interfaces. However, adoption and utilization of these standards have been sporadic or short-lived. Several factors produce difficulty in producing a unifying standard interface for spacecraft to launch vehicles that is universally (or at least, widely) adopted, represented in Figure 3.

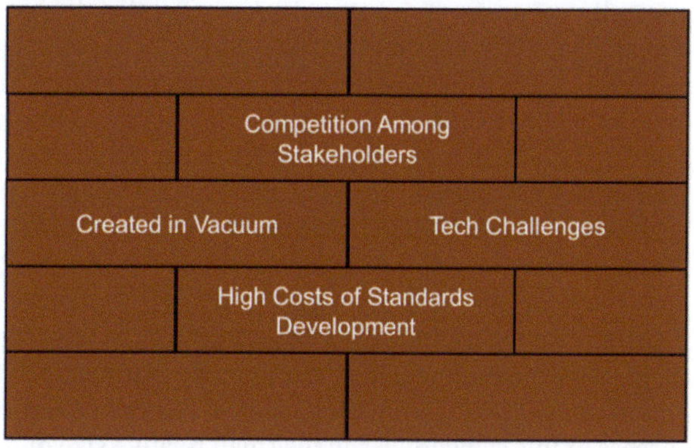

Figure 3. Factors That Hinder APL to LV Interface Standards.

Competition Among the Stakeholders

First of all, standards in general are difficult to generate because they are commonly developed by stakeholders that compete within the same market. Ganley writes that the parties that participate in the development of standards are "most often the result of market forces," that is, that oftentimes, competing companies and/or organizations often constitute the body that is called to define a particular standard.[16] This is problematic, he asserts, because "the primary end goal of the [standards development] process" is consensus with a "required mechanism for negative opinions and resolution." Futron's "Hosted Payloads Guidebook" also mentions the concerns of commercial stakeholders resisting a common standard because it potential creates "an alternative to proprietary interfaces, developed at great expense by each manufacturer, in favor of some common standard."[17] Specific to the smallsat industry, Ganley opines that the small satellite industry lacks the capability to create a "critical mass" of enough companies and organizations that would be willing to cooperate to develop common standards, stating bluntly that "cooperation in the space industry is all but impossible, with the true cooperation required for this level of standards development effort non-existent."[18] Moreover, when international governments are involved, the potential for conflicting goals and motivations bleeds beyond financial concerns into other factors, such as politics and nationalism. Therefore, this "tyranny of consensus" leaves many standards development efforts stifled and/or impotent.

Requirements Developed in a Vacuum

Another challenge for standards in general is that they are sometimes generated in a partial vacuum by stakeholders slightly removed from the international space industry as a whole. These types of standards are usually generated by representatives from governments or pure academia, but not often between government, academia *and* the commercial space industry *all together*. Therefore, these standard requirements end up failing to obtain advocacy by enough of the market, greatly reducing the probability of adoption. Since standards generation from these entities typically take a long time to develop, the requirements created are often can be outdated or irrelevant to the dynamic nature of the space technology environment. This seems to be the case for the aforementioned ISO standards for common spacecraft to LV interfaces. Furthermore, standards that are generated in this fashion, especially by governments, provide the sense of being imposed rather than encouraged. In their presentation entitled "Creating Standards for the Small Satellite Industry", Herrell, Primpikar, Hines, and Quintero seem to imply that the highly competitive smallsat industry would to be more receptive to standards that are "commonly accepted, not imposed."[19] Further, this author reported in the paper "Hosted Payloads or Dedicated Rideshare: What's the Best Way to Orbit?" that "many in the industry do not desire formalized standards, such as military standards (MIL-STD), which could be over- restrictive.[20]

Technical Challenges that Complicate Standardization

The development of standards is further frustrated by several technical challenges associated with the smallsat to LV interface. First of all, as the size of the small satellite goes up or the number of smallsats at a particular location increases, the technical complexity of the interface increases, as represented in Figure 4. The most blatant example of this is for coupled loads and dynamics phenomena between the satellites and the LV. Single smaller satellites, especially in the Picosatellite class of smallsats that generally range from 1 kg to 10 kg generally do not affect the overall loads and dynamics of the entire group of satellites with respect to the LV because they are so small relative to the overall mass of all spacecraft on a particular manifest. Therefore, industry standards for this size of spacecraft have been developed with widespread adoption across the international government, commercial and academic industries through the CubeSat standard. The CubeSat construct was developed by California Polytechnic University at San Luis Obispo, California in conjunction with Stanford University. The CubeSat, along with its associated dispenser system, the Poly Picosatellite Orbital Deployer (P-POD) system emerged as the industry standard for this class of smallsats. Figure 5 represents a CubeSat from Atkins' paper, and Figure 6 provides a representation of the P-POD from the P-POD Mark 1 ICD.[21, 22] Part of the success of this standard lies in the small relative size of the spacecraft, which presents negligible impacts to the overall coupled loads analyses (CLAs) associated with each mission.

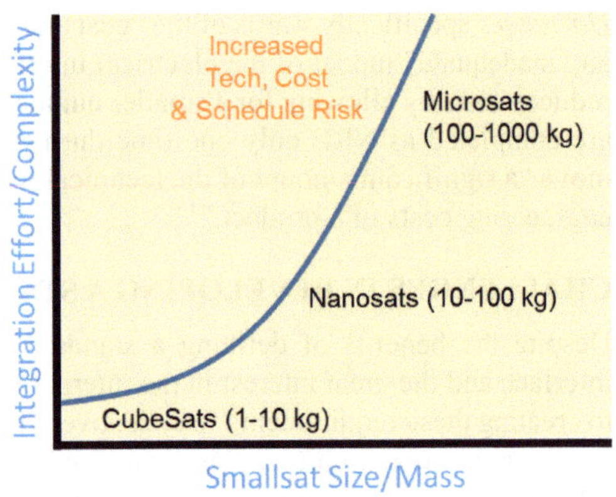

Figure 4. Qualitative Correlation between Smallsat Size and Mission Effort Required and Complexity.

However, as the size of the smallsats increases, the spacecraft create a more appreciable impact on the coupled interaction of all of the spacecraft with respect to the LV, requiring much more mission-unique analyses, as shown in Figure 4. This potentially introduces the technical risk of breaking components on the other spacecraft in the manifest, or even on the LV itself, due to the tremendously harsh loads and dynamics events associated with launching a rocket through the atmosphere to get to orbit. In order to mitigate this technical risk, SV and LV providers rely on at least two CLAs to characterize the coupled interaction between the primary spacecraft, rideshares, and LV. Subsequently, this introduces associated programmatic, schedule, and cost risks to the primary mission, which further increases reluctance to allow rideshare missions. For specific example, the publically released Launch Services User's Guide for the Delta IV booster requires that spacecraft providers deliver a dynamic model of their satellites a launch minus (L-) 24 months as input to the preliminary CLA for that mission, which is typically completed around L-18 months.[23] The European Space Agency's (ESA's) Ariane 5 User's Manual requires model delivery at L-20 months to complete the first CLA by L-18 months as well.[24] This is a particular challenge for smallsat providers on rideshare missions because smallsat providers typically do not begin development of their systems until between L-18 and L-12 months; thus they are typically not able to provide models with the necessary fidelity to input into the preliminary CLA. This introduces technical, schedule and cost risk to the overall mission because the mission integrators may discover loads exceedences late in the mission integration timeline, requiring costly additional analyses and possibly, even more expensive system re-designs. Even NASA GSFC's paper entitled "Primer on the Craig Bampton Method" admittedly describes the CLA as "long and costly."[25]

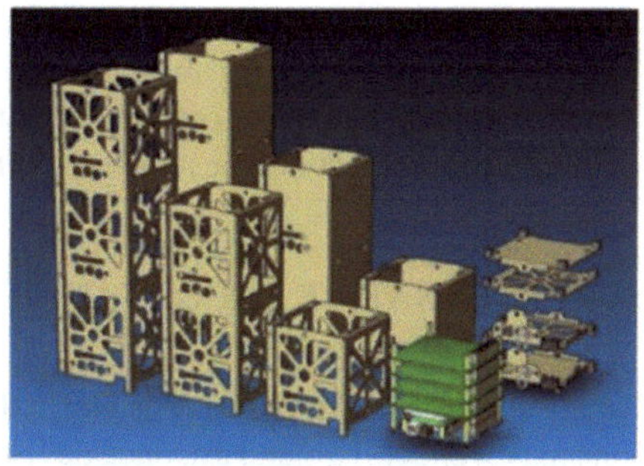

Figure 5. CubeSat Variants [From 21].

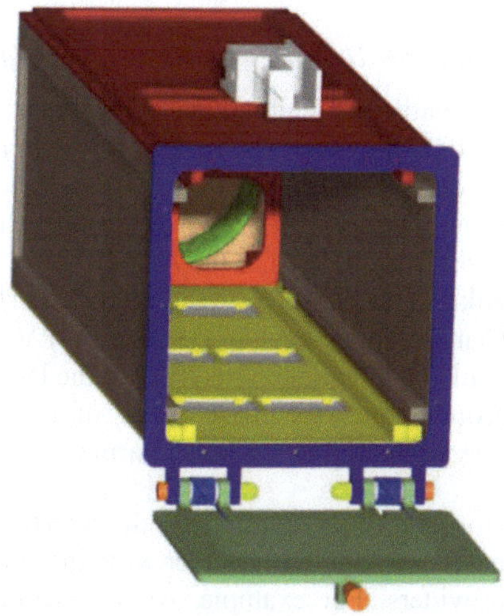

Figure 6. Representation of a P-POD [From 22].

Coupled loads issues significantly hinder the establishment of a common smallsat interface because the wide variability in spacecraft size, volume, overall mass and center of gravity of smallsats larger than the Picosatellite class of spacecraft make it extremely difficult to create a standard that envelopes the coupled loads interactions. To many primary SV and LV providers and their associated leadership, the coupled loads risk itself is enough to deter them from allowing rideshares on sRS missions, aside from the additional technical and programmatic risks associated with ridesharing. Further, as more smallsats are aggregated onto a mission to decrease the overall launch costs, the combined mass of the rideshares, even if they are CubeSats in P-PODs, impact the overall loads and dynamics of the coupled total payload to LV interaction. Therefore, naturally, there are much more numerous launch opportunities in the current international launch landscape for very small groups of CubeSats than for larger clusters of CubeSats and/or smallsats of larger form factor.

In addition to the coupled loads issue, other technical factors cause a reluctance to establish and/or adopt and apply standards across the international smallsat community. As alluded to earlier, standards are a proverbial double-edged sword insomuch as they can overly constrict the smallsat spacecraft design. A strong example of this phenomenon is with the P-POD system mentioned earlier. The fact that there are increasingly more launch opportunities for CubeSats than other smallsats forces smallsat providers to cram as much capability into the small CubeSat form factor as possible. This author has previously stated that "smallsat providers spend more time and dollars miniaturizing technologies to fit the Cubesat standard in order to maximize their chances for launch. The smallsat provider's time and resources should be focused on the specific capability of the overall spacecraft instead of diverting much effort into miniaturizing components."[26] Many smallsat developers labor under the challenge of meeting the volume and mass restrictions of the CubeSat and P-POD standards.

High Cost of Standards Development

In addition to the factors already discussed, significant costs are associated with the coordination of the technical and programmatic requirements for standards development. Ganley succinctly states, "Standards development is expensive, costing at minimum $1 million [USD], and easily ranging into the $100's of millions for large efforts."[27] This is one of the major reasons why government institutions are often the creators of formalized standards documentation. The high costs and long timelines required for non-government entities to meet together to establish requirements for a common standard interface is virtually impossible for shallow-pocketed academic and commercial smallsat companies.

AN ALTERNATIVE FOCUS CAN GET US THERE

After digesting the previous discussion, one may wonder if the effort behind finding a standard interface between smallsat rideshares and the LV is worth the time and resources it would take to define the requirements. Furthermore, one may wonder if a standard, once developed, could be widely adopted and applied by the international launch vehicle and smallsat community. This author strongly asserts that a shift in focus can make possible the establishment of industry standards that would be attractive enough to encourage widespread embrace of its requirements. The focus should shift to allowing the adapter hardware providers and mission integrators to create "organic" industry standards.

Standards Defined with Respect to Adapters

Currently, the international space industry is already moving in the direction of defining smallsat interfaces with respect to launch adapter systems specifically designed to accommodate smallsat rideshares. Only in a few examples do LVs create direct interface locations for smallsats onto the booster itself, such as the Aft Bulkhead Carrier (ABC) location described in Atkins' thesis.[28] The ABC bolted interface is located in the aft end of the Centaur upper stage of the Atlas V LV, where a smallsat or set of smallsats could be integrated directly onto the launch vehicle. The associated interface requirements for these types of direct LV interfaces tend to be very restrictive to prevent direct harm to the booster itself.

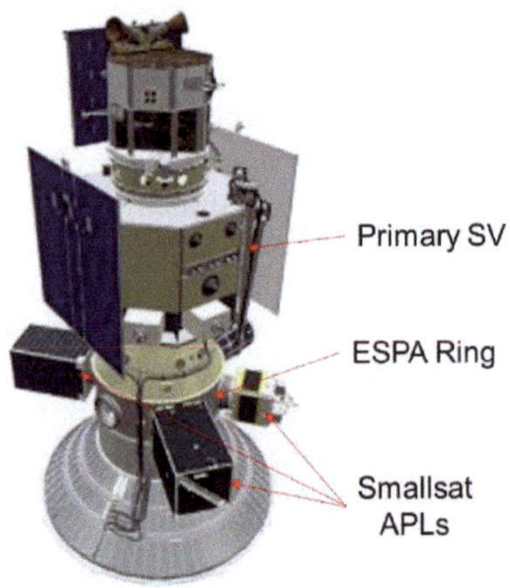

Figure 7. Example ESPA Rideshare Mission [Adapted from 29].

However, more commonly, the interface between the LV and smallsats are based on smallsat adapter systems that are vetted and approved by LV providers. For example, Atkins mentions the EELV Standard Rideshare Adapter (ESPA) ring and Arianespace's Ariane Structure for Auxiliary Payloads (ASAP) 5 adapter.[29] As shown in the representation of the ESPA ring in Figure 7, the launch adapters mate to directly to the LV and/or primary spacecraft provider, and then provide mating locations for APL rideshares. The ESPA ring, provided by Moog CSA Engineering, creates a standard mechanical interface between the LV, the ring, and the primary SV that remains static regardless of the various types of rideshare spacecraft that can bolt radially from the ring's center. Assoicated with the ESPA ring is a detailed set of interface requirements laid out in the ESPA Rideshare User's Guide (RUG), such as mission requirements, launch environments, APL to ESPA interfaces, and operational restrictions. Therefore, the LV and/or primary SV providers, as well as the APL smallsat providers have a strong understanding of the requirements to utilizing the ESPA system for rideshare missions, alleviating many of the inhibitions to rideshare.

In essence, the ESPA system is a strong example of the advantages of allowing rideshare hardware and mission integrators to create standards for the APL to LV interface. The ESPA RUG has become a well received set of standard requirements that has been embraced by many launch vehicle and spacecraft providers in the government, commercial and academic space industry around the world. The development of the ESPA RUG coalesced inputs from the US Air Force (USAF), their Federally Funded Research and Development Center (FFRDC), the Aerospace Corporation, and the commercial industry, and synthesized these inputs into a comprehesive set of requirements with enough detail to both assuage many of the concerns of the LV providers as well as provide sufficient APL spacecraft design boundaries to maximize the probability of clean acceptance for launch on rideshare missions. Atkins' entire thesis is centered around the strengths and capabilities of the ESPA system to provide a non-interfering, "nearly transparent" presence to the primary spacecraft.[30]

Although led by the government, this industry standard was devised by mission integration experts that were very well acquainted with both the LV and smallsat rideshare perspectives. Therefore, the contributors were the most congnizant of the particular challenges and issues that increase or decrease mission complexity and risk due to rideshare spacecraft. This is precisely why mission integrators are the most suited to lead the development of interface standards. Standards devised by mainly the LV community are most likely biased towards more severe requirements on rideshare spacecraft because the LV providers are wholly focused on the overall mission success of the primary spacecraft the safety of the rocket itself. Therefore, these standards tend to overly constrict the developmental trade space for the smallsat rideshare spacecraft. Conversely, common standard requirements developed by the smallsat APL providers will tend to allow the greatest amount of flexibility in the rideshare spacecraft design in terms of technical requirements, developmental timelines, and mission operations and orbital parameters. Thus, common interface standards should be devised by the entity that must balance the desires and requirements of the two sides of the interface: The mission success and safety of the LV and the other spacecraft in the manifest versus the greatest amount of mission and design flexiblity for the smallsat rideshare APLs. This neutral third party is the rideshare mission integrator of the single rideshare spacecraft or the aggregate of several APLs on a single manifest.

Despite the many strengths of the ESPA system and its associated RUG that serves as the industry standard for smallsats up to 180 kgs (400 lbs), the ESPA system by itself does not assuage the coupled loads concerns discussed at length earlier. However, enabling technologies are emerging from the interface hardware and mission integration industry to complement the ESPA system to address issues like the CLA alignment issue between the rideshare spacecraft and the primary mission.

As a specific example, TriSept Corporation, in conjunction with Moog CSA Engineering is developing the FANTM-RiDE™ dispenser system to specifically address all of the mission integration and technical concerns that dissuade LV and primary spacecraft providers from allowing rideshare spacecraft on their missions, to include the coupled loads dillema. The FANTM-RiDE system, with the dispenser depicted on an ESPA ring in Figure 8, addresses the key concerns of the LV and primary SV communities with respect to rideshare by providing four major features.

Figure 8. FANTM-RiDE with Representative APL Rideshare Spacecraft on an ESPA Ring.

First, it is comprised of a containerized dispenser that provides the assurance that in the unlikely case that a smallsat APL component breaks off of the spacecraft during launch, the debris would be captured within the dispenser. Currently, the FANTM-RiDE system allows for one or several spacecraft to be loaded within an available volume of 24 in by 24 in by 32 in space with a mass availability of 240 lbs (108 kg) for one or more APLs. Containerization also enables the integration of potentially more than one spacecraft per dispenser, further dividing up the launch costs among the ridesharers.

Additionally, the dispenser's most salient feature is that it can be "mass-tuned" so that the mass properties of the dispenser can be manipulated so that every FANTM-RiDE dispenser will have the same overall mass and center of gravity (CG), regardless of the number, shape and size of the spacecraft that fit within the dispenser. This specifically addresses the coupled loads concerns mentioned earlier because a LV provider can execute their preliminary CLA once for the dispenser as NRE, and then never have to re-execute final verification CLAs for that dispenser configuration again because the mass dynamics properties will remain unchanged despite the configuration or number of spacecraft within the dispenser.

This mass tuning feature is a perfect example of allowing the adapter hardware and mission integrators define a set of standard requirements instead of defining them with respect to the direct LV interface to the APLs. FANTM-RiDE's mass tuning capability allows a static configuration to the LV provider, i.e., the overall mass and dynamics properties will not change for each dispenser, even between mission to mission. Therefore, it inadvertantly creates a psuedo-standard interface to the LV, in complement of the ESPA RUG requirements. FANTM-RiDE's capability to become a transparent "phantom" to the LV and other spacecraft in the manifest perfectly complements the ESPA system.

Likewise, the requirements that will be defined in the FANTM-RiDE user's guide and associated ICD will provide adequate bounds for smallsat rideshare spacecraft designs, while allowing a much greater flexiblity in design flexibilty. Unlike the P-POD or similar CubeSat dispenser systems, FANTM-RiDE's size, containerization, and mass tuning capabilites allow greater flexiblity in the physical dimensions and mass properties of the contained rideshare spacecraft, especially since the overall mass properties will be tuned to the standard mass and CG. For instance, this allows for more allowances for protrusions for deployables and antennas, as well as the potential to allow propellants and pressure vessels to be carried without further risk the other spacecraft in the manifest because of containerization.

FANTM-RiDE's third feature of a stand alone architecture further enables transparency to the LV and primary SV. Basically, FANTM-RiDE's dispenser contains all of the necessary commodities to sustain and deploy one or numerous smallsats from each dispenser. These include independent battery power for trickle charging of spacecraft, and an independent sequencer system that can trigger numerous events and deployments for a single dispenser. This allows for a single, standard electrical harness for the dispenser, common to all LVs, that will receive a dual redundant separation signal from the LV when they are allowed to begin deployement of the APLs. Further, the mechanical interface is a 15 in circular bolted interface that has become standard throughout the international space industry because of the ESPA system. Therefore, the FANTM-RiDE dispenser can be integrated on any interface that can accommodate the mass and CG of the dispenser, not exclusively on an ESPA ring.

The final FANTM-RiDE system feature is total mission integration. This provides a comprehensive "concept to orbit" turnkey rideshare solution. With FANTM-RiDE, TriSept mission integrator will guide all APL rideshare provider through the entire mission integration process, from launch opportunity selection and manifesting, contracting, documentation development, verification, validation, test, integration, launch and on-orbit operations.

All in all, the FANTM-RiDE system will create a set of standard requirements that appeal to LV and primary SV providers due to the invariability of CLA contributions and the non-interference, transparent nature of the system. Basically, FANTM-RiDE allows for a decoupling of the mission integration timeline of the primary mission from the rideshare spacecraft, enabling smallsats to enter and exit the manifest without perturbing the overall mission. Additionally, the requirements should widely appeal to smallsat providers because it provides a greater level of technical design, programmatic, operational, and schedule flexiblity. Other technological and methodology innovations akin to FANTM-RiDE, P-POD, and ESPA will further enable more frequent and lower cost launch opportunities for smallsats.

Promulgating Organic Industry Standards

In many of the examples provided thus far, many of the APL spacecraft to LV interface requirements have "unintentionally" become standards for much of the international space industry. For instance, the ESPA RUG requirements have organically propagated as the industry standard not as a result of a body of government representatives in a vacuum, purposefully intending to create a universally accepted common standard interface. Instead, the government, with interaction with the commercial space industry, worked to define a set of requirements that would enable the widespread use of the ESPA system on US Government launches. Organically, these requirements have become the "rule of thumb" for spacecraft design and interface requirements for applications outside of merely ESPA and US Government missions.

Therefore, the smallsat industry should look to the most effective and widely accepted interface hardware systems and mission integration requirements to realize common interface standards. This provides the greatest potential for wider adoption and utilization than a top-down approach of a standards organization pushing down standards for the international community to adhere to. This also ensures that the requirements are more contemporary to the most current technologies and practices employed by the smallsat industry.

A strong analogous example of this is Apple's iOS construct and its associated hardware systems. The popularity of Apple's intuitive user interface, hardware features, ergonomics, aesthetics and reliability has resulted in over 800 million iOS devices sold as of June 2, 2014, according to an article on Mashable.com.[31] Although highly proprietary, these iOS devices' interfaces with other hardware and software systems and applications have driven numerous industry standard requirements and protocols. For instance, Apple's proprietary Cocoa Touch programming framework has been utilized to drive the development of over 1.3 million applications (apps) and third-party software programs as of September 9, 2014, according to the Wikipedia entry for iOS.[32] Furthermore, the devices' proprietary hardware interface requirements have driven a multi-billion US dollar third-party market of hardware devices and products, such as protective cases, keyboards, docking interfaces with audio systems, and even devices like home thermostats and lighting systems, and golf swing analyzers. Additionally, Apple's signals protocols, such as their AirPlay feature, allows for iOS devices to stream music and video to a large number of third-party audio and visual systems that must comply with the AirPlay design protocol requirements. All of these devices and apps must strictly adhere to Apple's proprietary interface requirements to work with iOS device. However, the consumer electronics industry has widely accepted and adopted Apple's interface requirements to capitalize on the high popularity of Apple's iOS devices. Apple did not intend to directly codify a set of interface requirements for all mobile devices. Instead, they focused on creating strong interface requirements for only their products, which have dominated the consumer electronics industry around the world.

Similarly, the CubeSat and associated P-POD's simple, low-cost, effective, and low-risk form factor and interfaces have played a large role in the international assumption of the standard interfaces of the CubeSat paradigm. Again, the CubeSat program did not directly intend to create a common standard interface when it devised its requirements. Instead, it defined the requirements for maximum acceptability by the LV and Picosatellite provider community, which then organically evolved into the industry standard for 1 to 10 kg spacecraft. As a result, there are numerous launch opportunities for CubeSats, to include launches from the International Space Station and standard locations directly on boosters themselves.

SUMMARY AND SPECIFIC RECOMMENDATIONS

The international space industry needs reform to break free from the bonds of developing extremely expensive satellite systems that require the highest levels of reliability, which drives excessively high launch costs with highly conservative reliability requirements to deliver these costly systems to orbit. Smallsat technologies could potentially propel the industry to a healthier state of lower complexity, less expensive, yet mission-effective space architectures and more frequent and lower cost launch opportunities for spacecraft of all sizes. As asserted earlier, the smallsat industry is a the cusp of helping to break the destructive cycle of the current space development and acquisition paradigm if allowed more numbers of launch opportunities at a more affordable price point made possible from rideshare missions. Additionally, if the smallsat industry can recognize and widely adopt interface requirements from strong interface hardware systems and mission integration methodologies to realize a set of organically derived industry standards, ridesharing opportunities will only increase for missions around the global market.

The smallsat industry should look to the phenomena surrounding the development of the industry standards brought about by CubeSats, P-PODs, and ESPA to glean lessons on how to recognize interface standards for rideshare spacecraft to LVs from strong interface hardware solutions and mission integration practices. Furthermore, the industry would benefit from a careful study of analogous interface standard developments of non-smallsat systems, such as Apple's iOS devices. Finally, international smallsat providers should seek out and explore the strongest interface hardware systems and mission integration providers to find more lower cost launch options that will a set of common industry standard requirements for the interface between the rideshare APLs and the LV to organically emerge. New technological and integration methodology innovations that enable smallsat rideshare access to space will highly benefit the entire international space industry. Change must occur now to exploit the current opportunities to significantly change the international space industry for the better.

REFERENCES

1. Lim, daniel. (2014, June 9). *Shaping Our Future: Practical Actions to Bring Real Industry Change.* Paper presented at the 2014 Small Payload Rideshare Association Rideshare Conference, Pasadena, CA.
2. Atkins, Robert. (2011, September) *EELV Secondary Payload Adapter (ESPA) Ring: Overcoming Challenges to Enable Responsive Space* (Master's thesis). Naval Post Graduate School, Monterey, CA.
3. Lim, daniel. (2014, May 21). *Defining a Roadmap to Bringing the US Space Industry Back to Health.* Paper presented at the 30th Space Symposium Technical Track, Colorado Springs, CO.
4. Lim, daniel. (2014, May). *Practical Knowledge on Opening Up Low-Cost US Launch Opportunities for International Smallsats.* Paper prepared for the 2014 4S Symposium, Majorica, Spain.
5. Atkins, pp.68-69.
6. Ganley, Jeff. (2004). *Small Satellite Standards Development*, p. 1.
7. Atkins, p.5.
8. Ibid., p.6.
9. Reese, Kenneth, Alex Martin, and David Acton. (2014). *STPSat-3: The Benefits of a Multiple- Build, Standard Payload Interface Spacecraft Bus.* Paper presented as the 2014 Small Satellite Conference, Utah State University, Logan UT.

10. Futron Corporation. (2010, August). *Hosted Payloads Guidebook*. Guidebook prepared for NASA Langley Research Center, Bethesda, MD.

11. Boland, Phillipe. (2005, October). The launch business: Standard formats for launch vehicle - spacecraft interface documents. *International Standards Organization (ISO) Focus Magazine*.

12. Boland, Phillipe. (2004, March). Developments and Initiatives: Interfaces between spacecraft and launch vehicles. *ISO Focus Magazine*.

13. International Standards Organization. (2012). *ISO 26869:2012: Space systems -- Small-auxiliary-spacecraft (SASC)-to- launch- vehicle interface control document*. Abstract retrieved from http://www.iso.org/iso/home/store/catalogue_ics/catalogue_detail_ics.htm?ics1=49&ics2=140&ics3=&csnumber=57530

14. NASA Goddard Space Flight Center. (2003, July). *Design and Manufacturing Standard for Electrical Harnesses, Rev. C*. Retrieved from http://eed.gsfc.nasa.gov/docs/harness/harness.html

15. Ganley, p.2.

16. Ibid., pp.2-3.

17. Futron, p.8.

18. Ganley, p.3.

19. Herrell, Linda, Milind Pimprikar, John Hines, and Andy Quintero. (2009). *Creating Standards for the Small Satellite Industry*. Presentation for the CANEUS Small Satellite Sector Consortium (SSSC).

20. Lim, daniel. (2013, October). *Hosted Payloads or Dedicated Rideshare: What's the Best Way to Orbit?* Paper presented for the 2013 Reinventing Space Conference in Los Angeles.

21. Atkins, p.52.

22. Nason, Isaac, Michelle Creedon, and Nick Johansen. (2002, May). *CubeSat P-Pod Deployer Requirements, Mark 1*. California Polytechnic University, San Luis Obispo, CA.

23. United Launch Alliance. (2013, June). *Delta IV Launch Services User's Guide*. Centennial, CO.

24. Arianespace. (2008, July). *Ariane 5 User's Manual, Issue 5, Revision 0*. Evry-Courcouronnes Cedex, France.

25. NASA Goddard Space Flight Center. (2000, October 4). *Primer on the Craig Bampton Method*. Retrieved from http://femci.gsfc.nasa.gov/craig_bampton/Primer_on_the_Craig-Bampton_Method.pdf

26. Lim, *Hosted Payloads or Dedicated Rideshare: What's the Best Way to Orbit?*, p.15.

27. Ganley, p.6.

28. Atkins, p.53.

29. Ibid., pp.16 and 33.

30. Atkins, p.6.

31. Figerman, Seth. (2014, June 2). Apple Has Sold More Than 800 Million iOS Devices to Date. *Mashable.com*. Retrieved from http://mashable.com/2014/06/02/apple-wwdc-stats/

32. Wikipedia.com. (nd.). *App Store (iOS)*. Retrieved from http://en.wikipedia.org/wiki/App_Store_(iOS)

BIS-RS-2014-23

MSR IOD Study: Low Cost End-to-End In Orbit Demonstration of Key Technologies for the MSR Mission

Emanuele Monchieri
Airbus Defence & Space
Gunnels Wood Road, SG1 2 AS, Stevenage, UK; +44 (0)143877 3843
emanuele.monchieri@astrium.eads.net

Benjamin Dobke
Airbus Defence & Space
Gunnels Wood Road, SG1 2 AS, Stevenage, UK; +44 (0)143877 3146
benjamin.dobke@astrium.eads.net

ABSTRACT

The Mars Sample Return (MSR) mission is the number one goal for Mars planetary science and will present a major milestone in the exploration of the solar system. Despite the scientific instrumentation brought to the surface of Mars by previous and planned missions, studying unaltered Martian materials using the huge array of sensitive scientific equipment on Earth may result in a paradigm shift in planetary science, helping to answer questions about the nature of Mars, its formation, and the possibility of life on another planet. Various mission architectures are under consideration, and are evolving. All such mission concepts have a common fundamental requirement - the launch of a sphere - containing the collected soil samples and of the size of a football - into Mars orbit and its retrieval performed by a pre-existing orbiter. This sphere, Sample Container (SC), must be found thousands of kilometres away, approached, and then captured to be further packaged before its return to Earth. This complex sequence of operations must be done almost completely in an autonomous way, due to the typical delay in Earth-Mars communications, and the whole success of the MSR Mission relies on its successful performance. For this reason, this represents a major risk that must be mitigated before the mission actually begins. Previous studies have investigated some of the key aspects relevant to this rendezvous and capturing sequence, but they have also immediately showed the limitations in what can be actually tested/replicated on-ground and/or during parabolic flight/drop tower campaigns.

This paper describes how a low cost In-Orbit Demonstrator (IOD) can be used to validate this mission critical development, and pave the way for future autonomous in-orbit rendezvous missions. The defined mission was aimed at providing a cost efficient demonstration of key technologies such as long range optical detection, autonomous GNC for rendezvous & capture of uncooperative target, capture and securing of the free-flying Sample Container, and on understanding their applications on similar missions, e.g. debris removal.

A clear definition of the main mission objectives and of the associated benefits in terms of risk mitigation and TRL improvement has been derived, together with an overview of the mission architecture and of the full operations concept to be implemented. One of the most critical aspects that have been assessed is to verify the representativeness of a low-cost IOD with respect to the actual Mars Sample Return mission (which will likely be a multi-agency endeavour due to the complexity and overall cost). This has been demonstrated by identifying the main commonalities and differences with the MSR mission and environmental parameters, and how those shall be addressed in order to guarantee useful and reliable results. In support to these analyses, a phase-0 design of the sample capture payload has been performed, including a preliminary selection of potential platforms, launchers, and ground segment strategy, and a preliminary estimation of the main engineering budgets. This allowed demonstrating the feasibility not only from the technical point of view but also by ensuring to implement a low cost mission.

KEYWORDS: [In-orbit demonstration, Payload and platform technologies, International programmes and cooperation]

S. Hatton (ed.), *Proceedings of the 12th Reinventing Space Conference*, DOI 10.1007/978-3-319-34024-1_9

The development of space assets is nowadays experiencing a significant need for the introduction of novel and disruptive technologies aimed at improving the achievable performance. Telecommunications satellites are designed to improve the surface coverage and the type, quality and amount of data that can be transmitted from the orbit. Earth Observation satellites need to accommodate advanced payload for gather more and better information about the Earth and monitor its special features (winds, ocean currents, etc.). Exploration spacecraft are designed to gather information about the solar system and the universe, every time facing new challenges requiring the development of specific hardware/technology sometimes even existing only as a concept/idea. However, the novelty/complexity of such new technologies can require the implementation of expensive test campaigns, which may significantly affect the budget of a mission. Therefore, it becomes mandatory to develop the capability to perform test as early as possible in the mission, with the aim of de-risking those technologies and reducing their potential impacts on technical feasibility, costs and schedule.

In such a frame, In-Orbit Demonstrations (IODs) have represented a valuable mean for enabling this capability. IODs are nowadays used to validate technologies and products, allowing them climbing in the TRL ladder and getting flight heritage for use in main stream projects usually remaining within reasonable costs. Important examples are represented by the PROBA-1 and PROBA-2 missions that, despite their limited budget, allowed the demonstration of several novel platform/payload experiments now flying on operational missions.

Although this advantage is quite clear and straightforward for the development of technologies for Earth orbit, it is also true that to perform an in-orbit demonstration of a technology for an exploration mission poses a new challenge: the environment representativeness. In fact, it is difficult to justify that a technology aimed at operating in Mars orbit can be fully demonstrated with a mission in Earth orbit. Nevertheless, there are cases where this can be not only made possible, but also provide additional advantages.

A strong example is provided by the ESA "Mars Sample Return In-Orbit Demonstration" Study. This Study has been performed in the frame of the ESA's General Studies Programme (GSP), and was aimed at investigating the opportunity to demonstrate in Earth orbit some of the key critical technologies of the actual Mars Sample Return (MSR) Mission foreseen for the next decade: autonomous rendezvous and capture of an uncooperative target. The demonstration of such technologies was already discussed in previous studies of the MSR Mission, but the actual demonstration was planned to occur in MSR orbit. This would guarantee to be very representative of the mission environment, but it would significantly increase not only the risks (i.e. it is still a mission occurring in Mars orbit, with limited control from Ground), but also the effort in terms of development time and cost.

Successfully completed in the last May 2014, the MSR IOD Study led by Airbus DS UK provided a potential alternative low cost solution to this problem, demonstrating that those technologies can be validated up to some extent also with an Earth orbit mission. This is because the key parameters that drive the performance of the investigated technologies have been thoroughly understood and their value is fully in line with the one expected in the actual MSR Mission.

The resulting advantages to perform such a demonstration in Earth orbit are:

- Close capability to control/monitor the mission data and performance → some of the post processing operations can be performed on Ground, in a cheaper and quicker way than in the Mars orbit case, this reducing operations cost and the impacts on the onboard HW/SW without affecting the representativeness,
- Low mission cost → an Earth's orbit represents a less challenging environment with respect to a mission to Mars. Standard platforms can be used, as well as off-the-shelf/existing technologies. Moreover, several launch opportunities (i.e. piggy-back or secondary payload) are available, which helps in reducing the launch cost,
- Synergy with other similar missions → rendezvous and capture technologies are applicable to various other missions (i.e. debris removal), which would significantly beneficiate of this demonstration,
- Adoption of lean and novel approaches for hardware manufacturing/ software development may allow for cost reductions and shorter schedules.

The following sections of the paper will provide an insight of the major outcomes of the Study, clarifying the benefits and the importance in promoting such kind of missions.

II. BACKGROUND: THE MSR MISSION

As explained in the previous section, the Mars Sample Return Mission has been largely investigated during the last two decades, with several architecture concepts that have been proposed and analysed in-depth. During those studies, several critical technologies have been identified, some of which are currently under development. However, there are technological areas that, despite the efforts, still lack available data for calibrating/validating the developed models/simulators. Those technologies require an additional step for reaching a level of maturity/risk that is acceptable enough to make the main mission feasible.

The main objective of the MSR Mission is to collect samples of Martian soil and to safely carry them back for enabling in-depth analyses by the more powerful and advanced instruments equipping terrestrial laboratories.

In the most up to date architecture concept, the MSR mission can be divided in three main consecutive stages. Each of them is separated from the previous one by a gap of two years, allowing for a specific relative position/alignment between Earth and Mars for reducing transfer time and relax propulsion requirements. Briefly, the three stages envisage the following:

- Delivery of a Sample Caching Rover (SCR) aimed at collecting samples on Martian surface in a selected site,
- Delivery of an Orbiter in Mars Orbit aimed at providing communication/navigation services to surface assets, and hosting the Earth Re-Entry Capsule as well as the rendezvous and capture equipment,
- Delivery of a spacecraft on Martian surface including a lander, an ascent vehicle (MAV), and a Sample Fetching Rover (SFR). This latter is aimed at retrieving the SCR, taking the cache containing the samples and carrying it back to the lander site. Here, the cache is accommodated inside a Sample Canister (SC) that is stored inside the MAV. Once the SC is filled and sealed, it is delivered in Mars Orbit by the MAV. Therefore, the Orbiter detects the SC and then performs a rendezvous trajectory to capture it. Once captured, the SC is stored in the ERC and the return journey to Earth is performed.

In such complex sequence of operations that spans within a timeframe of about 6 years, one of the most critical operations is the "detection, rendezvous and capture" of the Sample Canister (see the following figure). This is a high risk manoeuvre affected by large uncertainties on:

- The altitude (and up to some extent the inclination) at which the Sample Canister is (successfully) released by the MAV after departure from Mars surface,
- The capability of the Orbiter to detect a free-flying, non-cooperative, very faint (magnitude is estimated to be about 13, TBC), and distant (>1500km) object as the Sample Canister while in Mars Orbit.

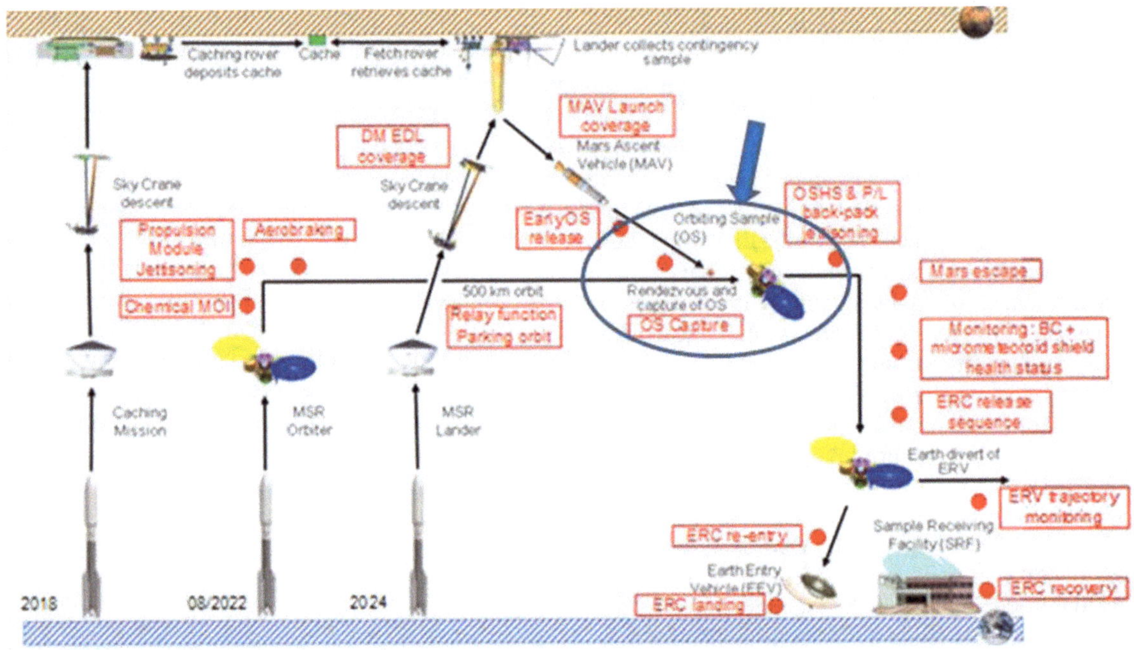

Figure 1: The actual MSR Mission

Those uncertainties and the risk of failure in capturing the unique Sample Canister, which would mean to fail the whole 6-year mission, require a robust and safe approach to increase the success probabilities.

III. THE TECHNOLOGIES TO BE DEMONSTRATED

Taking into account the described background, the MSR IOD Study investigated the opportunity to raise the TRL of the following technologies:

- Long-range optical detection
- Autonomous GNC for rendezvous & capture of an uncooperative target
- Capture mechanism

Long-range optical detection

The SC optical detection represents one of the main challenges of the overall MSR rendezvous operations. Detecting a very small and faint object from up to 1500km and discriminate this object against other bodies (e.g. stars, moons of Mars, etc.) requires a dedicated camera as well as state-of-the-art image processing algorithms. In the far range, the Sample Canister represents a discrete point object similar to a star. Its magnitude (up to 13 TBC, depending on the actual detection distance, Sun phase angle and Sample Canister design) makes it difficult to be detected because of the poor signal to noise ratio and because of the great number of objects of the same size and signal (typically about 2,000 stars in a 5 degree field of view, far more in worst cases) appearing within the camera Field of View (FoV). In the actual MSR mission, the Sample Canister is designed to house an RF beacon, while an RF sensor accommodated on the Orbiter is used to support the detection phase and to provide RF-based measurement for the whole rendezvous. Moreover, a Narrow Angle Camera (NAC) is also required in order to cope with the visual detection requirements. This allows for a single failure tolerant system.

The mission proposed in the MSR IOD Study envisages implementing the same type of sensors, i.e. a NAC and a RF sensor, in order to provide the same functional redundancy given by the two different methods and to guarantee the representativeness with the real mission.

At the same time, the image processing algorithms currently used for detection of very faint objects are all based on the *a priori* knowledge of the type of movement of the object to be detected. There are mainly two types:

- Algorithms based on the difference of angular rate between the target and the other objects: if the spacecraft is pointed towards the target, all moving objects are removed from the image. Alternatively, if the spacecraft pointing is stable with respect to the celestial vault, all fixed objects are rejected.

- Algorithm based on the trajectory of the object to be detected: if the trajectory of the object is known (position in the FoV = f(time)), only objects with the expected trajectory are selected.

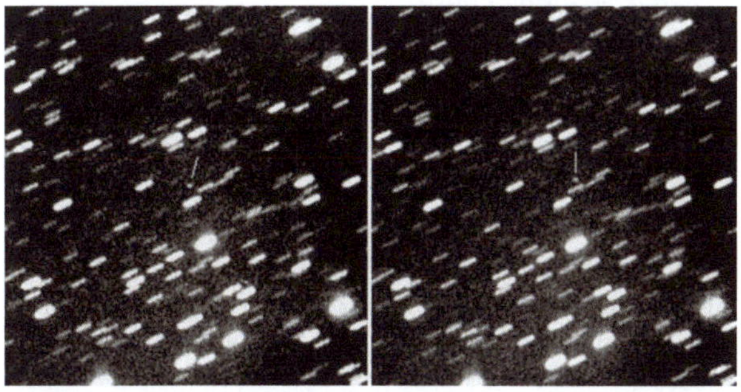

Figure 2: Example of faint object detection in a motion-blurred star field. Here, detection of an asteroid with a ground-based telescope.

Previous studies allowed the consolidation of those image processing algorithms taking into account various effects as star background, radiation environment, straylight, parasitic objects (e.g. Mars moons), detector and camera optics modelling. However, all the performed analyses relied on simulated images and experimental data, while the actual in-orbit images are still lacking.

This can be fulfilled with in-orbit acquisition through the IOD mission

The actual approach would consist in releasing the Sample Canister from the Chaser and detecting it with suitable algorithms. Robustness and performances of detection algorithms for large relative distances are demonstrated and validated also by using GPS data acquired onboard the Sample Canister and then downloaded after capture. In this way, the collected data would allow reconstructing the SC trajectory and validate the trajectory "simulated" by the algorithms. The post processing for the long range optical detection is foreseen to be performed on ground. This is allowed because of the time scale of this operation (days of acquisition), because the same approach is foreseen in the actual MSR mission, and because it allows reducing the requirements on the on-board processing capability.

Autonomous GNC for rendezvous & capture of uncooperative target

In the frame of the MSR mission, the rendezvous & capture phase requires an unprecedented level of autonomy. In particular, during the terminal phase of the rendezvous manoeuvre, Ground close interaction is no longer possible due to the large distance from Earth, which causes significant delays in the communications (~30 minutes). It becomes then mandatory to provide the Chaser of the capability to operate autonomously, which can be achieved thanks to the contemporaneous use of on-board sensors and algorithms. Particularly critical is the "last meter" operation. In fact, this is performed in "blind" condition, completely relying on the predictions of the SC trajectory provided by the on-board GNC algorithms (i.e. in the actual MSR mission, the accommodation of the GNC sensors is such that in its last meters before entering the "capture cone" the SC is not anymore in the sensors' field of view). During this final approach, it is also important to take into account that a switch to safe mode may cause a Delta-V that, depending on the mission thruster configuration, may set the Chaser on a collision trajectory with the SC, seriously decreasing the chances of mission success. For these reasons, the rendezvous scenario must be tested and made robust to safe-mode occurrence. In the following figure, an alternative collision-free scenario proposed during the past Mars NEXT Study is showed.

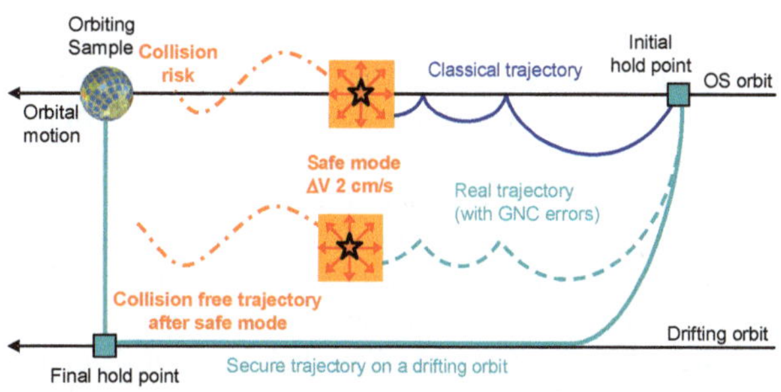

Figure 3: Collision-free scenario proposed in Mars NEXT Study

Finally, in addition to the above contingencies, the capability to resume the rendezvous scenario must be ensured to allow re-attempting in case the capture fails. Nowadays, in orbit autonomous rendezvous & docking but with a co-operative target has been already performed (e.g. ATV mission) in Europe. On ground, autonomous GNC rendezvous has been instead demonstrated in the frame of the HARVD study by performing software and hardware-in-the-loop simulations.

The relative navigation represents then a critical step for the successful performance of the actual MSR mission, in particular for such level of autonomy and accuracy, and it is of primary importance to demonstrate the robustness of the proposed algorithms/approach with respect to the conditions of the space environment.

The MSR IOD Study proposed to demonstrate and validate the image processing required to track the Sample Canister and the navigation filter able to fuse vision-based and inertial measurements (e.g. IMU and Star Trackers). At the same time, FDIR and capability to compute anti-collision manoeuvres in real-time will be also demonstrated.

The autonomous GNC for rendezvous & capture of an uncooperative target therefore encompasses several challenges, all to be addressed in this IOD mission:

- Autonomous GNC during the final approach phase, encompassing hop manoeuvres and forced translation, until start of the free-drift phase
- Autonomous collision avoidance
- Free-drift and capture

Capture mechanism

The last critical technology of the MSR mission that has been here considered for the MSR IOD Study is the Capture Mechanism. This element is aimed at actually "capturing" the Sample Canister when this latter is flying close to the Chaser. The capture phase is critical because, as explained above, it is performed in "blind" condition, so only on the basis of the predictions of the Sample Canister trajectory provided by the GNC algorithms. In fact, during the terminal phase, the SC is so close to the Chaser that it is not in the sensors' field of view and it is detected only once it is inside the capture cone and it is passed through a set of detection sensors. After the SC detection, a device (i.e. a robotic arm) is used for retaining the Sample Canister bouncing against the capture cone walls and preventing its loss in space. As for the rendezvous, this operation has to be performed in an autonomous way, without any Ground intervention, due to its very short duration with respect to the communications delay. At the moment, in Europe the development of such a mechanism is ongoing, including ground and parabolic flight testing. However, the SC dynamics during capture represents a critical aspect that cannot be fully simulated on-ground (because of the need of a 6 DOF SC) or on a plane (because the time for the test is not compatible with the whole duration of the capture). Finally, while a number of docking mechanisms have been used in space, none has been operated yet on a non-corporative target of such a small scale as the intended Sample Canister (~10 to 23cm).

The MSR IOD mission is designed to allow testing the "last meter" in space, focusing in particular to the coupling between the design of the capture device and the achievable GNC performances at the end of the forced translation phase.

Taking the above into account, the MSR IOD Mission is aimed at replicating, demonstrating, and validating up to some extent the full End-To-End Rendezvous and Capture Scenario of the actual MSR Mission, allowing increasing the maturity of some of its most challenging technologies/operations. A successful IOD will significantly reduce the high risks associated with such a manoeuvre and increase the probability of mission success.

IV. MISSION DESIGN

The MSR IOD Mission lasts about 2 years and envisages as primary demonstration the performance in Earth orbit of the end-to-end rendezvous and capture sequence. Therefore, a secondary demonstration is performed with the goal to demonstrate the capability to perform rendezvous and, potentially, approach and capture a piece of debris having a size similar to the Sample Canister. Finally, the last period of the mission is dedicated to conclude some guest experiments hosted on the platform or to actually enable their performance, depending on the resources still available.

A summary of the sequence relevant to the primary demonstration is represented in the following figure.

After launch, insertion in a Sun-Synchronous Orbit at around 700km altitude, and LEOP, the actual primary demonstration starts with the release of the Sample Canister. Therefore, the Chaser moves reaching a higher altitude (about 40km higher) in order to get in a condition similar to the one expected during the actual MSR Mission. Once there and once the desired distance between the Chaser and the Sample Canister has been reached, the Chaser uses its RF beacon receiver and the Narrow Angle Camera to detect the Sample Canister. The duration of the search phase can vary depending on various parameters, among which the altitude difference between the Chaser and the Sample Canister, and the orbit altitude.

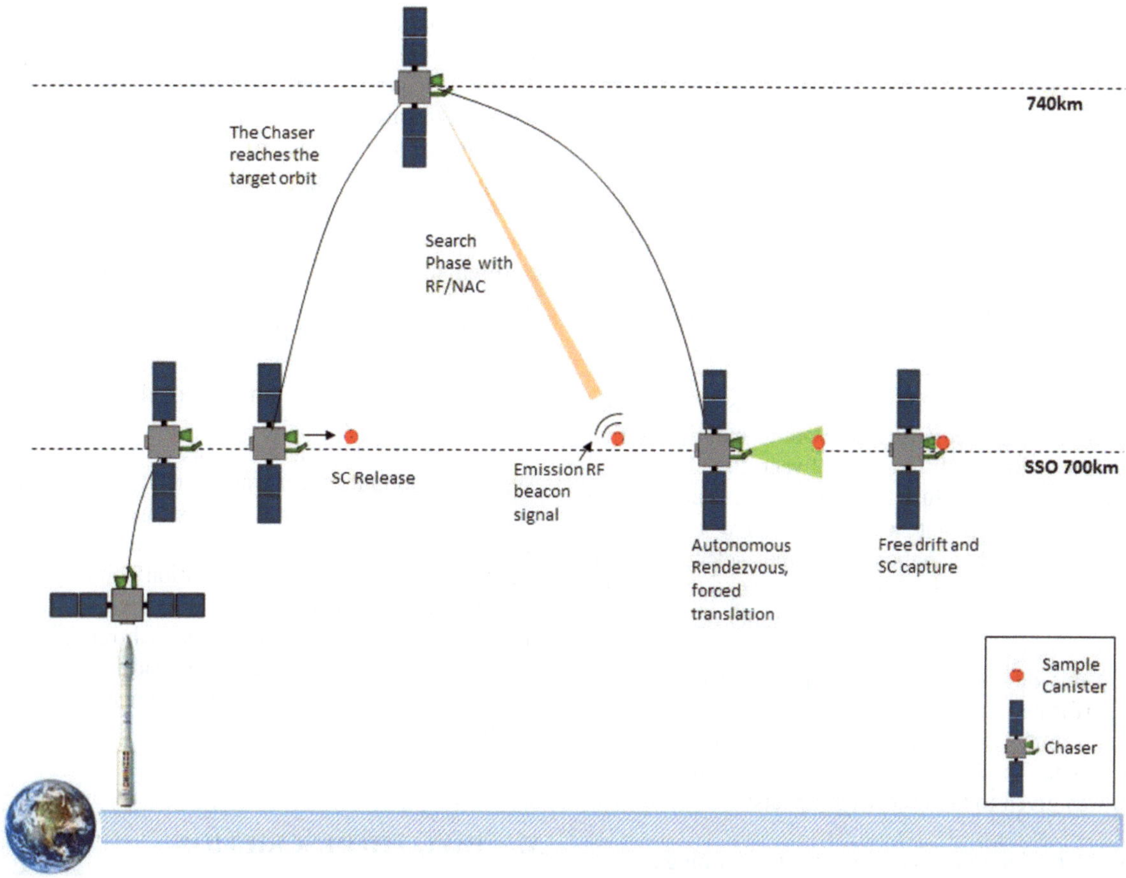

Figure 4: The MSR IOD Primary Demonstration Scenario

After at least two to three detections, the Chaser progressively reduces its orbit getting closer to the SC until the short range sensors, i.e. Wide Angle Camera and LiDAR, can be used. While during the performed search phase the ground provide support for what it concerns the image processing capability, during the approach phase this can be done autonomously on-board. This is called "forced translation" phase, and it ends once the Sample Canister is too close and it is practically not anymore in the sensors' field of view". In this case, a free drift phase is allowed until the SC is inside the cone where it is captured and secured.

Representativeness vs. actual MSR Mission

A major issue in planning such a mission was to guarantee the highest representativeness of the obtained results.

The following table reports a brief summary of the main MSR Phases and of how the MSR IOD Mission matches them during the actual demonstration.

Table 1: MSR Phases and those that are to be demonstrated via the MSR IOD mission

MSR Phase	MSR IOD	Notes
Search Phase	✓	Smaller difference between Chaser and SC orbits
Approach Phase	✓	Chaser and SC are on orbits with the same inclination (this mainly impacts the delta V requirements)
Re-Acquisition Phase	✓	Reduced impulse manoeuvre

MSR Phase	MSR IOD	Notes
VBar Hops Phase and Terminal Phase	✓	Fully demonstrated
Free Drift Phase	✓	Fully demonstrated
Capture	✓	Fully demonstrated

The above table shows that several aspects of the actual MSR Mission are demonstrated, with as main limitation the reduced altitude difference between the Chaser and the Sample Canister. Such a difference was necessary in order to be compliant with the limited amount of propellant available onboard of the selected small platforms. However, this is still representative of the overall system behaviour (just with smaller delta V).

Taking this into account, an in depth investigation was then performed during the study to understand how the chosen orbit(s) was impacting the representativeness with respect to the MSR Mission. In particular, the focus was given to the following aspects:

- Detection of a very faint object at far range
- Illumination conditions different from Mars Orbit

Regarding the first aspect, three main factors were taken into account:

1. *Sample Canister Signal To Noise Ratio (P1)*: the Sample Canister can be detected if the signal received at detector level is sufficiently high with respect to noise sources

2. *Sample Canister motion (P2)*: the sample canister is more easily detected when its apparent motion in the image is slow, allowing accumulation of the signal over several images.

3. *Background stars distribution and motion (P3)*: the canister is expected to have a magnitude between 10 and 14. The density of stars of comparable magnitude in the field of view is high, causing additional noise, possible confusion and even possibly masking the canister signal.

Assuming to perform a mission in a Low Earth Orbit, and investigating the sensitiveness of the above defined parameters with respect to the changed environment, two scenarios were proposed for which the representativeness was guaranteed. One called "true scaling", requiring no modification on the sensors design/performance, and one called "low delta-V", ensuring the performance of the mission with a reduced spend of delta-V but implying to modify the used sensors. It has to be said that the modification on the sensors was not considered as impacting the representativeness of the mission as this is instead driven by keeping constant the three discussed parameters: P1, P2, and P3. Taking also into account the constraint on the available delta-V, the second scenario was finally chosen to be implemented in the MSR IOD mission.

Another important aspect was related to the illumination conditions to be considered. In this case, the idea was to enable the possibility of testing the sensors in the highest variability of the illumination conditions on the Sample Canister. This, together with the need to choose an orbit characterised by a high launch rate (since IOD is expected to be secondary payload), has driven to choose a Sun-Synchronous orbit with a local time of 9:30h.

V. PAYLOAD DESCRIPTION

The Payload of the MSR IOD Mission is composed by two main elements:

- the Capture Mechanism
- the Sample Canister (x2 for redundancy)

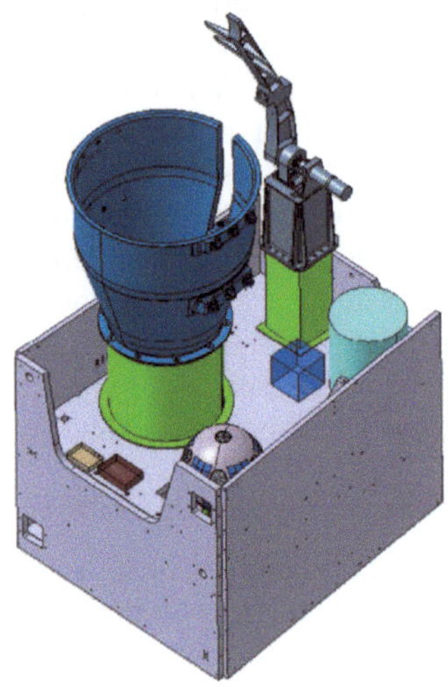

Figure 5: The Payload of the MSR IOD Mission accommodated on one of the investigated small

platforms (i.e. SSTL-150 Platform). One of the two Sample Canisters is not visible as it is accommodated inside the Capture Cone

The Capture Mechanism is accommodated on top of the Platform and is equipped with a specific suite of instruments required in order to achieve the defined demonstrations. Its main functions are:

- to release the Sample Canister
- to detect the Sample Canister during the long range demonstration
- to reconstruct the Sample Canister trajectory in order to get closer to it
- to perform the final rendezvous with the Sample Canister
- to capture and secure the Sample Canister.

The Sample Canister instead has the size of a football sphere, i.e. 23cm diameter, and it is a mock-up of the spacecraft that in the actual MSR mission will host the Martian soil samples and will be delivered in Martian orbit by the MAV for its capture. In the MSR IOD Mission, it is released from the Platform/Capture Mechanism once the demonstration begins. Its main functions are to provide a RF beacon signal to support the long range detection and to survive in space until capture.

An important aspect that has been driving for the design of those elements was the definition of the required GNC sensors. In fact, their selection was made bearing in mind the level of representativeness to be achieved with respect to the actual MSR Mission, the cost associated with their development, and the current uncertainties relevant to their design. Moreover, the share of the AOCS/GNC functions among the Platform, the Payload, and the Ground Segment was also defined in support of the final choice. As explained above, the same functionalities requested to the GNC sensors in/algorithms in MSR Mission were ensured also for the MSR IOD Mission. In particular, the decision to perform during the search phase image processing on ground allowed saving onboard computing resources and reduced impacts at the platform level.

On such a basis, the following set of AOCS/GNC components was identified:

- A *Narrow-Angle Camera* (NAC) for search phase, far range navigation. Its design is specific to the IOD Mission, but still close to the expected MSR design. Based on the current VISNAV design,
- A *Wide-Angle Camera* (WAC) for short range navigation. Several cameras are already available, including COTS cameras requiring space qualification to already space-qualified cameras.
- *GPS receivers* on both Platform/Capture Mechanism and Sample Canister; processed on ground (no real-time) to provide reference trajectory. E.g. the Phoenix receiver provided by DLR represents a valuable candidate,
- A *LiDAR sensor*. Two different technologies are available: scanning LiDAR and flash LiDAR. Scanning LiDARs are representative of the full MSR sensor chain (as foreseen today, but still subject to change), but imply high mass and cost. An interesting alternative is to use a short range Flash imaging LiDAR. Although the operating range of such LiDAR for space applications in Europe is currently much shorter, such a sensor would still allow demonstrating short range navigation and collision avoidance, for a lower mass and power penalty.

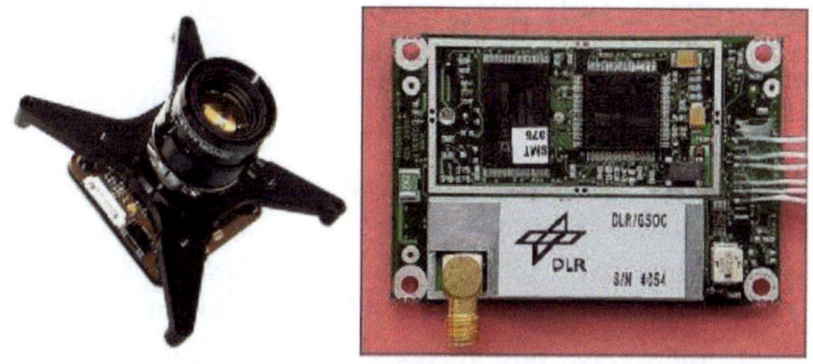

Figure 6: Some of the potential GNC Sensors: WAC GomSpace NanoCam (on the left), Phoenix GPS receiver (on the right)

During the Study, various configurations were investigated, assuming an increasing level of complexity and of cost. Such sensors were then distributed between the Platform hosting the Capture Mechanism and the Sample Canister.

For the *Capture Mechanism*, the reference design has been based on the current development performed at OHB-CGS in the frame of the ESA "Sample Canister Capture Mechanism" (SCCM) Study. However, this may change in later studies, in case additional requirements are provided and investigations are carried on. It includes the following main components:

- The *Capture Cone Assembly*. This component is aimed at providing an envelope volume within to capture the incoming Sample Canister, at enabling the detection of the Sample Canister once located inside the Funnel, and at providing a "hold" point to stow the Robotic Arm during launch. It is mainly composed by a rigid funnel, by a set of LED sensors for SC detection and accommodated on the funnel, and by an HDRM to hold the robotic arm in stowed configuration,

- The *Robotic Arm Assembly*. This component is aimed at retaining, transferring, and securing the Sample Canister during capture. It is mainly composed by a 1DOF robotic arm able to perform multiple operations,
- The *SC Trap Assembly*. It is aimed at accommodating the first Sample Canister during launch and, then, at hosting it during capture. The SC Trap is made of an aluminium/CFRP cylinder, attached at the bottom of the Funnel. The SC release could be done by using a dedicated simple mechanism, ideally a spring, or could be guarantee just by thrusting backwards the Platform, this guaranteeing the required separation distance

In order to control the above described mechanisms, to support the search, detection and rendezvous operations, additional equipment is required:

- The *C&DH/Comms System*. It includes all the electronics and the boards that are required to support the described elements, together with the OBC Software,
- The *GNC Sensors*. These sensors are aimed at supporting the overall demonstration and include one NAC, one WAC, one Flash LIDAR, and an RF Beacon receiver,
- The *Thermal Components*. These are standard MLI, heaters, and thermal sensors (thermostats, thermocouples, etc.) that are required to guarantee the survival at the space environment conditions.

The Sample Canister is a football sphere size of 23cm, and its design has been based on the one for the Mars NEXT Study, although with some modifications. First of all, the overall mass and the external surface were kept similar to the SC of the actual MSR Mission, in order to guarantee representativeness and do not impact the GNC sensors performance. However, due to the different environment, Low Earth Orbit vs. Mars Orbit, the core electronics has been selected in order to ensure the same functionalities but at a reduced cost. For this reason, the core electronics box has been derived from Cubesat applications.

The SC includes the following main components:

- The *Structure*, represented by an external shell inside which a set of frames/rails are accommodated. This shell is 23 cm diameter faceted sphere made in Aluminium or CFRP and at its turn is composed of two hemispheres, which are sealed together. The external surface provides the accommodation for the solar cells, the microstrip antennas, and for the retroreflectors (in case the use of LiDAR technology is also envisaged),
- The *Power System*. Solar cells are spread over the surface in order to provide enough power to guarantee survival and RF beacon transmission. Cubesat EPS and batteries are envisaged for power distribution and storage,
- The *C&DH/Comms System*. The OBC is derived from Cubesat applications, while a memory is provided to store housekeeping data and GPS data,
- The *GNC Sensors*. These include 1 RF beacon signal transmitter, a set of microstrip antennas, which provide omnidirectional coverage, and a GPS device keeping track of the position of the Sample Canister while free-floating,
- The *Thermal Components*. These are standard MLI, heaters, and thermal sensors (thermostats, thermocouples, etc.) that are required to guarantee the survival at the space environment conditions.

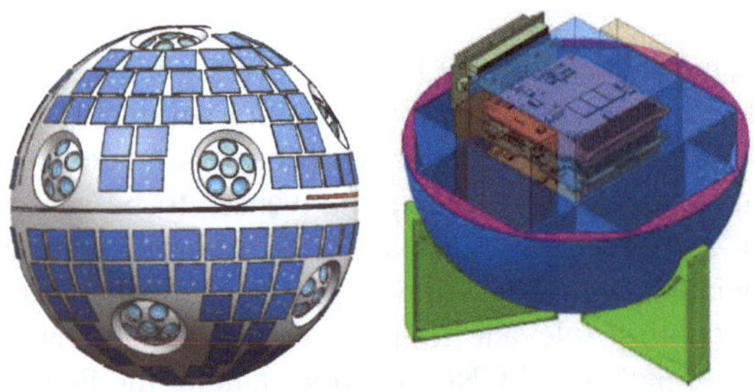

Figure 7: External and internal views of the Sample Canister. The external surface of the Sample Canister is representative of the actual MSR Sample Canister

VI. PLATFORM IDENTIFICATION

The Platform was selected through a trade-off process involving several platform suppliers. The list of platforms to be traded-off was made taking into account the following main aspects:

- suitability with the rendezvous mission
- suitability for accommodating payload mass and envelope

- suitability for supporting payload needs in terms of power and delta-V requirements
- availability within a constrained timeframe
- low mass and cost

Each supplier was asked to provide a specific set of information, and therefore two platforms were selected for investigating more in depth the potential accommodation and integration of the designed payload on the platform itself during dedicated workshops. The set of considered platforms included:

- The PRISMA MANGO Platform developed by OHB Sweden,
- The PROBA NEXT Platform developed by Qinetiq Belgium,
- The Myriade SPIRALE/Myriade Evolution Platforms developed by Airbus DS France in collaboration with CNES,
- The SSTL-150 and SSTL-300 Platforms developed by SSTL Ltd,
- The TET-X Platform developed by Kayser-Threde,
- The CHEOPS Based Platform developed by Airbus DS Spain (former EADS CASA).

The resulting selected platforms were the PRISMA MANGO Platform, which is still operational, and the PROBA NEXT Platform. Those platforms were selected because they provided a good compromise with respect to the above listed aspects. However, the payload was designed with particular attention to maximise the compatibility with most of the investigated platforms. In this way, a higher flexibility was ensured, together with the possibility to select another platform in case new constraints and/or requirements occur.

VII. LAUNCHER OPPORTUNITIES

For completing the overall investigation, potential launcher opportunities have been also identified. In this case, the goal was to find a launch at a reasonable cost for an in-orbit demonstration but still providing the amount of mass and envelope required by the mission. In such analysis, the main issue was to find a slow cost solution guaranteeing enough envelope to host the whole spacecraft (platform + payload), which has a considerable size due to the specific features of some of its component (e.g. the funnel and the robotic arm).

For such a purpose, several launchers have been considered, including:

- Small Class (e.g. VEGA, PSLV, EuRockot, DNEPR, SOYUZ-1),
- Medium Class (e.g. SOYUZ-ST, Cyclone-4).

Figure 8: Launchers providing potential launch opportunities

Various options have been analysed, among which the possibility to perform:

- A dedicated launch with a small launcher, exploiting the full launcher capability,
- Shared as a primary/secondary payload or Cluster launch (e.g. as auxiliary payload),
- Piggy Back launch, launched together with another satellite which drives orbit/time of launch.

From this analysis, two launchers were preliminary selected: VEGA in its VESPA configuration, and SOYUZ being an auxiliary payload. These were selected not only for their technical performance but also because of the requirement for a European launcher imposed by ESA. However, the analysis shown that the IOD Spacecraft could be launched also by other investigated launchers and potentially at lower cost.

VIII. OPERATIONS SEQUENCE

Regarding the main operations sequence, this can be split in a set of operations/demonstrations that are performed within the MSR IOD Mission. This includes four main phases:

- Launch and Early Operations (LEO) Phase
- Primary MSR Rendezvous and Capture Phase
- Secondary Debris Rendezvous and Capture Phase

- Guest Experiment and Long-Duration Testing Phase

(1) The *LEO Phase*: this phase includes the initial orbital insertion, standard separation and initial acquisition steps, and trim-manoeuvres to set the platform in the approximately correct attitude for initial testing and calibration, as well as solar array deployment, dependant on the final platform chosen for the MSR IOD. Telemetry and data relay can be checked during this time, along with preliminary activation and testing of the platform OBC. Subsystems calibration such as gyro and pointing calibration, guide star locating, battery conditioning, propulsion system and thruster testing, etc., will then be performed to ensure correct and satisfactory operation of the overall platform. Any major issues can be mitigated or addressed at this point. The unique payload of the MSR IOD will require switching on and confirming activation of the main rendezvous sensors, including the Wide and Narrow Angle Cameras (WAC, NAC), LiDAR, and RF receiver. The capture mechanism arm will be activated and perform an initial open-and-close cycle to both confirm operation of the arm HDRM and motor, while simultaneously confirming the correct operation of the detection sensors accommodated on the capture cone. It will also be possible to calibrate some of the camera algorithms at this stage using bright point sources such as known nearby satellites, the moon, and/or bright guide stars. The Sample Canister is in effect a separate spacecraft, and as such will require an initial systems check. Primary systems include the on-board EPS, the RF transmitter, and the basic OBC and system clock, and batteries. Testing and conditioning of the secondary batteries can begin at this point, while still connected to the platform. This will depend on charge levels and the predicted time-lapse until the first demonstration. Once disconnected from the platform, charging is not foreseen to be possible without adding additional complexity (e.g. charge-docking mechanism, etc.). It is anticipated that the early phase testing and calibration can take between two and four weeks, depending on progress and the occurrence of any unplanned issues.

(2) The *Primary MSR Rendezvous and Capture Phase*: - This phase is aimed at enabling the demonstration of the identified critical technologies. It therefore represents the most important phase of the MSR IOD. The primary demonstration requires the implementation of a complex sequence of events incorporating the careful operation and monitoring of two spacecraft, the Chaser and the Sample Canister, the second one being non-cooperative, over large and short range orbital distances. The performance of those operations may require in some cases on-ground in-the-loop control, as well as autonomous guidance, navigation and control, depending on the kind of operation is performed. A high-level, step-by-step process of the operational stages involved in the primary MSR demonstration phase is as follows:

- ✓ Stage 1: Initial release of SC
- ✓ Stage 2: Demonstrate forced translation, free drift and capture
- ✓ Stage 3: Demonstrate final approach
- ✓ Stage 4: Demonstrate far-range detection
- ✓ Stage 5: Demonstrate end-to-end rendezvous, from far-range detection down to capture

Within each stage, the achievement of safe-hold points is always ensured in order to provide reference points for starting/ending each demonstration.

The primary MSR capture demonstration is expected to last between 6 – 12 months depending on Delta-V allowance for repeatability and unforeseen technical issues. The baseline envisages repeating the end-to-end demonstration for three times, and it includes tests under varying conditions (illumination conditions, attitude pointing law), and a demonstration of autonomous collision avoidance.

The above described stages are mapped below versus the achieved demonstrations.

Table 2: The Primary MSR Rendezvous and Capture Phase Synthesis

Demonstration	St 1	St 2	St 3	St 4	St 5
Data acquisition	✓	✓	✓	✓	✓
Relative station-keeping	✓	✓	✓	✓	✓
Capture Strategy		✓	(✓)		✓
Forced Translation		✓	✓		✓
R-Bar/V-Bar hops			✓		✓
Autonomous Collision		(✓)			✓

Demonstration	St 1	St 2	St 3	St 4	St 5
Avoidance					
Sensors Hand-offs			✓		✓
Visual Detection and Orbit Determination				✓	✓
RF Detection and Orbit Determination				✓	✓

(3) The Secondary Debris Rendezvous and Capture Phase: - Once the primary MSR IOD rendezvous and capture phase has been completed, the secondary debris capture can be demonstrated. The operational sequence will be similar to the one planned for the capture of the Sample Canister, except that the RF beacon will remain inactive throughout the entire capture demonstration (this sensor cannot be used because the generic debris cannot transmit any RF beacon signal). The operations will begin by commanding the platform to being manoeuvred in order to get closer to the orbit of a pre-determined item of suitable space debris. At this point the Wide- and Narrow Angle Cameras will be switched on, and along with the detection algorithms, search and isolate the targeted debris. Autonomous collision avoidance will be tested again here, since the debris will be irregular in shape and spinning.

The secondary debris rendezvous and capture phase is predicted to last ~ one month depending on the distances involved to suitable debris, and the use of the most efficient transfer orbits to reach the target, as well as contingency for unforeseen technical issues.

(4) Guest Experiment and Long-Duration-Testing Phase: - The final operational phase is the activation of the guest experiments accommodated on-board, and the long term performance testing of both the guidance sensors and capture mechanism. Depending on the final guest experiment/s chosen, and particularly the peak power required, it may be possible to begin the experiments earlier in the mission. In the actual MSR mission the Capture Mechanism will stay in orbit for 12 to 24 months before performing the critical capture operation. Therefore, testing the long-term performance of the mechanism and of the sensors, and comparing the results against the ones obtained through ground-testing and initial space-based performance, will be very beneficial for understanding their robustness with respect to the environment. Once the primary phases of the MSR IOD have been completed, it will also be possible to test the long-term survivability of the Sample Canister by releasing for an extended period of time, and simulate the full free-floating mission of the Sample Canister in Mars Orbit as in the actual MSR mission.

At the end of life, the platform will perform its de-orbiting manoeuvre. The selected platforms already foresee a specific amount of delta-V dedicated for it.

IX. BENEFITS TO OTHER SIMILAR MISSIONS

The implementation of the MSR IOD Mission represents then an alternative solution to a Mars In-Orbit Demonstration for the development of the described technologies. The achieved in-orbit validation would allow gaining sufficient confidence before their implementation on the actual MSR Mission.

However, the benefit of their development would not be limited to the increase of their TRL, but it would be also spread to other missions that have strong synergies with the MSR rendezvous and capture operations. An example is given by the Active Debris Removal (ADR) missions, where the GNC technologies identified for the MSR IOD can be largely reused applying similar operations strategy. In fact, while long-range optical detection is not a required technology for ADR as debris orbits are already roughly known from the Ground observations, the image processing and navigation filter may still be used to refine the knowledge of these orbits thanks to the accuracy reached through optical means. Furthermore, although a reduced level of autonomy may be accepted for ADR missions, autonomous GNC for rendezvous of uncooperative target remains fully applicable to ADR missions.

There are at least five main categories of missions where the technologies advanced thanks to the implementation of the MSR IOD mission will provide significant benefits:

1. Near Earth Asteroid (NEO) sample retrieval
2. LEO debris removal
3. Lunar/Mars-satellite sample return
4. Full Mars sample return mission
5. Automated docking of small-medium satellites

Further investigations should be made to understand other potential synergies and to create future mission opportunities.

X. ACHIEVEMENTS AND CONCLUSIONS

The above sections reported the main outcomes of the MSR IOD Study. The Study has successfully demonstrated the feasibility of a low cost demonstration in Earth orbit of some of the most critical technologies of the MSR Mission. The resulting spacecraft has a mass ranging between 200kg and 300kg, including a payload of about 50kg, and can be launched as secondary payload into a Sun-Synchronous Earth Orbit.

During the Study, both design and operational aspects of the actual MSR Mission and of the MSR IOD Mission have been investigated and compared in order to not only better understanding them, but also to provide solutions for their mitigation and for increasing the mission success probability. Moreover, particular attention has been given to the achievable representativeness and level of fidelity with respect to the actual MSR Mission scenario to guarantee the validation of the results.

The depicted scenario may change depending on the imposed requirements, and taking into account also the changes that the MSR Mission itself may have in the future, but it is still applicable also to the demonstration/performance of other similar missions. In particular, an interesting outcome is the possibility to setup a low cost "flying" (IOD) testing facility, which could support the development of future applications

Summarising:

- The Assessment of a Mars Sample Return In Orbit Demonstration has been performed successfully demonstrating its Feasibility
- The Technologies to be demonstrated have been identified and analysed in–depth, taking into account
  - ✓ Benefits of demonstrating them in orbit vs. on ground
  - ✓ Representativeness vs. actual MSR
- The Overall Mission Architecture and Operations Concept have been defined, also identifying
  - ✓ Compatible Platforms
  - ✓ Launcher Opportunities
- The Phase-0 Design of Main Payload Elements has been performed, including a first assessment of the Programmatic & Cost Aspects

As future step of this study, a follow on (Phase A Study) would be now necessary to further define the mission and provide better cost figure (here estimated as well below <100M€, including launcher).

Acknowledgments

✓ Jaime Reed (Airbus Defence & Space UK)

✓ Thomas Chabot & Nöela Despré (Airbus Defence & Space Toulouse)

✓ Stefano Santandrea (ESA) – ESA Technical Officer of the MSR IOD Study

BIS-RS-2014-14

The Small Satellite Integrated Communication Environment (ICE)

John E. Ploschnitznig
Riverside Research
15059 Conference Center Dr. Chantilly VA 20151; 703-908-8373
jploschnitznig@riversideresearch.org

ABSTRACT

As the SMALLSAT and NANOSAT communities advance space technology, a communications evolution must also begin. The infrastructure to effectively maintain, monitor, command, and communicate with a satellite requires resources. Historically, this communication link was accomplished by a dedicated ground site with a dish antenna, which provides limited daily access, requires dish operation, and is limited to a single satellite at a time. Although satisfactory for the traditional large satellite architecture, small satellites will break this paradigm as their numbers increase. The value in small satellites may not be in their individual performance, but rather in their numbers. As the number of SMALLSATS and NANOSATS increase from hundreds to thousands of satellites, dedicated ground sites will be rendered ineffective. This paper offers a concept that utilizes existing cell phone technology and a modified cellular infrastructure to revolutionize communications for all satellites. This novel approach will significantly advance satellite technology by providing access to space for academia and entrepreneurs, while off ering the military unconventional access to perform its mission.

KEYWORDS: **[Satellite ICE Communication Cellular NANOSAT]**

PROBLEM

With the proliferation of SMALLSAT and NANOSAT satellites, the usual method of communicating with satellite systems, that is, using dedicated ground stations with dish antennas is rendered ineffective. This method provides only limited daily access, requires dish operation, and is limited to a single satellite at a time.

SOLUTION

The solution is very simple: modify existing cellular towers to provide a fixed upward pointing narrow beam antenna. A single cell tower provides limited communication access, but distributing these antennas over appropriately spaced cell towers provides overlapping coverage spanning large areas of arbitrary size (Figure 1).

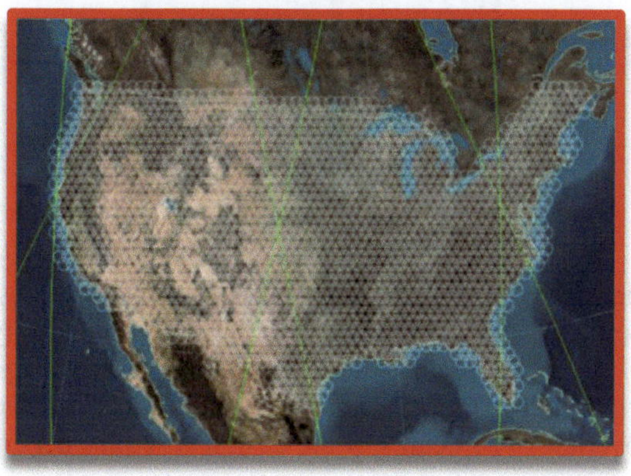

Figure 1 – Illustrates Overlapping Coverage

The goal is to integrate satellite communications into the existing cellular networks and break free from the legacy ground station concept, offering a new Integrated Communication Environment (ICE) for all future satellite programs.

S. Hatton (ed.), *Proceedings of the 12th Reinventing Space Conference*, DOI 10.1007/978-3-319-34024-1_10

Satellite ICE Module

Many of today's SMALLSAT and NANOSAT satellite systems use smartphone technology for onboard data processing based on size, sophistication, and needed capabilities. ICE leverages the communication capability of these powerful devices to transmit and receive data to and from the satellite. As the ICE communication technology advances, waiting hours to access the next available ground site will become history and watching live streaming video from space will become the norm.

Smartphone technology is currently being considered for the operating system on many of these new small satellite applications (Figure 2).

Figure 2 – DROID™ Integration into Satellite Technology

Although this technology has greatly evolved over the past decade for terrestrial use, some modifications are required for a satellite communication implementation. The link budget analysis points to increasing transmit power and adding a downward pointing directional antenna as two variables that can be adjusted in order to secure the link.

ICE Cell Tower

Current cellular towers are dedicated to providing the necessary coverage for terrestrial communication through antennas aimed at the horizon. ICE requires that a directional upward pointing antenna be integrated on existing cell towers (Figure 3).

Figure 3: Zenith Antenna Integration

Based on the particular cellular network and available transmit/receive frequencies, an antenna configuration can be designed to provide the required orbital footprint at Low Earth Orbit. For this analysis, we assume a 300 km altitude orbit. A 4G network operating at 4 GHz requires a 25 cm dish (slightly larger to minimize interference with terrestrial antennas) to create an 18 degree footprint in space. This footprint requires an ICE cell tower antenna every 60 km to provide pattern overlap. The details of the design will vary depending on the selected network and available frequencies.

What makes this approach interesting is a network of many cell towers, appropriately spaced, in order to provide overlapping orbital coverage at Low Earth Orbit (LEO) satellite altitudes. Only a fraction of the many available cell towers are required to provide continual communication over large areas of land.

A thorough link budget analysis is needed to ensure dedicated communications can be established with any satellite passing through the ICE antenna beams. The balance of the analysis defines various factors that can be adjusted to help define the ICE network size, number of antennas, transmit power, receiver sensitivity, and cost to build and install equipment. Ultimately, a solution that provides the largest effective individual footprint will drive the ICE cell tower antenna design since this will result in a lower ICE network cost. A business case must be made which demonstrates the return on investment for populating an ICE cellular network.

The Virtual Ground Site

The ICE approach relieves the satellite community from depending on traditional dedicated ground sites, and instead a virtual satellite monitoring system will be established. A generic approach will be used to send data to and from the satellite in logically-small transfer packages (ICE Pacs). Based on cell tower access statistics (worst-case short-duration access), packages would be designed to identify standardized data types: Command and Control (C2) uplink; Status of Health (SoH) downlink; and Data Packages (Puzzle Pieces in Figure 4), etc.

Software will be designed to monitor SoH, manage customer needs, perform satellite sensor tasking, create satellite commands and manage satellite operations, operate the ICE Network (under certain operational demands such as Multiple-Cell Tower Operations), and manage the ICE Pacs into logical ICE Trays (Figure 4).

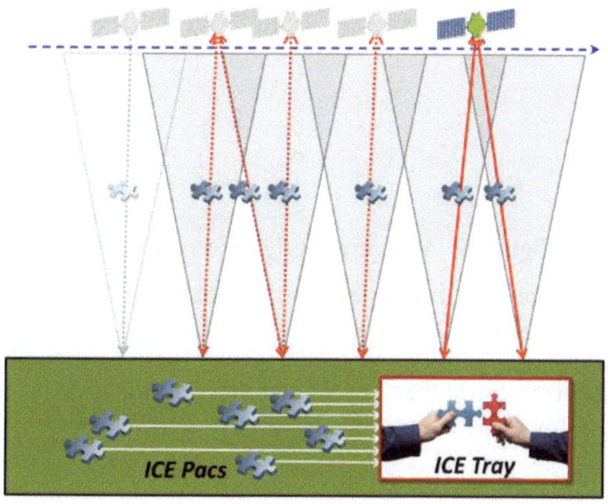

Figure 4: The Virtual Ground Site

Although the existing telecommunication infrastructure will be used to seamlessly integrate the ICE network, there are refinements that can be made to minimize typical cellular communication operational overhead. For example, the fact that we know the satellite trajectory allows us to configure the appropriate sequential cell towers to act as one continuous connection and offer seamless video transmission.

TECHNICAL ISSUES

Although this concept leverages the success and maturity of existing cellular network technology, integration of these pieces into the ICE system presents several hurdles: cell phone communication hardware on satellites, satellite antenna and transmitter power requirements, cell tower antenna and associated transmit/receive hardware, selecting the appropriate cellular phone network, software to manage the system, etc. The goal is to make communications achievable and affordable by simply incorporating the necessary cellular equipment into the satellites. Despite the many technical issues that must be resolved, the most difficult challenge is likely to be political: "How do we convince industry to invest in the necessary cellular infrastructure to make the ICE concept feasible?

Link Budget Analysis

A link budget analysis is needed to define the ICE configuration parameters (antenna gain, transmit power, receive sensitivity, etc.) of the satellite phone and cell tower system. Currently used cell phone parameters may be modified since they will not be operated in the conventional fashion and will not be restricted by human safety requirements. Likewise, ICE cell tower antennas will be pointing upwards and can incorporate low sidelobe and backlobe antennas. The goal is to ensure feasibility of an executable communication system that provides seamless satellite-to-cell tower communications while operating within FCC approved restrictions.

Antenna Patterns

The cell tower antenna pattern will set the pace. The antenna will be mounted rigidly on top of the cell tower and be pointing upwards. This will minimize interference with the many other antennas in the vicinity of the cell tower. The antenna beam width is a critical operations and cost driving factor. It defines the orbital intercept area in space and the maximum distance between ICE antenna installations.

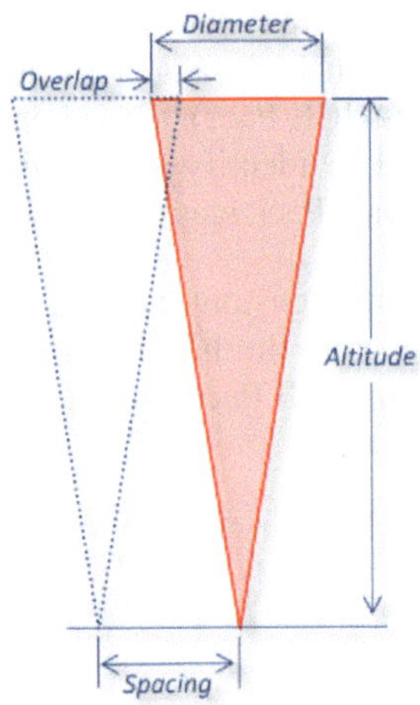

Figure 5: Antenna Design Considerations

These physical settings establish the satellite operating environment (access time per cell), which impact communication operations (ICE Pac size, etc.).

The link budget analysis will determine whether any candidate configuration is feasible. The desire will be to minimize the number of ICE cell tower antennas needed to perform the mission, thus minimizing ICE hardware installation costs (Figure 6).

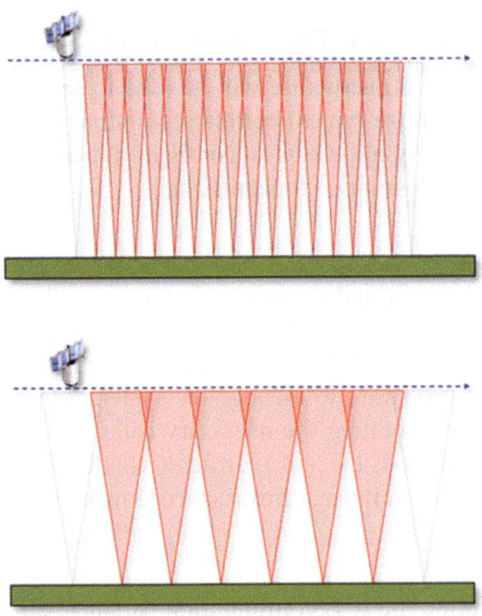

Figure 6: Antenna Gain Performance vs Spacing

Likewise, the satellite ICE communication antenna must be designed to support simultaneous coverage over neighboring cell towers to ensure uninterrupted communications (Figure 7).

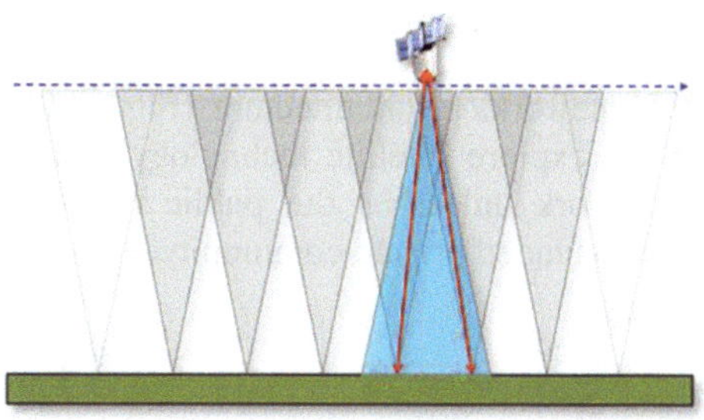

Figure 7: Simultaneous Coverage

In short, the link budget analysis will help define minimum equipment requirements and performance levels to ensure a robust communication capability.

Cellular Networks

Cell phone services have significantly evolved over the past decade. Advances in telecommunication technology have transformed a simple voice phone into a media storefront. Networks are now able to stream video into the palm of your hand. These advances should also benefit the small satellite community.

What service is the best: AT&T™, Verizon™, Sprint™, 3G, 4G, GSM? Each has benefits and shortcomings, but the key factors that will determine the initial choice will be the service that offers a programmable interface, can be used globally, and offers low operating costs. Since many of the satellites that will use ICE are for research and academia, the cell network of choice will be the one which offers the most flexibility for the satellite owner/operator. Open architectures allow academia and commercial industry to explore various communications concepts for their satellite applications. The cost of the ICE cell tower equipment (antenna, receiver, transmitter, cell tower integration, etc.) must be minimized. Non-recurring costs associated with the initial design and fabrication should be subsidized by the federal government to reduce the upfront investment (similar to government's role in developing the internet and GPS).

The ultimate goal is to allow a satellite vendor to purchase a satellite-enabled phone that can be integrated into their satellite hardware. To reduce weight, these phones would offer no video interface or antenna, and these phone circuit cards could be easily integrated into a NANOSAT or SMALLSAT chassis. A monthly service fee or data-package fee could be charged, and at least initially, a usage fee may apply.

ICE Software

This is a major R&D growth area. Firmware is necessary to translate remote commands into satellite operations. Software is necessary to manage satellite tasking, the communication network, and information flow to/from the satellite (Figure 8).

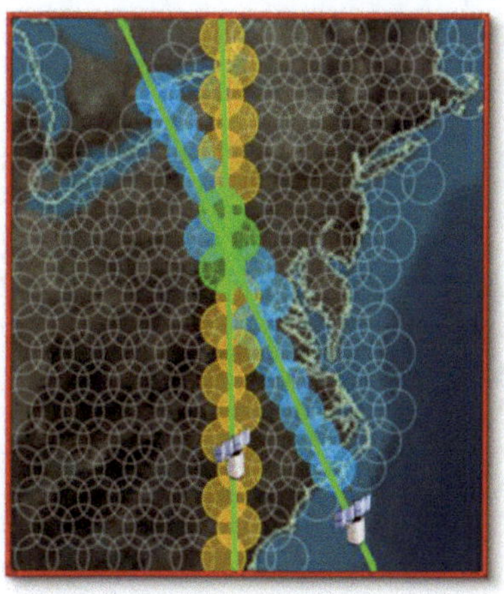

Figure 8: ICE Operations Management

Applications will be developed for handheld devices to request collection tasking and to view products.

Ultimately, the goal is to provide a framework, an architecture (Figure 9), from which academia, commercial industry, or even the ordinary citizen can explore satellite technology. Eliminating the physical ground site and offering a virtual communication framework unleashes the public to explore and invent in an area that was previously controlled by a limited and well-funded private community.

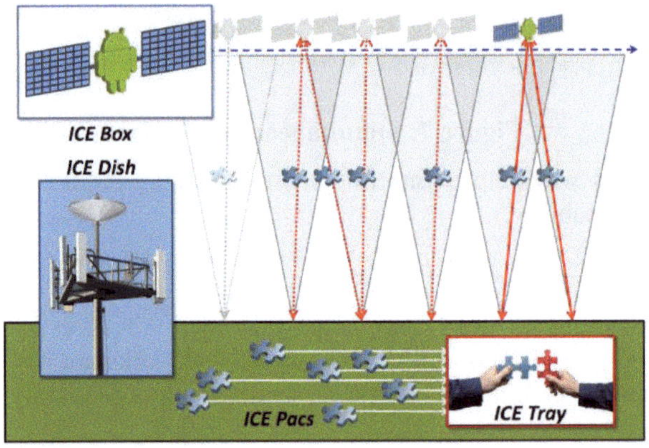

Figure 9: ICE Architecture

APPLICATIONS

The applications are many and include homeland security, emergency management, natural disaster, monitoring, communication, atmospheric and ionospheric research, sensor calibration, forest fires, weather, land survey, etc. There are many concepts for SMALLSATs, NANOSATs, and PICOSATs, from single satellites to constellations of tens to hundreds of satellites. Legacy communication systems and single pedestal antennas would be overwhelmed by the increased demands associated with managing thousands of satellites and systems, monitoring their health, commanding each satellite, and managing the flow of data/information to and from each satellite. This is the problem that the ICE concept addresses. The lessons learned from today's cellular communication networks in simultaneously managing thousands of phone calls will be used to manage the information to and from these many satellites.

"Look at me"

Let's assume we have a constellation of 100+ satellites (multiple sun-synchronous orbital planes with many satellites in each plane). Such a constellation would offer on-demand coverage over large areas of land. These satellites could be configured in many ways (video, Electro Optical, Thermal, Hyper/Multi Spectral, etc). With the click of a button on a handheld phone application – "Look at me" – the phone geo-location would be sent to the task manager and would designate the appropriate satellite(s) to point towards the uploaded latitude and longitude (Figure 10).

Figure 10: "Look at me" Application

The application for this could be used many ways: natural disaster, news, tornado sightings, etc. Anyone could request snapshots and videos through a simple application or even on the computer by simply clicking the map. Immediately, the ICE Tasking Management System would task available satellites supporting the "Look at me" service.

The Dynamic Earth

Videos could be painted onto any global web service: Google Earth™, NASA World Wind, CESIUM, etc. These form the digital canvas upon which to paint information from hundreds and thousands of data sources. Imagine 100 satellites feeding streaming video onto this global canvas. This application would allow users to log in and task a satellite via a simple point-and-click interface.

Military Application

The government may choose to encrypt their data (eICE Pac or Black ICE) to protect the content of all uplink and downlink data. In certain cases, the government may utilize ICE as a way to transmit satellite live data to designated receive sites, with this data then forwarded to appropriate person(s) through established secure means. USSOCOM would be a likely customer in such cases.

The US State Department may utilize such a system to monitor international activities in such global hotspots as Libya, Egypt, Syria, Yemen, etc. This would provide real-time situational awareness on large scale international activities. This effectively becomes an eye in the sky (albeit with limited resolution).

There are many different applications that can be attempted at very low cost. Space Situational Awareness, signals monitoring, hyper-spectral, weather, etc. Whatever the technology, ICE offers a robust communication infrastructure to get the data down quickly.

CONCLUSION

The overall ICE benefit is to provide large area unrestricted satellite communications using relatively simple, inexpensive, redundant, and scalable technology. Utilizing the existing cellular network leverages the infrastructure of a massive data communications network to disseminate the information. As the numbers of small satellites and constellations increases, this approach offers unlimited access. It can be applied globally on cell towers or at any broadband internet access point. The legacy dish pedestal approach can only handle a single satellite within its field of view, whereas ICE can simultaneously handle many satellites with a single cell tower; no pointing, no tracking, no prioritizing, no man-in-the-loop. Just as the internet and cell phone technology offered people unrestricted access to each other, ICE extends the reach of these two technologies and offers unrestricted access to space.

Riverside Research has a rich history of bringing technology-based solutions to our customers in the most cost effective manner, especially in the satellite collection tasking and mission management arena. As a not-for-profit corporation, Riverside Research has supported the US Government for over four decades and looks forward to meeting future challenges.

Further details regarding the ICE system and technology are disclosed in U.S. Patent No. 8,751,064 B2 entitled "Methods and Systems for Satellite Integrated Communications," which issued June 10, 2014.

BIS-RS-2014-33

SPARTAN: Scramjet Powered Accelerator for Reusable Technology AdvaNcement

Dawid Preller
University of Queensland
Australia
d.preller@uq.edu.au

Professor Michael Smart
University of Queensland
Australia
m.smart@uq.edu.au

ABSTRACT

The global economic environment combined with the rapid pace of technology advancement is placing importance on reducing the cost and increasing the responsiveness of access to space systems. Based on decades of practical experience with rocket-only launch vehicles, current technology is operated close to theoretical limits and only marginal further efficiency improvement is achievable. In order to further improve the efficiency of access-to-space vehicles, new propulsion systems will be required. Airbreathing engines, and scramjets in particular, are considered the most promising alternative. Scramjets have an advantage over rocket propulsion in terms of a significantly higher specific impulse; other benefits of airbreathing propulsion for access-to-space are increased launch flexibility, such as shorter time to rendezvous with a target spacecraft, and increased launch window duration and number of opportunities. This project investigates the use of a three-stage rocket-scramjet-rocket system for transporting payloads of approximately 500 kg to a Sun Synchronous Orbit. It is believed that this mission profile meets the requirements of many missions, such as responsive surveillance of man-made and natural disasters and several earth science missions.

The first stage is being developed as an advanced academic programme in Australia, South Africa and France, and is named the Austral Launch Vehicle (ALV). The ALV is a re-usable liquid rocket stage used to accelerate the stack to the point of Scramjet ignition at Mach 6, after which it is recovered by flying back to the launch site. The reusable second stage named the Scramjet Powered Accelerator for Reusable Technology AdvaNcement (SPARTAN), is based on a winged-cone vehicle initially developed for the US National Aerospace Plane program. It is powered by liquid hydrogen fuelled Scramjets and it is intended to provide acceleration until the net specific impulse drops below useful levels. The final stage is deployed upon scramjet shutdown, using a conventional expendable liquid-fuelled rocket motor to place the payload in the desired orbit. Preliminary analysis of the complete three-stage system indicates a better overall performance, in terms of the payload mass fraction to orbit, than current rocket-only systems of this scale. In order to explore this concept further the project has advanced to the next phase of development which includes the design, manufacturing and testing of scaled down demonstrators of the ALV and SPARTAN vehicles.

1. INTRODUCTION

The global economic environment combined with the rapid pace of technology advancement is changing the requirements of current and future space-access systems. Both reduced scale and increased responsiveness will be the future drivers of access-to-space. This is in contrast to the late 20th century where large scale multi-experiment satellites with long lifespans led to the development of large launch systems. Due to the rapid development of micro-scale, low power electronics, satellites that were many thousands of kilograms, now weigh just hundreds of kilograms. There is also a movement to even smaller satellites (< 100 kg) with limited life.[1] The current costs for launching small satellites are at least a factor of three greater than large satellites on a per kilogram basis, so while the reduced mass is a significant advantage, the lower structural efficiency of smaller rockets inhibits its usefulness. A technology shift to reusable systems with an air-breathing propulsion component may be the best solution for the launch of small satellites.[2]

Rocket based expendable launch systems are expensive and lack the responsiveness that will be desired by the space community in the future. Based

S. Hatton (ed.), *Proceedings of the 12th Reinventing Space Conference*, DOI 10.1007/978-3-319-34024-1_11

on decades of practical experience with rocket-only launch vehicles, current technology is operated close to theoretical limits and only marginal further efficiency improvement is achievable. The introduction of new commercial players such as Space-X has increased the organisational efficiency of space launch, with a flow-through reduction in costs. But the fact remains; throwing away a significant portion of your launch system each time you fly will always be inherently expensive. Two key changes in technology are needed to improve the cost and responsiveness of space access. Firstly, a substantially re-usable system will decrease costs if the technology is right. Secondly, introduction of an air-breathing facet to space launch, and therefore aircraft-like operations, will remove the rigidity of fully rocket based systems in terms of orbital inclination, time to rendezvous and safe abort.

Many concepts involving air-launch have been proposed in recent years.[3] These launchers take-off from a runway, typically fly to high subsonic speed and approximately 10 km altitude, and release a multi-stage rocket in the desired direction. They can also return to base if problems occur, or manoeuvre around bad weather. The main advantages of introducing air launch are, (i) lifting the rocket stage above the lower atmosphere (reducing drag and heating), and (ii) increasing launch flexibility. In terms of energy, the subsonic launch speed and 10 km altitude do not substantially contribute to the required delta-V of the system, so the overall payload mass fraction is relatively unchanged by air launch. An alternative to this is to use a 3-stage rocket-scramjet-rocket system. In this system the 2nd stage scramjet supplies the flexibility of aircraft-like operations, but also contributes to an increase in payload mass-fraction by using airbreathing propulsion over a meaningful proportion of the launch trajectory. Furthermore, the scramjet powered vehicle is inherently reusable, and its high Lift-to-Drag ratio enable it to return to base for the next launch.

For significant reductions in launch costs, the first stage booster of such a rocket-scramjet-rocket system should also be reusable. To this end, UQ has teamed up with Heliaq Engineering to combine a fly-back booster first stage with a second stage scramjet powered vehicle. The small third stage can be a standard expendable upper-stage rocket, as this is a relatively small proportion of the overall mass at take-off. The third stage will be deployed at high dynamic pressure, so it will require thermal protection. Preliminary studies of this system have shown real promise in accomplishing the combined goals of reusability and flexibility.

In this article some background on the status of scramjet development will first be described, along with details of the proposed configuration of the fly-back booster. The 3-stage rocket-scramjet-rocket system will then be described in the context of a system for delivery of approximately 500 kg to Sun Synchronous Orbit (SSO). Finally, a technology road-map will be presented, including sub-scale flight tests.

2. CURRENT STATUS OF SCRAMJET PROPULSION

Based on decades of practical experience with rocket-only launch vehicles, current technology is operated close to theoretical limits and only marginal further efficiency improvement is achievable. To further improve the efficiency of access-to-space vehicles, new propulsion systems will be required. Airbreathing engines, and scramjets in particular, are considered the most promising alternative.[4] In contrast to conventional rockets that carry separate tanks for both fuel and oxidizer, scramjet-powered systems carry the fuel only, using atmospheric oxygen for combustion. Scramjets have an advantage over rocket propulsion in terms of a significantly higher specific impulse (Isp).[5] However, scramjet operation requires an engine-airframe integrated vehicle design, which is inherently more complex than a rocket propulsion system.[6] A further benefit of airbreathing propulsion for access to space is increased launch flexibility, such as shorter time to rendezvous with a target spacecraft, increased launch window duration, and number of opportunities.[7] These benefits are obtained through airbreathing propulsion throttling and aerodynamic turning and pitch control available from a high L/D vehicle. The importance of these aircraft-like operational characteristics for space access missions has only recently begun to be explored.

Recent scramjet flights, such as NASA's Hyper-X[8] and the US Air Force X-51[9] have given impetus to the practical application of airbreathing propulsion to hypersonic flight. Furthermore, the HIFiRE Program, a joint Australian/US program for the development of sustained hypersonic flight involves 9 flights of increasing complexity.[10] The final flight (scheduled for 2016) of an autonomous hypersonic vehicle that will fly at Mach 7 for more than 30 seconds (Fig. 1) will include significant use of high temperature composite materials, enabling the vehicle to fly with a "hot" structure. All these flight tests have

Figure 1. HIFiRE 8 Autonomous vehicle powered by a scramjet at Mach 7

indicated that the theoretical advantages of scramjets over rockets can be realised. A series of studies have been conducted at The University of Queensland to examine the use of a three-stage rocket-scramjet-rocket system for transporting payloads of approximately 100 kg to Low Earth Orbit (LEO).[11-13] In these studies a conventional rocket 1st stage was used to boost the vehicle to the point of scramjet ignition at Mach 6. The reusable second stage was based on a winged-cone vehicle (WCV) and powered by a near-term Mach 6-12 3-D scramjet using hydrogen fuel.[14] The final stage was deployed upon scramjet shutdown, using a conventional liquid-fuelled rocket motor to place the payload in LEO. The second stage scramjet powered accelerator was the subject of a series of Multi-Disciplinary Optimisations (MDO) of increasing fidelity which involved flying complete trajectories. The resulting vehicle is shown in Fig. 2. Key aspects of the vehicle that led to high net specific impulse over a wide Mach range were a tight integration of the vehicle and the propulsion system, relatively large engine capture and low trim requirements. Payload mass fractions of the order of 2% were obtained; a value that compares very favourably with rocket based systems of the same scale.[12] High temperature Ceramic Matrix Composite (CMC) materials were critical to enabling the reusability of the vehicle and also to minimise its structural mass fraction. Two options for third stage separation have been considered for such a system; (i) high dynamic pressure, horizontal flight, and (ii) a pull-up of the scramjet vehicle to a low dynamic pressure at a relatively large flight path angle. It has been found that the aerodynamic losses of the pull-up far outweigh any advantages of this method, so a high dynamic pressure separation of the third stage is a key part of the system.

Figure 2. Optimised scramjet powered accelerator

3. ALV: A FLY-BACK 1ST STAGE BOOSTER

The Austral Launch Vehicle (ALV) project is an international effort to develop a cost-optimized partially Re-usable Launch Vehicle. Now in its fourth year, the ALV project originated as an investigation into how re-usability can provide real launch cost reduction. Requirements of modularity, flexibility and simplicity were identified as most important.[15] Modularity is required to increase the vehicle flight rate (through larger payload range and more module flights per launch) and to reduce development cost (through reduction of the number and size of newly developed elements). Flexibility is required to ensure a wide market can be captured, while simplicity leads to critical reductions in development and operational costs as well as increased reliability. Based on a response to these requirements, the ALV includes a re-usable liquid rocket booster stage that takes off vertically, deploys upper stages to continue to orbit, and fly's back to the launch site (Fig. 3).

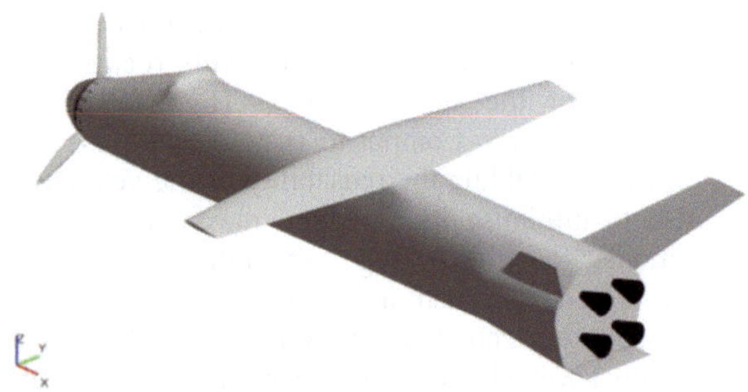

Figure 3. ALV fly-back booster in aircraft mode

This is achieved by the booster performing a high angle-of-attack re-entry (from approximately 1800 m/s using a body-flap and all-moving fins for control), followed by deployment of wings and propeller at subsonic speed. The booster then travels back to the launch site and lands like a conventional aircraft under autonomous control.

4. SPARTAN SYSTEM ARCHITECTURE FOR SUN SYNCHRONOUS ORBIT

The reusable second stage named the Scramjet Powered Accelerator for Reusable Technology AdvaNcement (SPARTAN), is based on a winged-cone vehicle initially developed for the US National Aerospace Plane program.[16] It is based around a conical fuselage with tightly integrated scramjet engines. Delta wings with a diamond cross section are included to provide the required lift for return flights and landings. Three views of the vehicle are shown in Figure 4. The vehicle has a length of 25m and a gross mass of 11.5t which includes the 4.5t third stage booster. The expendable third stage is positioned on SPARTAN in a piggy back configuration. It has a liquid propellant rocket engine. A full cladding heat shield is included for thermal protection during the high dynamic pressure stagging event and during the atmospheric acceleration period. The 500kg payload is positioned in the front section of the 3rd stage.

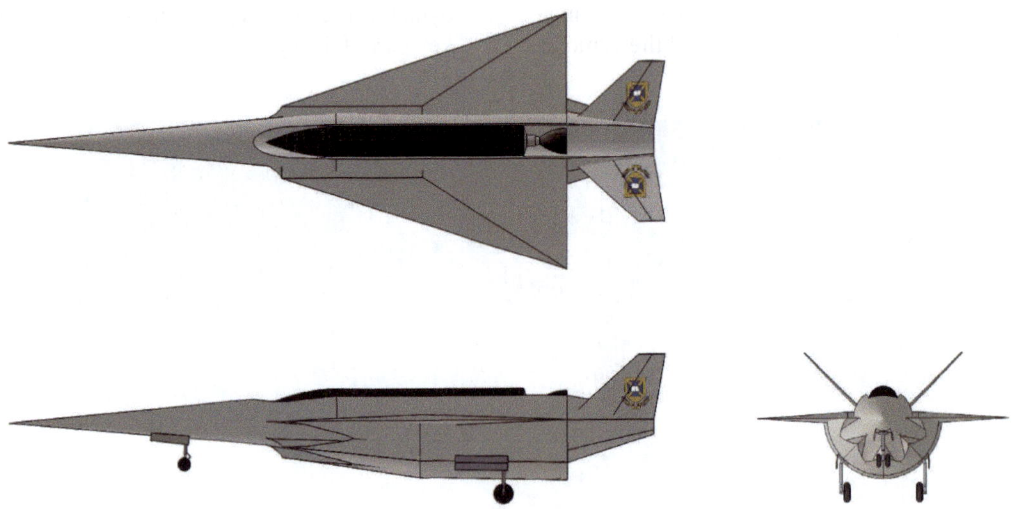

Figure 4. Views of SPARTAN

SPARTAN was designed around a mission profile of delivering a 500kg payload into a SSO of 566km altitude. It is believed that this requirement combined with a small launch capability focusing on rapid deployment, one mission only satellites, and short lifespans of 2-4 years will meet the needs of many missions, such as responsive surveillance of man-made and natural disasters and several earth science missions. The SSO is a popular orbit used for earth science missions, because it provides numerous desirable characteristics that satisfy many key mission requirements. The key feature is that the orbit crosses over the equator at the same local time each day as shown in Figure 5. This orbit allows for consistent scientific observations with fixed lighting conditions along the mission ground track. Another benefit is a global coverage due to the high inclination angles. [17]

The launch trajectory for the chosen SSO mission starting from Cape York (-12.25° Latitude, 143.1° Longitude) in Northern Australia is shown in Figure 6. Cape York is an attractive potential launch site because of its remoteness and close proximity to the equator. The launch stack will travel in a Northern direction at a 97° retrograde resulting in a SSO with an altitude of 566 km. This gives 15 revolutions per day at an orbital period of 5760 seconds, resulting in a view of 2671.7km between adjacent ground tracks; Characteristics that are believed to be very favorable for the intended earth science missions.

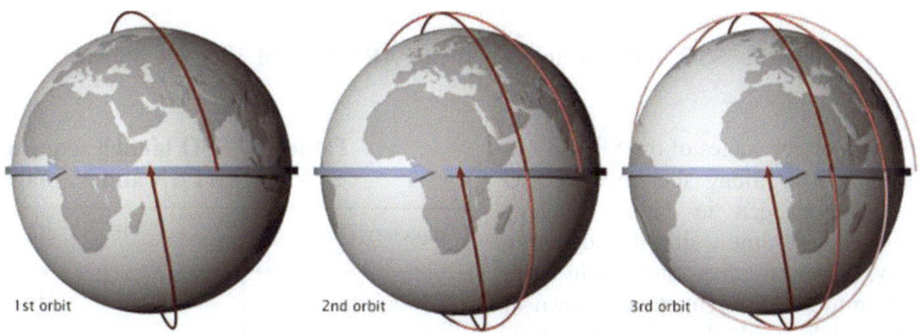

Figure 5. Sun Synchronous Orbit showing the ground tracks at fixed lighting conditions

Source: http://earthobservatory.nasa.gov/Features/OrbitsCatalog/page2.php

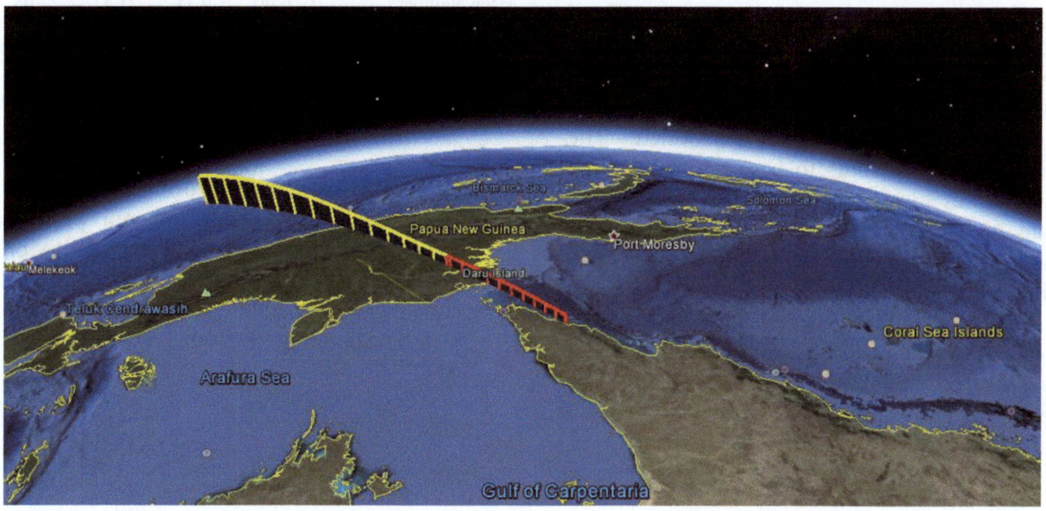

Figure 6. Launch trajectory from Cape York Australia. SPARTAN flight shown in red and 3rd stage trajectory shown in yellow.

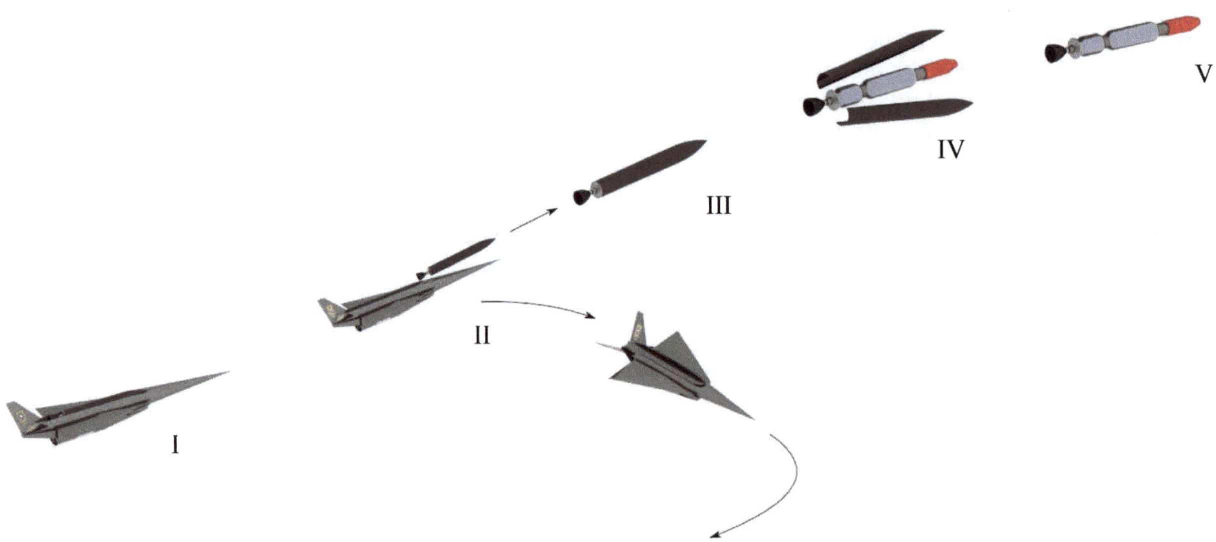

Figure 7. SSO launch stages for the SPARTAN and 3rd stage booster

The different trajectory stages of the SPARTAN and the 3rd stage booster along the chosen SSO launch mission can be seen in Figure 7. Key trajectory data such as time, altitude, dynamic pressure, velocity, Mach number and flight path angle corresponding to the different stages shown in Figure 6 are reported in Table 1. A mission starts with the system launching from Cape York. The 1st stage rocket booster accelerates the SPARTAN to favourable operating conditions for the scramjet, typically Mach 6 at a dynamic pressure of 50kPa. At this time (Fig. 6I), the scramjet engines ignite and SPARTAN separates from the booster. The reusable 1st stage booster now returns to base. SPARTAN accelerates until the performance, in terms of net ISP, drops below useful levels. At this point (Fig. 6II) the liquid rocket on the final stage ignites and a high dynamic pressure separation is performed. The reusable SPARTAN returns to base while the final stage accelerates away. The 3rd stage continues to accelerate up to a point where it can coast to a dynamic pressure at or below 10Pa (Fig. 6III). At this point (Fig. 6IV) the thermal protection system is no longer required and is discarded while the rocket continues to coasts towards the apogee of its orbit. At the apogee (Fig. 6V) the 3rd stage performs a Hohmann transfer to insert the payload into the SSO at 566km altitude.

Table 1. SSO launch sequence data for SPARTAN and 3rd stage booster

Trajectory Phase	I	II	III	IV	V
Time, *s*	0	206	348	414	503
Altitude, *m*	26665	32593	67445	96352	110418
Dynamic pressure, *Pa*	50000	50000	1000	10	1
Velocity, *m/s*	1798	2839	6327	6206	6183
Mach Number	6.00	9.34	21.47	21.06	20.98
Flight path angle, *deg.*	1.36	0.27	5.07	2.93	0

Preliminary analysis of the complete three-stage system indicates a better overall performance, in terms of the payload mass fraction to orbit, than current rocket-only systems of this scale. The presented system achieved a payload mass fraction of 2% corresponding to 500kg into a 566km altitude SSO.

5. TECHNOLOGY PLAN

The next steps needed to progress the concept involve sub-scale flight testing of key technologies that are particular to the launch system, and have not been done before. Table 2 lists all the key technologies needed to be developed for the overall concept to be successful. There are many other necessary technologies, such as those related to the 1st stage booster rockets, but these are considered to be typical of the current space industry.

Those technologies that are new and un-proven have been assigned a high risk level in Table 1. The planned sub-scale flight tests have been particularly designed to prove the viability of these for the first time. Technologies that are currently being developed in other programs and are more generic (such as sustained scramjet propulsion), have been assigned a moderate risk level. Those technologies that are particular to the concept, but are slight variations on current space practice, have been assigned a low risk level. It is expected that once the high risk technologies have been proven, confidence in the voracity of the overall concept would be high.

On the fly-back booster, the subsonic wing deployment is assigned a high risk level. This type of deployment has not previously been accomplished, and is a key differentiating feature of the ALV. On the scramjet accelerator, the high-q deployment of the upper stage is assigned a high risk level, as it a complex procedure that is very dependent on the particular vehicle configurations. There are no technologies on the upper stage that have been assigned a high risk level.

Two subscale flight tests are planned to provide proof that solutions can be found for the high risk items. These flights will be at reduced scale, and designed solely to prove the high risk technologies. To keep costs to a minimum, they will not be sub-scale versions of the complete system. The characteristics of the planned test vehicles are as follows:

ALV-1: Subscale fly-back booster

- Boost to supersonic speed using solid propellant motor
- Re-enter under body-flap and all moving fin control
- Deploy wings and propeller at subsonic speed
- Fly-back and land

SPARTAN-1: Subscale Scramjet 2nd Stage

- Boost to scramjet take-over (using an ALV or other booster)
- 30 second scramjet operation
- Deploy upper-stage at high dynamic pressure with dummy payload (upper stage does not go to orbit)
- Re-enter, glide-back and land

Table 2. Technology plan with associated risk levels

Stage	Technology	Risk Level	Comment
Booster	2nd stage deployment	low	In-line separation
	Controlled re-entry	medium	Lifting body with body flap and all moving fin control (X37-B)
	Subsonic wing deployment	high	Not done before
	Propeller deployment and fly back and land	medium	Similar technology exists
Scramjet Accelerator	Sustained scramjet operation	medium	based on HIFiRE technology
	High-q upper stage deployment	high	Not done before
	Re-enter, fly/glide back and land	medium	High L/D vehicle
Upper Stage	High-q motor ignition	medium	Not standard for liquid fuel rockets
	TPS drop-off on way to orbit	low	Similar to fairing drop-off at low q

6. CONCLUSIONS

The current economic environment and the rapid pace of technology advancement is driving a need for reduced scale and increased responsiveness in access to space systems. This article showed that one alternative to rocket only systems is the application of airbreathing technology. To this end an overview of the current status of scramjet propulsion was given to highlight the performance benefits such as improved specific impulse. Mission benefits include increase flexibility and increased launch window duration. A three stage to orbit rocket-scramjet-rocket system was introduced. The first stage is a reusable rocket booster focused around modularity, flexibility and simplicity in order to provide real launch cost reduction. The second stage vehicle is a winged cone vehicle powered by Scramjet engines, and the final stage is an expendable liquid propellant rocket booster. A preliminary study found that this configuration can deliver a payload mass fraction of 2% (around 500kg) into a 566km altitude Sun Synchronous Orbit. In addition to gaining two reusable systems the performance results compares very favourably to expendable systems of the same scale. The next phase of the project involves the demonstration of key technologies of the system. A plan was outlined identifying technologies needed for the success of the project. The key, as-yet un-proven technologies were, i) subsonic wing deployment in the first stage booster, and ii) the high dynamic pressure upper stage separation from the second stage airbreather. Future work will involve two subscale flight tests to provide proof that solutions can be found for these items.

7. REFERENCES

1. Baker, D. N., and S. P. Worden, "The Large Benefits of Small-Satellite Missions", *Eos Trans. AGU*, 89(33), doi:10.1029/2008EO330001, (2008).

2. Bowcutt, K.G., and Smith, T.R., "Responsive and affordable launch of small satellites: Areusable air-breathing concept". AIAA paper RS-2012-5001, 2012.

3. Chen, T.T., Ferguson, P.W., Dreamer, D.A., and Hensley, J., "Responsive Air Launch using F-15 Global strike Eagle",4th Responsive Space Conference, 24-27 April, 2006, Los Angeles, CA, AIAA-RS6-2006-2001.

4. Cook, S. and Hueter, U. (2003). "NASA's Integrated Space Transportation Plan - 3rd Generation Reusable Launch Vehicle Technology Update," *Acta Astronautica*, vol. 53, no. 4-10, pp. 719–728.

5. Kerrebrock, J.K. (1992). *Aircraft Engines and Gas Turbines*. MIT Press, London.

6. Heiser, W.H. and Pratt, D.T. (1994). *Hypersonic Airbreathing Propulsion*. American Institute of Aeronautics and Astronautics, Washington.

7. Flaherty, K.W., Andrews, K.M. and Liston, G.W. (2010). "Operability Benefits of Airbreathing Hypersonic Propulsion for flexible Access to Space," *Journal of Spacecraft and Rockets*, vol. 47, no. 2, pp. 280–287.

8. McClinton, C. R., "X-43–Scramjet Power Breaks the Hypersonic Barrier Dryden Lectureship in Research for 2006," AIAA, 44rth Aerospace Sciences Meeting and Exhibit, AIAA-2006-1, Reno, NE., USA, January 2006.

9. Hank, J. M., Murphy, J. S., Mutzman, R. C., "TheX-51A Scramjet Engine Flight DemonstrationProgram," AIAA 15th International Space Planes and Hypersonic Systems and Technologies Conference, AIAA 2008-2540, Dayton, OH, April2008.

10. Bowcutt, K., Paull, A., Dolvin, D., Smart, M.K., (2012). "Hifire: An International Collaboration To Advance The Science And Technology Of Hypersonic Flight", ICAS 2012 CD-ROM Proceedings. 28th International Congress of The Aeronautical Sciences, Brisbane, Australia, September 24-28, 2012.

11. Smart, M.K. and Tetlow, M.R., "Orbital Delivery of Small Payloads using Hypersonic Airbreathing Propulsion", AIAA Journal of Spacecraft and Rockets, Vol. 46, No. 1, pp117-125, 2009.

12. Jazra, T, Preller, D and Smart, M.K., "Design of an Airbreathing Second Stage for a Rocket-Scramjet-Rocket Launch Vehicle", AIAA Journal of Spacecraft and Rockets, Vol. 50, No. 2, 411-422, 2013.

13. Preller, D. and Smart, M.K. (2012) "Design of a minimum trim hypersonic airbreathing accelerator vehicle". *In: Proceedings of the 18th AIAA/3AF International Space Planes and Hypersonics Systems and Technologies Conference.* Tours, France, 24-28 September 2012, AIAA-2012-5825.

14. Suraweera, M.V., and Smart, M.K., "Shock tunnel experiments with a Mach 12 Rectangular-to-Elliptical Transition Scramjet at off-design conditions", AIAA Journal of Propulsion and Power, Vol. 25, No. 3, pp 555-564, 2009.

15. Schutte, A. and Tolyarenko, N., "The AUSTRAL launch vehicle: Reducing space transportation cost through reusability, modularity and simplicity", 64th International Astronautical Congress, Beijing, China, 2013, IAC-13-D2.4.11.

16. Shaughnessy, J.D., Pinckey, S.Z., McMinn, J.D., Cruz, C.I. and Kelley, M.L. (1990). "Hypersonic Vehicle Simulation Model: Winged-Cone Configuration". NASA TM102610.

17. Boain, R.J., "A_B_Cs of Sun-Synchronous Orbit Mission Design", 14th AAS/AIAA Space Flight Mechanics Conference, Maui, Hawaii, 8-12 February 2004, AAS 04-108.

BIS-RS-2014-28

The Disruptive Potential of Subsonic Air-Launch

David J. Salt
Telespazio VEGA Deutschland GmbH
Europaplatz 5, D-64293 Darmstadt, Germany; +49 (0)6151 8257-0
dave.salt@telespazio-vega.de

ABSTRACT

This paper tries to show that big launch vehicles may not be required to enable big space operations. It first highlight the essential role of space launch in enabling all space operations and indicates the dominant role that 'customer' demand has played in both enabling and constraining its development. It then discusses the advantages and drawbacks of a subsonic air-launched reusable launch vehicle (RLV); comparing and contrasting them against a wide range of other possible launcher concepts. In doing this it highlights the unique evolutionary opportunities that this concept has to offer and provides some insight as to how these may be realised and enhanced via existing technologies. Finally, the paper highlights the radical improvements in operational architectures afforded by such a vehicle. It shows how an RLV with a relatively modest launch performance of less than 5t to low Earth orbit could be capable of supporting almost all current and future launch demand by forming the key element of a fully reusable space transportation infrastructure.

KEYWORDS:

ACES	=	**Air Collection & Enrichment System**	**LEO**	=	**Low Earth Orbit**
CR	=	**Collection Ratio**	**Mg**	=	**Metric tonne**
ELV	=	**Expendable Launch Vehicle**	**Mn**	=	**Mach number**
GEO	=	**Geosynchronous Earth Orbit**	**SSTO**	=	**Single Stage to Orbit**
GTO	=	**Geosynchronous Earth Orbit**	**TSTO**	=	**Two Stage to Orbit**
IRR	=	**Internal Rate of Return**	**RLV**	=	**Reusable Launch Vehicle**
kg	=	**kilogram**			

1. THE LIMITS TO GROWTH

THE frequency and complexity of space operations has evolved relatively slowly over the last three decades and is certainly much reduced in comparison with the first two decades of the Space Age, which began more than half a century ago. The reasons for this 'phase change' can be easily understood when one considers the changes in both political and economic drivers that control most space activities, especially those involving human spaceflight.

The past five decades of space activity have, to a large degree, been driven by a few specific issues such as national security and conservation of the industrial base. In contrast, the slow growth of commercial ventures has been due mainly to market and financial constraints, rather than any basic limitation of the available technology. As a consequence, the diversity and intensity of spaceflight operations have also been paced by these trends, though the manner in which they are performed, on both the ground and in space, has been radically improved by the phenomenal advances in computing and software over this same period.

1.1 The Current Space Paradigm & Potential of NewSpace

We first consider future possibilities to identify the factors that may either prevent or severely restrain their realisation. Given the importance of markets in the development of commercial activities, this assessment also considers how such factors may also influence their growth and sustainability.

i) Current Constraints

Current space activities range from pure science missions through to civil and military applications like communication, navigation and observation systems. Nevertheless, growth and evolution in all these areas is limited by a few key factors:

- government priorities and constraints;
- competition from terrestrial alternatives;
- low market 'elasticity' (i.e. lower prices stimulate only limited market growth);
- launcher cost/availability/reliability.

The first factor is important because the growth of space activities is still dominated by government programmes, both civil and military. Communication satellites represent the nearest thing to a truly commercial market sector, but government funding still underpins much of their basic R&D while the second and third factors have placed significant restraints on their growth and evolution, as witnessed by the problems of commercial ventures like Iridium, Globalstar, ICO, SkyBridge and Teledesic.

S. Hatton (ed.), *Proceedings of the 12th Reinventing Space Conference*, DOI 10.1007/978-3-319-34024-1_12

To put the situation into perspective, Figure 1 shows a breakdown of the global space industry's annual revenue, which was $304 billion in 2012. However, this was still less than the annual turn-over of a single successful commercial company like Wal-Mart [RD.2], which was founded in 1962 but has managed to outgrow the entire world space industry by servicing vastly bigger and established markets to give a turnover of $445 billion in 2011.

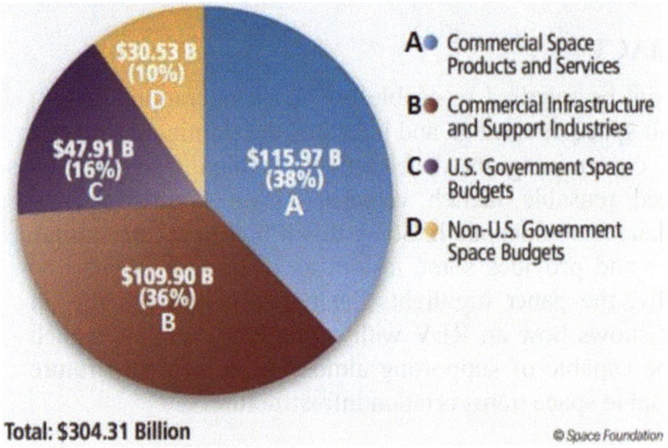

Figure 1. Global Space Activities, 2012 [RD.1]

ii) Future Potentials

A wide range of future space-based activities and associated business opportunities[1] have been discussed for many decades (e.g. space manufacturing facilities, solar power satellites) but their realisation has also been limited by a few key factors:

- large investment requirements;
- operation and utilization cost uncertainty;
- market demand and 'elasticity' uncertainty;
- launcher cost, availability and reliability.

Given these circumstances – and in the absence of a major government imperative, equivalent to that which justified Apollo (i.e. the Cold War) – it has become clear to many that the current paradigm will not lead to any significant growth of space activities in the foreseeable future. As a consequence, a number of NewSpace ventures have begun to emerge[2] that represent an attempt to change the paradigm by placing greater emphasis on entrepreneurial rather than government activities.

NewSpace ventures believe that the best way to change the paradigm is to stimulate existing and/or new markets in order to drive and sustain their growth, primarily through the power of commercial enterprise. Moreover, as launch issues are seen as the common factor that limits both current and future growth, most have chosen to address this issue first; their ultimate aim being to reduce specific launch costs by an order of magnitude to below about $1000/kg to LEO, the point where significant growth in all market sectors is expected to be triggered.

Nevertheless, it is important to realize that the NewSpace paradigm is not solely restricted to entrepreneurial start-up companies. A more thoughtful definition would also include groups working within established companies, such as Boeing, Lockheed-Martin and Orbital Sciences, who are also seeking to stimulate existing and new markets by applying novel technologies and commercial practices such as fixed-price, rather than cost-plus, contracts.

1.2 NewSpace and the Realities of Space Access

As launch services are one of the most significant constraints on the growth of future space operations, we now consider the significant improvements in vehicle design, operation and economics that will be required and the ways in which these could be realized.

It has long been recognised that the only way to achieve significant improvements in space access is via reusable launch vehicles (RLVs) instead of expendable launch vehicles (ELVs), because they offer:

- major reductions in marginal costs, as expensive components tend not to be discarded after use;
- better amortisation of investments, as costs can be spread across more users;
- higher reliability and safety, due to the intrinsic value of the vehicle.

Unfortunately government efforts to field such systems have, to date, either missed many of their original goals (i.e. Shuttle) or been outright failures (X-33/VentureStar, X-34, etc.). Moreover, commercial efforts to develop such systems have been hampered because their development costs are difficult to justify against potential markets, for example:

- many studies estimate it will cost $10-20 billion to field an operational system;
- the existing markets are insufficient to justify their development because they have limited growth and 'elasticity'[3] (i.e. lower prices stimulate only limited market growth);
- the new markets that could justify their development are far too uncertain and speculative.

Such factors show that both market and financial issues play just as important a role as the obvious technical ones. They also explain why NewSpace ventures have chosen to begin by developing RLVs to service sub-orbital markets, which demand significantly less of an initial investment, with many estimating that only $100m-$200 million will be required.

---

[1] For example, the Commercial Space Transportation Study (CSTS) performed a comprehensive review in 1994 of all current and foreseeable markets [RD.3]

[2] Summaries and links for all current ventures are at (http://www.space-frontier.org/commercialspace/)

[3] Space market elasticity is difficult to estimate due to the relatively small size and low diversity of current markets, though studies such as the CSTS [RD.3] and the NASA ASCENT Study [RD.4] have derived tentative estimates.

Nevertheless, it should be appreciated that cost isn't everything and that frequent flight availability and a timely and efficient integration process are just as important. A good example of this is NASA's Get Away Special (GAS) canisters [RD.5] that were priced on the order of $100/kg to LEO but, because of the long and complex Shuttle integration process, were undersubscribed so that many GAS canisters were filled with ballast and the service was eventually discontinued after the Columbia accident.

2. THE CASE FOR SUBSONIC AIR-LAUNCH

Having identified space launch as a fundamental enabler of future space operations, this section discusses the advantages and drawbacks of a subsonic air-launched fully reusable launch vehicle (RLV). In doing this it highlights the unique evolutionary opportunities that this concept has to offer and provides some insight as to how these may be realised and enhanced via existing technologies. It explains why subsonic air-launch is the only realistic way of enabling space launch from conventional airfields within the foreseeable future and discusses the other major operational advantages of this concept, such as: much enlarged and flexible launch windows; recovery of all flight elements to the same geographic location; increased contingency options for launch abort; the potential to harvest propellant during its cruise to the launch point.

2.1 Brief History of Air-Launch

The idea of air-launching a rocket has a long history that dates back to the early 1950's when *rockoons*, which were sounding rockets launched from helium balloons. These allowed the rocket to achieve a higher altitude so that it did not have to move under power through the lower and thicker layers of the atmosphere. Unfortunately, they had some serious disadvantages because the balloon could not be steered and so both the launch direction and the region where it fell was not easily to control. Possibly the most successful was the USAF's Project Farside, which launched six vehicles in late-1957 though only two reached their target altitude of just over 2000km.

The first aircraft launched rockets were primarily developed as anti-satellite (ASAT) weapons. The first of these was Project Pilot, which was an attempt by the Naval Ordnance Test Station (NOTS) at China Lake to orbit a 1kg payload in response to Sputnik. The vehicles, named NOTS EV-1 (NOTSNIK), were solid rockets launched by a Douglas F-4D1 Skyray and ten were flown in mid-1958, though none were successfully tracked to orbit. Similarly, a Bold Orion missile, which was air-launched from a B-47 Stratojet on 19th October, 1959, against the Explorer 6 satellite. However, this was a limited test and it was not until 13th September, 1985, that an F-15A launched an ASM-135 ASAT destroyed the Solwind P78-1 satellite flying at an altitude of 555 km. Since then, the only operational air-launched rocket has been Pegasus, which was developed by the Orbital Sciences Corporation as a commercial satellite launch vehicle and first flown on 5th March, 1990, with 42 launches to date.

Most air-launch concepts carry the rocket external to the launch vehicle, either on top or under the fuselage or wing. However, a few concepts have proposed carrying the rocket inside the fuselage and 'extracting' it during launch via drag chutes, which also provide stability during the subsequent free-fall phase, before igniting the rocket motor. The USAF tested air launching a Minuteman ICBM from a C-5A Galaxy transport aircraft on 24th October 1974, but this concept was never pursued. However, the AirLaunch LLC performed significant demonstration tests in 2006 of a very similar concept called QuickReach for the DARPA/USAF FALCON programme, which launch a liquid ELV from a Boeing C-17A. Similarly, the Air Launch Aerospace Corporation proposed an air-launched system capable of placing satellites into LEO using the AntonovAn-124 "Ruslan", though this was never developed.

Other concepts proposed in the mid-1990's have envisaged towing the launcher behind an aircraft (i.e. Astroliner, proposed by Kelly Space & Technology) while others have envisaged in-air fuelling of the launcher in order to reduce take-off mass (i.e. Black Horse, proposed by Pioneer Astronautics). Neither of these approaches were ever pursued beyond the conceptual design stage, though Kelly did perform tow tests of an F-106 jet behind a C-141 cargo aircraft in early-1998 under a NASA SBIR award.

The most recent air-launch concept to attract serious attention has been the Stratolaunch Systems proposal in 2011 to build a massive aircraft by combining the wings and fuselage of two Boeing 747 airliners. However, the exact nature of the launch vehicle was not specifically defined and initial speculation was that it would be a variant of the SpaceX Falcon. It now appears that OSC will build the rocket, called Pegasus II, using two solid-stages and a cryogenic upper stage, which will be capable of launching a 6.1t payload into LEO. Two other air-launch concepts have been proposed in recent times: the Lynx III from XCOR and LauncherOne from Virgin Galactic. Both are evolved from sub-orbital launch systems but plan to launch much small satellites than Stratolaunch, on the order of 100kg, by using an expendable rocket launched from the sub-orbital vehicle: XCOR's Lynx rocket plane, separating at a high supersonic speed of around 4Mn; Virgin Galactic's WhiteKnight 2 carrier aircraft, separating at a subsonic speed of around 0.9Mn.

2.2 RLV Design Factors, Issues & Trades

Before discussing the specific benefits of a subsonic air-launched RLV, this section provides some insight of the advantages and drawbacks of the myriad possible designs that have been considered to date. Although limited in technical detail, it is based upon a synthesis of RLV conceptual designs. The synthesis is presented in more detail within the Appendix of RD.6 and is based upon a large number of authoritative papers that reported the results of detailed design studies, performed mainly for/by NASA. Most were published between the late-1980s or mid-1990s; a period that covers the last serious effort by the US government and aerospace industry to build a fully reusable launch vehicle through initiatives like NASP, DC-X and both the X-33 and X-34 projects. Sadly, with the exception of the DC-X, all of these efforts were cancelled before any significant hardware could be flown or even tested and, as a consequence, all subsequent new launch vehicle initiatives have focused upon the development and/or evolution of expendable designs.

To enable a sensible comparison of the incredibly large if not infinite variety of RLV concepts, the synthesis classified and assessed them with respect to three basic design and operational characteristics.

- *Propulsion system*; pure rocket, or some combination of rocket and air-breather (a/b)
- *Configuration*; two-stage-to-orbit (TSTO), or single-stage-to-orbit (SSTO)
- *Launch and landing mode*; vertical take-off and vertical landing (VT/VL), or vertical take-off and horizontal landing (VT/HL), or horizontal take-off and horizontal landing (HL/HL).

In addition, the impact of several design and operational issues (*flight profile*, *payload size* and *technology assumptions*) was also assessed.

i) Propulsion

Pure rocket vehicles are always lighter (dry) than vehicles with equivalent payload performance that use some form of air-breathing propulsion due to the installed mass of air-breathing engines. Assuming dry mass relates directly to research and development (R&D) and production costs (which is a reasonable first order approximation), then pure rocket concepts will be cheaper to both develop and produce. The only exception is when an existing air-breathing vehicle can serve as a TSTO booster. Moreover, as all-rocket concepts will be mechanically less complex than one using some combination of air-breathing and rocket engines, then the pure rocket concepts will be easier to maintain and so be cheaper to operate.

Pure rocket vehicles with dual-fuel propulsion are lighter (dry) than equivalent vehicles which use only hydrogen, but the extra cost of developing and operating a hydrocarbon engine in addition to a hydrogen engine, or the development of tri-propellant engine technology, will make the life cycle costs of dual-fuel vehicles significantly more than the equivalent hydrogen only vehicles. Dual-fuel vehicles using propane are lighter (dry) than those using other hydrocarbons.

ii) Configuration

TSTO concepts are lighter (dry) than equivalent SSTO concepts, but the extra complexity of developing and operating essentially two distinct vehicles in parallel means that the SSTO life cycle costs are less than those of an equivalent TSTO. The one exception to this may be the Siamese concept, in which the orbiter and booster are designed to be as similar as possible in order to minimise, or even eliminate, duplicated effort and equipment during the development, production, and operational phases.

iii) Launch & Landing Modes

SSTO rocket concepts based upon HT/HL designs are lighter (dry) than equivalent VT/HL designs, due to lower T/W engines and a lifting ascent trajectory that reduces mission delta-v, but the added complexity of a launch assist device may make their life cycle costs more than equivalent VT/HL designs. Wet-wings (containing LOX) provide a significant way of reducing the mass of HT/HL designs, possibly enough to reduce their overall life cycle cost to below that of equivalent VT/HL designs.

iv) Flight Profile Impacts

If re-entry cross-range requirements are relaxed to around 100km, then pure rocket VT/VL vehicles using ballistic re-entry will have the lightest mass (dry) and the lowest life cycle costs.

If significant launch flexibility is required, such as a launch off-set capability (in both time and position) and/or a cruise/loiter capability (for ferry or reconnaissance purposes), concepts will have to use air-breathing propulsion to some degree in order to minimise the vehicle's mass (dry).

v) Payload Size Impacts

Small payloads, around 5t and less, will favour TSTO configurations because system scaling factors tend to reduce SSTO payload fractions as absolute size decreases, plus it also becomes more feasible to consider using an existing vehicle as a TSTO booster in order to significantly reduce the R&D costs.

Large payloads, around 60t and more, will favour VT/VL with ballistic re-entry because lifting re-entry vehicles have a far-aft centre of gravity problem that tends to increase with vehicle size.

vi) Technology Assumption Impacts

SSTO air-breathing concepts will be significantly lighter (dry) than equivalent TSTO air-breathing concepts if they can use the advanced technologies envisaged for NASP to provide more than a 40% reduction in structural and system mass relative to the Shuttle (e.g. titanium metal matrix composite fuselage/wings/frames, silicon carbide hot structures, graphite composite tanks, slush hydrogen, etc.).

2.2 *Benefits of Subsonic Air-Launch*

There have been numerous studies of air-launch concepts and Table 1 provides an overview of a very small but representative selection of them. Significant effort was invested in developing these concepts because air-launch provides some important benefit with respect to performance, operations and the potential for evolution and these are discussed in the following subsections.

Config.	*Concept Name*	*Designer/Year*	*Air-launch Vehicle*	*Propellant*	*Reusable*	*Payload*
Captive on Top	Boeing AirLaunch	USA/1999	747	Solid	No	3.4t
	Interim HOTOL	UK/1991	An-225	LH2/LOx	Fully	7.0t
	MAKS-M	USSR/1989	An-225	RP-1/LH2/LOx	Partly	5.5t
	MAKS-OS	USSR/1989	An-225	RP-1/LH2/LOx	Partly	8.3t
	Pegasus II	USA/2011	Stratolaunch	Solid+Cryo	No	6.1t
	Saenger II	Germany/1991	Mach 4.4 turbo-ramjet	LH2/LOx	Fully	9.0t
	Spiral 50-50	USSR/1965	Mach 6 turbo-ramjet	RP-1/LOx	Partly	10.0t
	Teledyne-Brown	USA/1986	747	LH2/LOx	Fully	6.7t
Captive on Bottom	Global Strike Eagle	USA/2006	F-15	Solid	No	0.3t
	Pegasus	USA/1990	L-1011	Solid	No	0.5t
	Yakovlev HAAL	USSR/1994	Tu-160	Solid	No	1.1t

Table 1. Selection of external carriage Air-Launch concepts (excludes towed or internal carriage)

vii) Performance Benefits

Rocket operations above the dense atmosphere reduce significantly both drag and gravity losses. It also allows for a significant increase in engine specific impulse (Isp) by allowing the use of a larger expansion ratio nozzle, which is constrained at lower altitudes because over-expanded nozzle flows suffer destructive instabilities. Theoretically, the latter problem can be overcome by using some sort of altitude compensating nozzle, though the additional mass and complexity tends to cancel out any performance benefit.

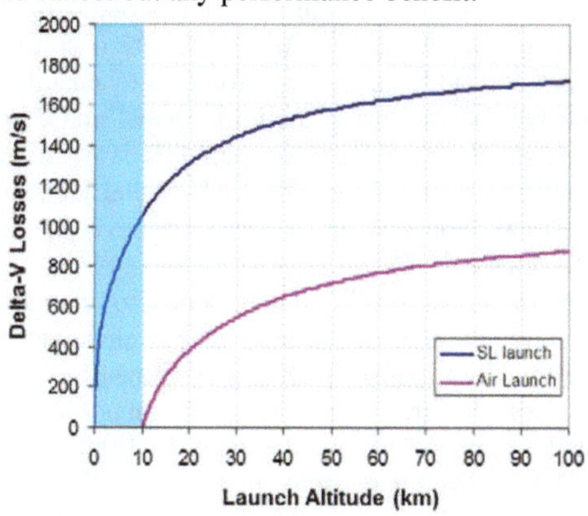

Figure 2. Delta-V Loss Comparison

Figure 2 illustrates the delta-V losses encountered by a rocket as a function of launch altitude for both sea-level and the 10km case, which represents a subsonic air-launch. It shows a major reduction in velocity losses and, more specifically, that these losses represent around 20% of the ideal ascent delta-V for a typical sea-level launch (i.e. 7.7km/s to LEO), but only around 10% of an air launch.

Another small but positive benefit of air-launch is that the launch point may be chosen to match the inclination of the target orbit. This not only allows for maximum exploitation of Earth's rotation (~400m/s for equatorial orbit), it also reduces trajectory losses by reducing or even removing the need for plane changes to achieve the target orbit.

viii) Operational Benefits

Air-launch offers the only realistic way to operate a space launch system from existing airfields, including the possibility of one day operation out of major civil airports. This is because the launch aircraft uses air-breathing propulsion as opposed to a pure rocket, which enables an enormous reduction in noise during take-off due to the reduced exhaust velocity. However, concepts that use a supersonic military jet will never be as 'quiet' as those that use a subsonic transport and will also be penalised because of their much reduced payload capacity, which will likely be at least one order of magnitude less.

Using an existing military or commercial aircraft also means that the air-launch system can build upon this vehicle's inherent safety, reliability, maintainability and availability. Moreover, these will be extremely valuable if rapid and/or frequent launch is one of the primary system requirements. In addition, it leads to a launch system whose elements are all processed and operated horizontally, which helps to streamline the maintenance and launch workflow as it simplifies access to the vehicle.

As already mentioned, air-launch offers the possibility to choose the launch point to match the inclination of the target orbit. An additional but extremely important benefit is that the launch point can be 'tracked' so that the launch window for rendezvous with an orbiting target can be widened significantly. This not only improves operational flexibility but, as already stated, also reduces the need for plane changes to achieve the target orbit and so has the potential to reduce the size of the upper stage by reducing the on-orbit propellant requirements.

Figure 3 presents a schematic of the operational profile of a generic subsonic air-launch RLV and also shows another operational advantages of this concept, which is that it can use the launch aircraft to ferry the rocket back to the launch site if it should have to land at an alternate. More importantly, it also highlights the potential to use the cruise phase to harvest liquid oxygen, which would require the aircraft to carry an Air Collection and Enrichment System (ACES) and is discussed in more detail within the next subsection. Additionally, it indicates the potential to refuel the aircraft in-flight in order to either reduce its take-off mass or extend its cruise and/or loiter capability.

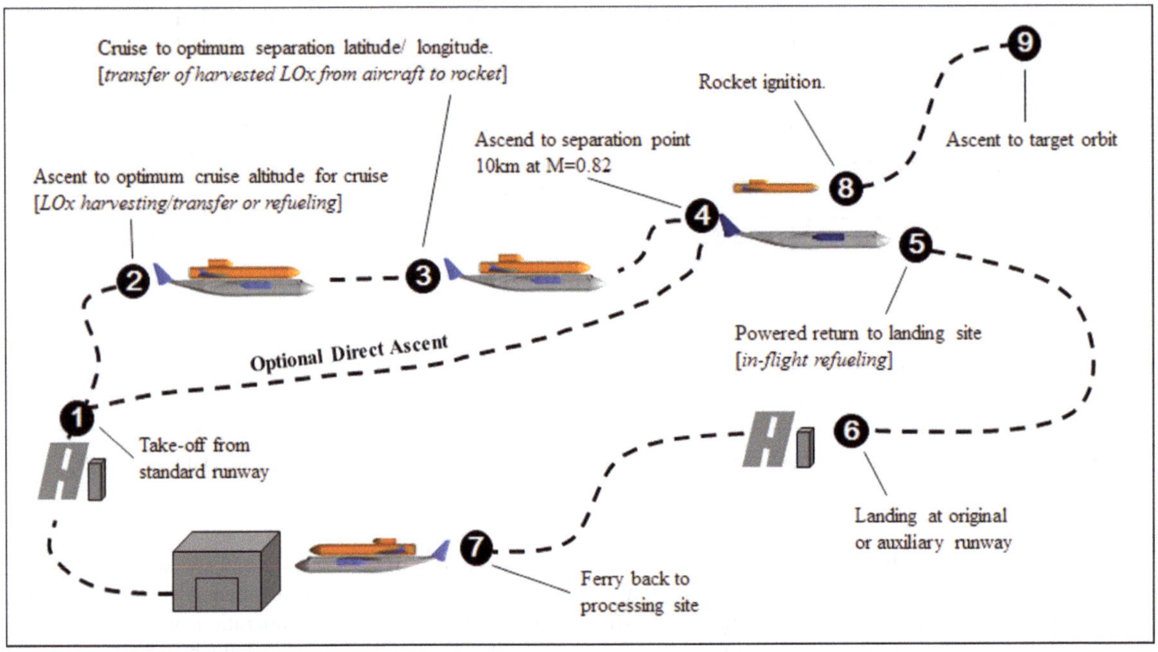

Figure 3. Subsonic Air-Launch Operations ***[optional in-flight LOx harvesting/transfer or in-flight refueling]***

Another capability that is not obvious from the figure but could have very important operational benefits is the ability to fly the launch vehicle up-range so that the 1st stage booster of any TSTO RLV can return directly to the launch site after staging, thus avoiding the need to fly or glide back up-range. Requiring the booster to fly-back up-range is a very constraining problem for ground launched TSTO RLVs because it either:

- limits the staging to around 3Mn at 30km altitude to ensure the booster has sufficient 'energy height' to glide back to the launch site;
- forces the booster to carry extra propellant in order to perform an up-range boost-back manoeuvre;
- forces the booster to carry an additional air-breathing propulsion system in order to fly back up-range;
- requires an additional landing site down-range of the launch site as part of the basic infrastructure.

For an SSTO RLV, it also means that an aborted launch could fly-back directly to the launch site should the abort occurred sufficiently early in the mission. Thus, air-launch also increases the number of abort options and so improves both safety and operational robustness.

ix) Evolutionary Benefits & ACES

Air-launch offers the ability to adapt an existing ground launched system and increase its performance by acting as a high altitude launch platform. As an example, the Pegasus system uses Orion solid rocket motors and adds an a wing structure to ensure a high flight path angle during the initial boost phase in order to maximise its performance. In this way, it may be possible to evolve an existing sub-orbital launcher into an orbital launcher, or at least improve its payload performance.

As increasing vehicle size tends to increase both development and operational costs, air-launch could offer an importance path for commercial ventures. However, the orbital payload performance of any air-launch concept is fundamentally limited by the aircraft's carrying capacity and, more specifically, its maximum take-off mass. Currently, the world's largest operational aircraft is Russia's An-225 but this is a one-off design, based upon a heavily modified An-124, which is likely to be an impractical option for an air-launch system. Commercially available options include the Airbus A380 and the Boeing 747-400, though the former is relatively new and so is very expensive. Two 747-100 aircraft (SCA-905 & SCA-911) were converted to carry the Space Shuttle Orbiter for both test and ferry flights. The mass of the drop/glide tests orbiter (OV-101 Enterprise) was 68Mg, though later orbiters had an empty mass of 78Mg, so there is good reason to believe that a second-hand 747-400 would be a good candidate for an air-launch concept. Table 2 gives an overview of the relevant performance of the most likely candidate aircraft and includes a rough estimate of the maximum payload mass, taken from RD.7, they could deliver to LEO if used as the basis for an air-launch system.

Candidate Aircraft	*External Mass (Mg)*	*Max. P/L to LEO (Mg)*
An-225	200	13.8
A380-800F	120	7.8
747-100 SCA -911	109	7.0
747-400F	140	9.1
Dual-fuselage C-5	350	23.7
Stratolaunch Carrier	120	6.1

Table 2. Candidate Aircraft for Air-Launch

The gross mass of any launch vehicle that uses liquid oxygen (LOx) as an oxidiser will be dominated by the amount of LOx it must carry. Typical oxidiser/fuel ratios of 5.2 and 2.3 respectively for liquid hydrogen (LOx/LH2) and kerosene (LOx/RP-1) fuelled rockets mean that the LOx will account for more than half the launch vehicle's gross mass at take-off. Therefore, it is reasonable to think any design approach that enables the LOx to be loaded *after take-off* should offer a number of significant advantages such as:

- increased payload performance for any given aircraft;
- improved safety during ground operations and take-off due to elimination of the LOx.

A cursory reflection on this idea may well lead one to think it illogical as, without LOx, the rocket cannot function and the mission will be futile. However, more thoughtful consideration shows the idea has some merit and that two approaches appear possible:

i) transfer the LOx in-flight from a 'tanker' aircraft;
ii) utilise the cruise phase to harvest the LOx from the atmosphere.

Though the first approach is the most obvious, it requires not only an additional aircraft but also the ability to transfer very large amounts of LOx in-flight between two independent vehicles, separated by many tens of meters – something that has never been attempted, to date. The second approach also requires technology that has yet to be fielded aboard an aircraft (i.e. the separation and liquefaction of LOx from the atmosphere), though it is a process that is performed routinely on-ground by industrial facilities. However, if an Air Collection and Enrichment System (ACES) could be 'miniaturised' sufficiently to fit inside an aircraft, it offers the possibility of mounting it within the carrier aircraft and so avoids the need for an additional vehicle.

There is actually a long history of RLV concepts that have used air collection as the basis for their propulsion cycle; the earliest being the USAF's Aerospaceplane programme of the late-1950s and early 1960s to develop a hypersonic airplane. Since then, studies of this approach have tended to focus upon SSTO RLVs using the Liquid Air Cycle Engine (LACE), though the resulting designs were always judged to be too complex and heavy, due to the mass of the LACE engine having to be carried all the way into orbit.

ACES can therefore be regarded as a variation of this general approach that avoids most of the performance penalties by placing the heavy machinery outside of the rocket (i.e. within the launch 'platform'). In fact, several patents have been issued for ACES designs [RD.8 & RD.9] and some work has been performed to develop and test representative hardware [RD.10 & RD.11]. Moreover, air-launch concepts based upon such devices have already been proposed that involve both subsonic [RD.10] and supersonic [RD.12] separation of the rocket stage.

A schematic of the ACES cycle and its key components is shown in Figure 4. The operating principle is that, while cruising to the launch point, the ACES device generates liquid oxygen by ingested air from the atmosphere and separating out the nitrogen component through a series of heat exchangers and a rotational fractional distillation unit. The heat exchangers use LH2 to super-cool incoming air, which is either tapped off the aircraft's main engines or drawn in by a dedicated compressor. The resulting LOx is then pumped from the ACES system on the carrier aircraft into the empty LOx tanks of the launch vehicle during flight. The tank holding the LH2 that super-cools the incoming air is sized by the volume of LOx required by the rocket, so this 'collection ratio' (CR) is an important performance characteristic of any ACES concept.

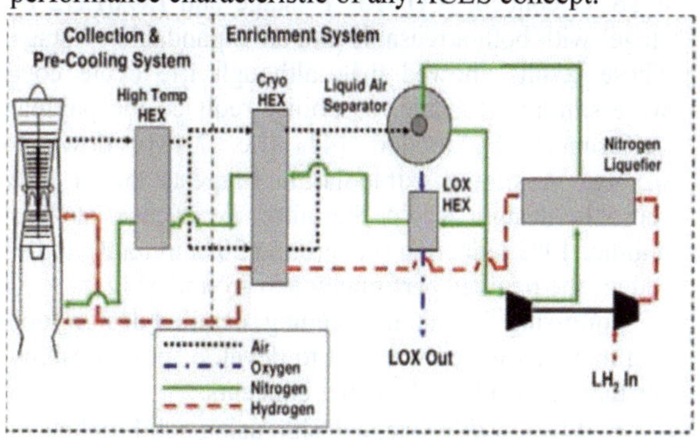

Figure 4. ACES Cycle Schematic [RD.12]

Perhaps because of its novelty, the ACES concept has rarely featured in launch vehicle design studies, though recently it was considered within a NASA-DARPA assessment of air-launch concepts [RD.7]. Although the study did not include ACES within any of its three vehicle point designs, its preliminary screening task did include a trade-off of ACES against others technologies (e.g. high-energy propellants) and found it to be the most promising option, in terms of cost/benefit, to enhance air-launch performance. The results of these findings are considered in more detail in the next subsection, which uses them as the basis for a more detailed assessment of the likely payload performance gains achievable by a design that includes ACES.

2.3 *Benefits of ACES*

The following high level assessments are presented merely to illustrate the potential benefits of augmenting an existing air-launch concepts with ACES. As such, the absolute magnitudes of the payload performance gains must be considered with a degree of caution and should in no way be thought of as definitive.

i) DARPA/NASA Study

In 2010, NASA and DARPA commissioned a joint study to assess horizontal launch concepts for military and civilian applications. Its goal was to recommend system concepts for subsonic and supersonic carrier aircraft options and to identify technology gaps for potential investments that included a near-term horizontal launch demonstration. The final results were published in October, 2011 [RD.7].

Payload market projections limited the detailed systems study to expendable concepts because the additional costs of reusability could not be justified for the assumed launch rate of 6 flights per year. Nevertheless, the initial screening analysis did compare the payload performance to LEO and lifecycle costs of a TSTO concept (LOx/RP 1st stage and LOx/LH2 2nd stage) with both a reusable and an expandable 1st stage. These results showed that, although life cycle costs were similar, 1st stage reusability reduced the payload performance by around 14% (i.e. 7.4Mg down to 6.5Mg). A simple extrapolation suggests that a fully reusable design would therefore experience at least another 14% reduction (i.e. around 30% in total) and so reduce the payload performance to around 5Mg.

Following the initial screening, detailed design tools and methods were then used to develop 'point designs' for three expendable launcher concepts:

- PD-1, a *three*-stage design using solid rockets on all stages;
- PD-2, a *two*-stage liquid design using LOx/RP on the 1st stage and LOx/LH2 on the 2nd stage;
- PD-3, a *two*-stage liquid design using LOx/LH2 on both stages.

Mass budgets were presented for each point design, along with trajectory, reliability and cost breakdowns that provide a very useful insight upon the impacts of propulsion choice. The point designs were also used as the basis for assessing cost/benefits of alternate technologies, which included both ACES and in-flight LOx transfer. Table 3 presents a short summary of the payload performance to LEO – 185km, due East – of four relevant designs and then estimates the impact of ACES, based upon the results of the technology trades.

Analysis Level	Screening	Screening	PD-2	PD-3
Stage Reusability (1st/2nd)	Yes/No	No/No	No/No	No/No
Stage Fuel Type (1st/2nd)	RP/LH2	RP/LH2	RP/LH2	LH2/LH2
Separation Mach Number (Mn)	---	---	8.4	11.7
Reported Payload Performance to LEO (Mg)				
Baseline	6.4	7.4	5.7	8.1
Baseline + ACES	---	---	7.7	9.9
Baseline + In-Flight LOx transfer	---	---	7.8	9.9
Impact of ACES	---	---	*+35%*	*+21%*
Impact of 1st Stage Reusability	---	*-14%*	---	---
Estimated Payload Performance to LEO (Mg)				
Reusable 1st stage	---	6.4	4.9	7.0
Reusable 1st & 2nd stages	---	5.5	4.3	6.1
Reusable 1st & 2nd stages + ACES	---	7.4	5.8	7.4

Table 3. Payload performance Summary [from RD.7] & Extrapolations

As would be expected, the reported performance values from the initial screening were rather optimistic with respect to those of the point designs (i.e. 7.4Mg compared to 5.7Mg). However, the impact of ACES, though clearly beneficial, was somewhat unexpected. The overall O/F ratio of the LH2/LH2 design (i.e. PD-3) means it should carry a larger proportion of LOx than the RP/LH2 design (i.e. PD-2) and so benefit more from ACES, but the results show the reverse (i.e. +35% for PD-2 and +21% for PD-3). Initial suspicions suggest that the density-volume impacts of the LH2 fuel in the 1st stage may have resulted in a heavier dry mass, which is also compounded by the higher separation velocity (11.7Mn for PD-3 and 8.4Mn for PD-2) that shifts the energy split between 1st and 2nd stages and so results in a larger booster.

ii) Parametric Assessments

In order to investigate these interesting results in a little more depth, a spread-sheet model was developed by the author that used vehicle mass and performance characteristics from both the DARPA/NASA study [RD.7] and the Future European Space Transportation Investigations Programme (FESTIP) [RD.13, RD.14, RD.15]. The baseline vehicle design and mission assumptions are listed in Table 4, which includes key performance characteristics and design factors used for assessing the impact of ACES.

The mass breakdown for the PD-2 and PD-3 designs were rescaled to account additional mass for reusability, which included wings and TPS as well as scaling of tanks and fuselage. As the FESTIP concept (FSSC-16) was a fully reusable TSTO design that used LOx/LH2 propulsion on both stages, the design re-scaling mainly accounted for performance effects relating to separation speed and the impact of using RP1 instead of LH2 in

RLV Design & Mission	
1	Baseline mission delta-v to 400km LEO = 7820 m/s
2	Delta-v loss: 1750 m/s from sea-level; 850 m/s from 10km
3	Existing rocket engines (e.g. Merlin 1C & RL10A-4-2)
4	Oxydised/Fuel ratio: 2.28 for LOx/RP; 5.24 for LOx/LH2
5	Isp: 450s @10km for LOx/LH2; 300s @10km for LOx/RP
6	Current available structural materials (i.e. TRL 6+)
7	TPS mass: 5% Booster dry mass; 20% Orbiter dry mass
8	Wings + Empennage + body flap: 7% dry mass
ACES Characteristics [RD.11]	
1	LOx collection plant (LCP) mass / volume = 4Mg / 6m^3
2	Collection Ratio (CR) = 2.0 (i.e. 1kg LH2 => 2.0kg LOx)
3	LOx collection purity = 90% (i.e. 10% N2)
4	LOx collection rate = 9 kg/sec
5	Isp = 292s @10km for LOx/RP with 90% purity LOx
6	Isp = 435s @10km for LOx/LH2 with 90% purity LOx

Table 4. Air-Launch Model – Assumptions

the 1st stage. The scaling rules applied to these models are outlined in Table 5.

The mass budget and payload performance of the vehicle was modelled for a range of separation speeds by splitting the baseline mission delta-v between the two stages but accounting the full delta-v loss only on the 1st stage. The resulting payload performance for each concept was then estimated for a range of separation speeds, crudely associated with Mach number by a simple linear interpolation, which gave specific point designs as shown in Table 6.

<u>Wing & TPS Mass</u>: Scales directly with materials factor (S) and the change, with respect to the baseline, in the sum of Fuselage, Tank, Systems, and Engine masses (Ms3 + Ms4 + Ms5 + Ms6).
<u>Fuselage Mass</u>: Scales directly with materials factor (S) and the change, with respect to the baseline, in the propellant tank mass (Ms4).
<u>Tank Mass</u>: Scales directly with materials factor (S) and change, with respect to baseline, in propellant mass (Mf) raised to the power of 2/3.
<u>Systems & Engine Mass</u>: Scales directly with the change in the propellant mass (Mf), with respect to the baseline.

Table 5. Air-Launch Model – Scaling Rules

When plotted together, these points produced curves that indicated an 'optimum' split for each concept and these are shown in Figure 5. In addition, each 'optimum' was then re-scaled to assess the impact of improved ACES performance with respect to the Collection Ratio and LOx purity. The impact of using a less capable carrier aircraft was also assessed, assuming the carrying capacity of a 747-100 instead of a 747-400 (see Table 3), while the impact of assuming more advanced structural materials (i.e. 10% lighter) was also investigate to show how far the payload performance from a 747-100 could be 'evolved'.

The limits of this relatively simple modelling are indicated by the payload performance differences between the PD-2/PD-3 and FSSC-16 results, which should be the same if the models were truly equivalent. Nevertheless, the curves are reasonably coherent and so their differences can be taken to indicate the level of modelling uncertainty and are shaded accordingly. More importantly, the models do show a consistent benefit of the ACES concept, which increases the baseline payload on the order of 40%. This value is in reasonable agreement with the DARPA/NASA study findings (Cf. Table 3.), although their result for the all-LH2 design (i.e. PD-3 with only a 21% increase) does seem rather low considering its much higher oxidizer/fuel ratio.

As mentioned before, the analysis was performed to simply illustrate the potential benefits of augmenting an existing air-launch concepts with ACES and should not be thought of as definitive. Though promising, these results have yet to include other design and operational issues that may have both positive and negative impacts, such as:

- the need to cruise for ~4 hour in order to harvest the required mass of LOx, based upon a nominal collection rate of 9kg/s;
- effects on carrier aircraft (e.g. lift, stability and cruise range) of increasing mass during LOx harvesting, which may reach as much as 20%;
- using sub-cooled hydrogen (e.g. stored at 16K instead of 20k, with a higher para-hydrogen fraction) to increase both its density and heat absorption capacity, which promises to reduce both LH2 tank size and LH2 boil-off during cruise while also increasing the ACES collection ratio (CR);

Separation Mach number (Mn) = 12	***Air-Launched***	***+ACES***
Materials density scaling factor (S) [%]	1.00	1.00
<u>TSTO Booster Details</u>		
Specific Impulse (Isp) [sec.]	450	435
Rocket equation factor (R=Exp(dV/Isp/g)	2.5968	2.6836
TSTO Gross Mass (MTg=MBp+MBs+MBf) [kg]	138576	200848
Booster Dry Mass (MBs=SUM(MBs1:MBs6)) [kg]	18508	25933
Wings Mass (MBs1) [kg]	*1414*	*1981*
TPS Mass (MBs2) [kg]	*1014*	*1421*
Fuselage Mass (MBs3) [kg]	*3390*	*4399*
Tank Mass (MBs4) [kg]	*3509*	*4555*
Systems Mass (MBs5) [kg]	*2020*	*2987*
Engines Mass (MBs6) [kg]	*7162*	*10590*
FSSC-16 Defined Propellant Mass (MBf) [kg]	85211	126006
Booster Payload (MBp=MOg, Orbiter Gross Mass) [kg]	34857	48908
Booster delta-V loss (LdV) [m/s]	850	850
Booster delta-V (BdV) [m/s]	3363	3363
<u>ACES Details</u>		
LOx fraction of TSTO gross mass	65%	66%
Total LOx propellant [kg]	90165	132436
LCP mass [kg]	---	4000
LH2 for ACES [kg]	---	66218
ACES 'kit' Mass [kg]	---	70218

	Air-Launched	***+ACES***
	1.00	1.00
<u>TSTO Orbiter Details</u>		
Specific Impulse (Isp) [sec.]	450	435
Rocket equation factor (R=Exp(dV/Isp/g)	2.7448	2.8421
Orbiter Gross Mass (M0g=MOp+MOs+MOf) [kg]	34857	48908
Orbiter Dry Mass (MOs=SUM(MOs1:MOs6)) [kg]	5379	7139
Wings Mass (MOs1) [kg]	*615*	*817*
TPS Mass (MOs2) [kg]	*1090*	*1446*
Fuselage Mass (MOs3) [kg]	*1251*	*1588*
Tank Mass (MOs4) [kg]	*1508*	*1914*
Systems Mass (MOs5) [kg]	*677*	*968*
Engines Mass (MOs6) [kg]	*854*	*1222*
FSSC-16 Defined Propellant Mass (MOf) [kg]	22158	31700
Resultant TSTO Payload (MOp) [kg]	7320	10070
Orbiter delta-V loss (LdV) [m/s]	---	---
Orbiter delta-V (OdV) [m/s]	4457	4457
<u>TSTO System Details</u>		
Total Mission Delta-V [m/s]	8670	8670
TSTO Dry Mass (MTs=MBs+MOs) [kg]	23887	33072
TSTO Gross Mass (MTg=MTs+MBf+MOf+MOp) [kg]	138576	200848
TSTO Gross Mass without LOx [kg]	---	68412
TSTO Gross Mass without LOx + ACES [kg]	---	**138630**

Table 6. TSTO Performance Analyses - FSSC-16 using 747-400

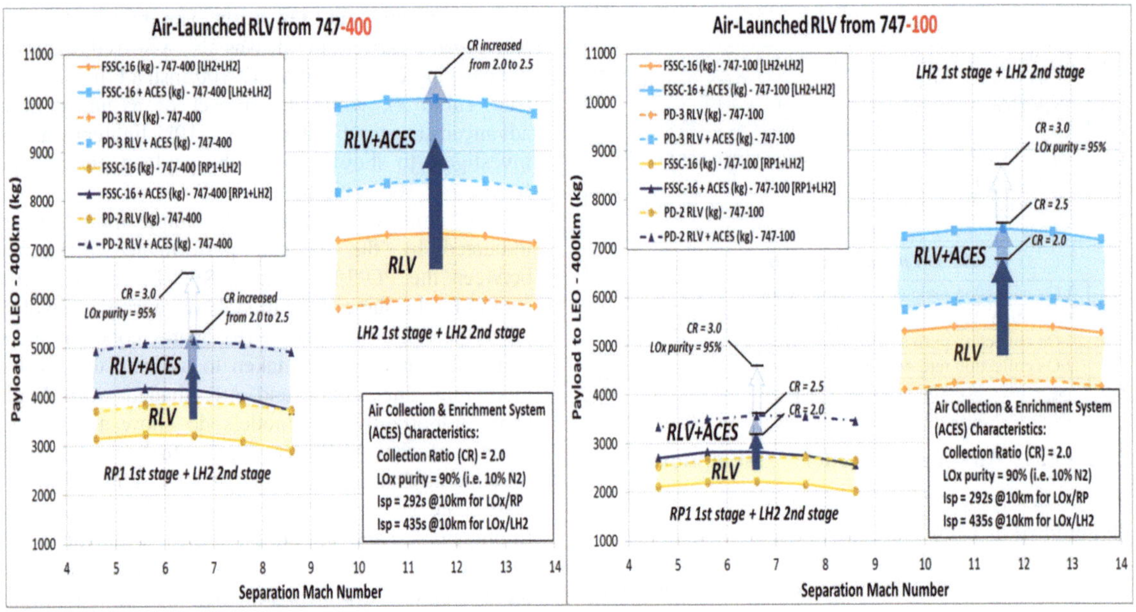

Figure 5. Air-Launched RLV – Payload Performance Sensitivity

- using hydrogen to fuel the aircraft's gas turbines [RD.10] and also to boost their thrust, via afterburning in the bypass duct, to increase flight-path angle at separation and so improve payload performance [RD.16].

Nevertheless, the current results give some confidence to the idea that an existing commercial aircraft could be used as a platform for air-launching a fully reusable TSTO rocket capable of placing 'commercially significant' payloads (i.e. ~4000kg) into low Earth orbit – a subject that is discussed in more detail within the next main section.

iii) Conclusions

The inherent operational principles the ACES concept (i.e. aircraft take-off without LOx) increases safety and makes use of the cruise phase the launch point to harvest LOx in a synergistic manner. More importantly, it offers a realistic way of 'evolving' an existing or planned air-launch RLV by increasing its payload performance to LEO by around 35% or more. In addition, it also holds out the potential to increase this to 80% or more if key ACES characteristics can be improved (i.e. Collection Ratio and LOx purity). Such a major performance boost could also widen the potential range of aircraft that could prove suitable for air-launch, which may be a very important factor for any future commercial venture – something that is discussed in more detail within the next section.

3. OPERATING BEYOND THE LIMITS

This final section highlights the radical improvements in operational architecture afforded by a subsonic air-launched RLV. It shows how such an RLV with a relatively modest launch performance of between 4-6t into low Earth orbit could be capable of supporting the majority of current and future launch demands by forming the key element of a fully reusable space transportation infrastructure. It identifies the RLV technologies and systems that are common to both orbital transfer vehicles and Lunar landers, as well as the synergistic way their development and production could be coupled in order to both reduce their costs and to increase their reliability, availability and safety. More importantly, it indicates how all these factors can be combined to radically improve the business case for pursuing these ventures as a commercial enterprises, funded almost entirely by private investment.

3.1 LEO Operations & Beyond

The science fiction writer Robert A. Heinlein is quoted as saying that "once you reach low orbit, you're halfway to anywhere in the Solar System", referring of course to mission energy rather than distance. Viewed in this way, it's clear to see why many regard Earth-to-orbit launch vehicles as *the* key enabler to opening up space for all humanity. However, simply reaching LEO is only part of the problem because most missions, both in and beyond LEO, are severely constrained by issues of both cost and schedule (i.e. operational factors such as the availability and frequency of launch are just as important as low cost). While reusing a launch vehicle may help reduce costs by eliminating the need to procure new hardware, the cost of maintaining both the vehicle and its associated ground infrastructure (i.e. facilities and people) may offset any savings if its flight rate is too low. Indeed, this was the critical factor that undermined the cost effectiveness of NASA's Space Shuttle, which was 'sold' on the basis that it could support 64 flights/year.

i) Availability & Launch Windows

Beyond launch frequency and the obvious requirements for reliability and safety, the availability to launch at short notice and to support as wide a range of launch azimuths and launch windows as possible are also important factors for missions requiring interception and/or rendezvous with an on-orbit target. Air-launch offers a very attractive and realistic solution to these requirements and Figure 6 illustrate this by showing how the cruise range increases the number of launch opportunities (i.e. from two per day from a fixed launch site to more than six for air-launch) and, by definition, a wider range of launch azimuths because it can move the launch to a point where the launch ascent ground track will not fly over populated regions (i.e. over open oceans). It also shows a typical 'dog-leg' manoeuvre that may be required when operating from a fixed launch site in order to enable injection into a specific orbital plane for rendezvous, which an air-launch can reduce significantly or even eliminate.

market must be sufficient to justify the initial investment while its 'elasticity' will be important to ensure growth and secure future investment.

ISS Servicing Vehicles	*LEO Mass (Mg)*
Soyuz (Government – Russian)	7200
Progress (Government – Russian)	7200
ATV (Government – European)	20200
HTV (Government – Japanese)	19000
Dragon (Commercial – SpaceX)	6000
Cygnus (Commercial – OSC)	4500

Table 7. ISS servicing vehicles mass in LEO

Beyond individual scientific satellites, primarily in polar orbits to support Earth observation missions, the International Space Station (ISS) currently represents the only significant market in LEO that needs frequent and routine transport services. They are currently supported by a fleet of both government and commercial vehicles, which are listed in Table 7. As can be seen, a number of them have an injected mass

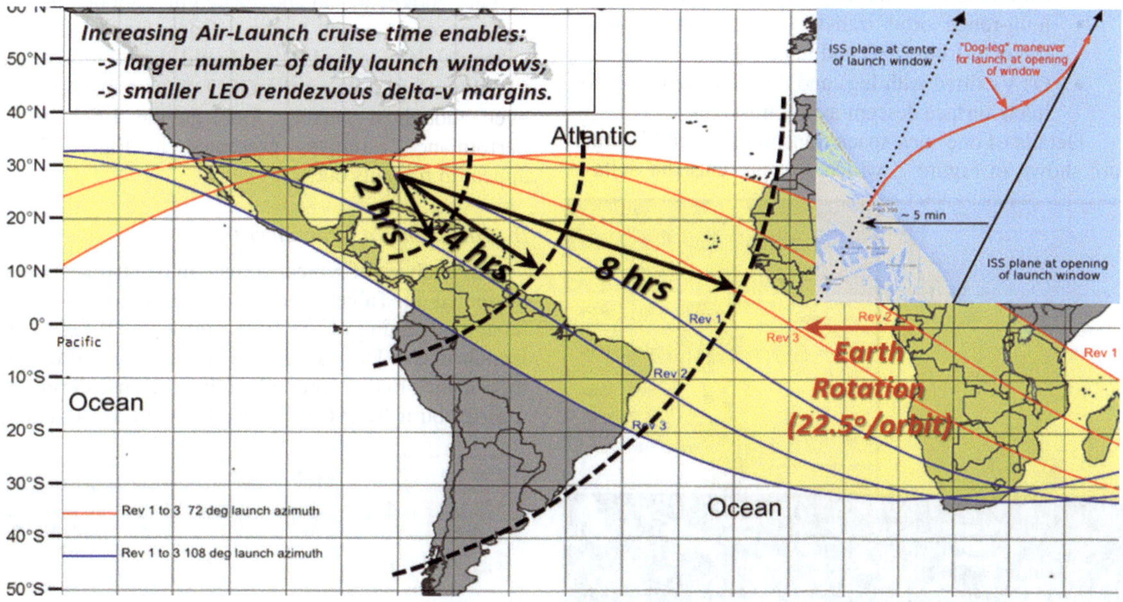

Figure 6. Impact of Air-Launch cruise time on launch LEO windows

These important operational benefits are why many concepts for rapid reaction launch vehicles have involved Air-launch. This is significant because military programmes like DARPA's XS-1, which aims to develop a reusable first stage of a space transport system that can reach Mach 10 or higher and fly 10 times in 10 days, see air-launch as a likely solution and so may represent an important stepping stone towards a viable air-launched RLV that could eventually be put into commercial service.

ii) Logistics & Crew Transportation

The success of any commercial venture requires both the capability to provide a service and a market that needs this service. More importantly, the size of the

into LEO that appears compatible with the payload performance of an air-launched RLV that uses ACES. This suggests that ISS logistics resupply may be potential market for any commercial venture, especially as two of these vehicles are already built and operated by commercial companies that have secured commercial resupply contracts with NASA. Note that the reusable X-37B, which has a mass of just under 5000kg, could also represent be another potential LEO payload though its military nature may make this possibility somewhat more unlikely.

Future commercial LEO space stations, like those planned by Bigelow Aerospace, represent another potentially lucrative market because they are predicated upon the availability of routine and frequent launch services. Like the ISS, they will also require the transportation of crew and so demand a demonstrated level of safety much greater than that needed for cargo re-supply. However, such levels should be more easily achievable via a fully reusable launch vehicle because its inherent value will demand better operational contingency options in addition to a crew escape system.

iii) Propellant Depot Resupply

A future LEO mission that may prove far more lucrative than those already mentioned is as the first leg of a space transportation infrastructure that consists of a set of operational nodes and transfer vehicles, namely:

- Space stations and human-tended experimental platforms;
- propellant depots to support missions both in, around and beyond LEO;
- short-range orbit maneuver vehicles (OMVs) to capture and transfer payload in and around LEO;
- long-range orbit transfer vehicles (OTVs) for travel to/from GEO and lunar orbits;
- OTVs fitted with legs and throttlable engines for lunar surface descent/ascent missions.

Details of one such space transportation architecture are shown in Figure 7, which also presents the delta-v

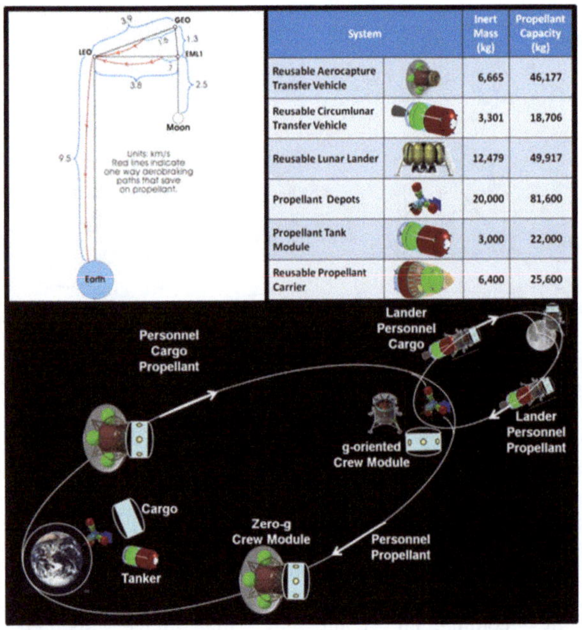

Figure 7. LEO-Lunar transport architecture [RD.17]

required to reach each node and the representative masses of each of the key elements. Interestingly, the dry mass of many of these elements falls within the payload launch performance of an air-launch RLV using ACES. However, a more important point to note is that the majority of each element's mass is propellant.

Analysis of the launch requirements for the build-up and operation of such an infrastructure [RD.17] show that the vast majority (~80%) of the mass launched into LEO is propellant. This is very significant because propellant can be infinitely subdivided and so would be the ideal payload for a small RLV capable of supporting both rapid and frequent launch and rendezvous missions. It therefore suggests that most of this architecture could be either launched and/or serviced by a subsonic air-launched RLV.

3.2 Commercial GEO Operations

Unfortunately, the markets identified so far are considered insufficient to justify the commercial development of a subsonic air-launched RLV because they are either too small or too speculative. Currently, the largest and most lucrative commercial launch market sector is the delivery of geostationary communications satellites (GEO comsats) into geosynchronous transfer orbit (GTO), with a perigee height ~200km and an apogee height ~ 36000km.

Any commercial business case for developing a subsonic air-launched RLV should therefore assess the viability of addressing the GEO comsat market sector, even though a cursory look at the LEO payload performance estimates presented in Figure 5 may appear to rule this out.

i) GEO Comsat Characteristics

An analysis of typical comsat mass characteristics is presented in Table 8 and indicates that the majority have a beginning of life (BoL) mass ~35% below their launch mass. This is because a significant fraction of their launch mass is propellant that they use during their transfer burn from GTO to GEO. More importantly, it

Payload Envelope (Stowed) ~ 7.3m x 3.8m x 3.4m

NOTE: Liquid apogee motor on all Boeing 702 & 601 series; Solid apogee motor on all Boeing 376 series

Spacecraft Name	Spacecraft Model	Launch Mass (lbs)	Launch Mass (kg)	BOL Mass (kg)	GTO to GEO Mass Fraction	RLV Flights (4t to LEO)
Anik F2	Boeing 702	13029	5910	3805	0.6438	5
	Generic	13000	5897	3796	0.6438	5
Anik F1	Boeing 702	10384	4710	3015	0.6401	4
Galaxy IIIC	Boeing 702	10604	4810	2835	0.5894	4
	Generic	12000	5443	3346	0.6148	4
Astra 1H	Boeing 601	8157	3700	2480	0.6703	3
Astra 2A	Boeing 601	7994	3626	2470	0.6812	3
Astra 2C	Boeing 601	8058	3655	2200	0.6019	3
	Generic	9000	4082	2658	0.6511	3
	Generic	6000	2722	1579	0.5800	2
Astra 2D	Boeing 376	3186	1445	824	0.5702	1
Astra 3A	Boeing 376	3318	1505	874	0.5807	1
	Generic	3000	1361	783	0.5755	1

Constraint

Table 8. Typical GEO ComSat mass characteristics

suggests that any vehicle capable of delivering a 4t payload into LEO could service the majority of currently planned GEO comsats if some sort of kick-stage were available on-orbit to perform the LEO to GEO transfer.

The key to servicing these markets with such a small reusable launcher is, therefore, the on-orbit assembly of a kick-stage capable of delivering the comsat directly into GEO, as illustrated in Figure 8. Such an operation would demand a rather special set of vehicle performance characteristics, namely the ability to perform:

- orbital rendezvous and docking;
- in-orbit propellant transfer or assembling sets of plug-in propellant modules;
- multiple launches within a short time period (e.g. a few days) to avoid effects of atmospheric drag, if low altitude orbits are used.

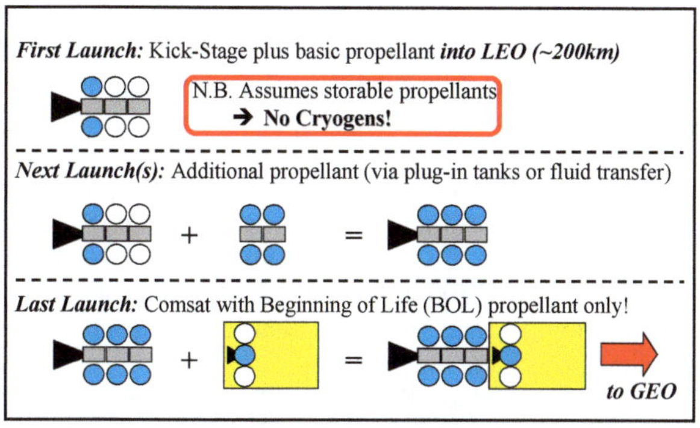

Figure 8. Orbital Assembly Scenario (Std. Comsat)

Such a vehicle would require an evolution of the basic orbital vehicle's capabilities but the upgrades to enable rendezvous and docking are not considered too major a technological challenge since they have already been demonstrated successfully by both Japanese and US spacecraft (i.e. ETS VII and Orbital Express). However, it is very unlikely that GTO customers would be willing to risk their satellites being launched in this manner until its operational complexity had been thoroughly proven, even if the launch price was half that of existing ELVs!

Nevertheless, GEO comsat missions are currently the most commercially viable market sector for any new launch vehicle and so this operational scenario is used as the basis for a brief business case analysis, which is described in more detail in the following sub-sections.

ii) Business Model Assumptions

Justifying the commercial development of a subsonic air-launched RLV requires more than just an assessment of the vehicle's design, operations and performance. It also requires an assessment of the associated costs and, more importantly, the revenue that it can be expected to generate from selling its services to commercial customers.

Development and operating costs can be based upon past estimates but will be highly uncertain. However, they can be used to bound the analysis and so indicate the range of values required to justify any investment.

Assessment of the potential market can be based upon data in RD.18, which gives annual projections for the number of GEO satellites within different mass groups and is summarized in Table 9. These market projections and estimated costs can then be used to construct a business model spreadsheet that generates an Income Statement and a Cash Flow Statement for any given scenario, which enables the performance of the venture to be assessed [RD.19].

Satellite Mass (kg)	Total No. (2013-2022)	Annual Average (2013-2022)	% of Total
Below 2200	29	2.9	13%
2200 to 4200	62	6.2	27%
4200 to 54000	46	4.6	20%
54000 and above	91	9.1	40%
Total Forecast	228	22.8	100%

Table 9. GEO ComSat size forecast [RD.18]

From the investors' point of view, the key is to get an acceptable return on any investment. A common yardstick to measure this is the internal rate of return (IRR), which is defined as *"the rate of return at which the present value of the cost of the investment and the present value of the future income stream equate"* – in simplistic terms, this is somewhat akin to the annual interest rate of a savings account. For high risk aerospace investments, the IRR has to be 20-30% for such projects to merit serious consideration. Another parameter of interest is the end-of-year (EOY) cash balance, which gives a good indication of the level of cash assets a company is generating and, more importantly, allows the payback period – the time needed to recoup the initial investment – to be assessed.

There are three or four fundamental parameters that drive the results: the available *market*; the *cost of services* (development and direct plus indirect operations cost); the revenue that can be generated by selling services at a given *price per flight*; and the *annual number or flights*. Other factors such as depreciation, taxes, amortisation and insurance generally have a relatively minor impact on the final result. Therefore, in order to simplify the analysis in the face of so many unknown or ill-defined values, a number of shortcuts or approximations were applied.

a) All up-front *investment* was expensed (i.e. put down as business expenses) in the same year it was applied. Strictly speaking, investments related to flight hardware and other capitalised equipment should be depreciated over their expected lifetime, however, as no useful breakdown is available here, they were expensed as they were incurred.
b) *Depreciation* was not accounted since it has only a marginal effect upon taxable income – it may change a 20% IRR into a 23% IRR, but not much more – and only occurs after the assets are paid for and in use.
c) Vehicle insurance, which could have been addressed by including at least one additional vehicle as an added expense (i.e. "self-insurance" against hull replacement), was simply taken as a nominal cost of $0.2M per flight against third party liability.

d) *Interest* was taken at a nominal annual rate of 10%, though this can vary and should be put to zero if the venture can be funded entirely by equity rather than debt, as assumed here. A more reasonable estimate for an all debt scenario could be 12-13%, which is essentially what it cost before taxes to borrow money at a corporate level in the US during the late-1999s, though this would have had minimal impact on the results.

e) *Tax*, which was accounted after interest and before the net income, was written-off when the venture incurred losses in the early years – in other words, it got a "tax credit" which could either be used to offset future gains or shared amongst the investors to offset gains in other investments. Therefore, assuming losses could be expensed against other gains, the net effect of taxes in the early years – especially during the development phase, which covers about three years – was to reduce the total out-of-pocket investment.

In addition, a set of financial and operational business parameters, shown in Table 10, was also developed in order to bound the business model and to investigate its sensitivity against changes in the baseline assumptions. A key point to note here is that the price per flight was only allowed to vary up to a maximum of $20M to ensure a reasonable margin against competing ELVs (e.g. Falcon 9 with a launch price of around $60 million for 4900kg GEO comsat, which would require 4 RLV launches). Also, as the kick-stage was assumed to be expendable, its cost were included within the overall variable cost and estimated to be around $2 million.

iii) The R LV Business Case

Business Parameter	*Value range*
Total R&D investment	$500-1000 million
Fleet size	3 operational vehicles
Price per flight	$10-20 million
Variable cost (per flight)	$2-10 million
Fixed annual operating cost	$40 million
Income tax rate	40%-60%
Interest rate	10% (for debt finance)
Annual flights (fleet max.)	100
First commercial launch	4 years after start

Table 10. RLV business model parameters

Assuming the maximum payload mass for the air-launched RLV is 4000kg, the number of flights needed for each class of GEO comsat are shown in the far right column of Table 8. This number was then uses to calculate the number of flights per year if 100% of the projected market was captured, which gave an average of 85 per year. However, a capture factor – nominally taken as 40% – was then applied to account for the fact that in the real-world a 100% market capture is considered as infeasible because at least one other competitor must be considered in any commercial scenario. The resulting annual flight rate, along with specific values for each of the business factors identified in Table 10, was then used to calculate the IRR and EOY cash balance over a ten year period from the venture's start.

This exercise was repeated for variations to the following key parameters in order to assess their overall impact:

- capture factor (25%, 40%, 55% & 70%);
- investment ($1000 million, $750 million & $500 million);
- price per flight ($10 million, $15 million & $20 million).

The results of this 'sensitivity' analysis are presented in Figure 9, which shows the evolution of IRR and EOY cash balance over the a ten year period from the venture's start with respect to a sub-set of the above values.

The plots show the impact of increasing market share (25% to 55%) and reducing launch price ($20M to $15M) for the $1000 million investment case, but also include one $500 million case ($15M price & 40% market) to illustrate the very significant impact of a reduced investment requirement.

Assuming that an IRR above 20% will be sufficient to justify investment in the venture, it is clear that an investment requirement of $1000 million *would not be acceptable* if the price per flight was $15 million (i.e. the price needed to be competitive with Falcon 9) and market capture was held at 40%. However, it *would become acceptable* if the market share could increase to 55% or the investment requirement was substantially reduced (e.g. down to $500 million).

iv) Observations on the Business Case

Clearly, this business case analysis is far too crude to judge the true commercial viability of such a venture. However, given these results, the general conclusion is that there are some good reasons for thinking that a fully commercial air-launched RLV venture may prove to be successful, particularly if its investment requirements can be kept around the $500 million mark and its launch price can be kept below $15 million. ***The major caveat here is that a development cost of $500 million appears extremely low for vehicles with such a payload performance, based upon current launcher development experience.***

As a point of comparison, the estimated development cost of the PD-2 concept [RD.7] was $940 million. However, being expendable, its per flight cost was $120 million, of which $112 million was for the production of each new vehicle and $8 million was for launch operations. Obviously the development costs for an RLV will be somewhat higher but the launch operations costs should be similar or better, which lends come credibility to the results of this rather simplified business case assessment.

One important observation here is that the business case can be improved significantly if some degree of leverage can be applied to reduce the initial investment.

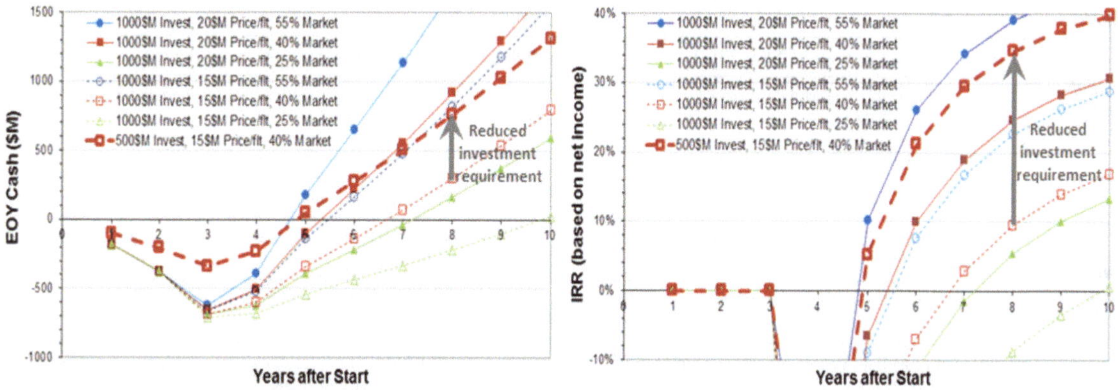

Figure 9. Air-Launched RLV business base sensitivity analysis

One obvious way to achieve such leverage would be to develop key elements of the system through a separate venture or business phase. DARPA's XS-1 initiative may be provide just such a leverage, while a business venture to service the nascent sub-orbital market may represent another. Whether these would be practical or sufficient to leverage development of an orbital RLV has yet to be determined. However, there are a number of real-world examples, both current and past, that may justify this approach, for example:

- SpaceX leveraging their NASA contracts to support development of the Dragon capsule;
- Boeing leveraging their USAF contracts for the KC-135 to support development of the 707.

Whatever the form of the leverage, this analysis serves to underscore the value of building up any space launch business in a series of small steps rather than one giant leap.

4. CONCLUSIONS

This paper has identified the key reasons why space activities have so far failed to achieve the great expectations set out at the dawn of the space age, over half a century ago. It has also described the ways in which small groups of people are attempting to change the current paradigm but, in doing so, has tried to indicate the enormity of the challenges they must overcome in order to realize their ultimate goal.

Having identified access to LEO (i.e. launch vehicles) as one of the main constraining factors for in-space developments and operations, it has assessed one very promising launch vehicle concept (i.e. the air-launched RLV) and identified potential technologies that could produce significant improvements in both its safety, operability and payload performance.

Based upon these insights, it has then shown how a relatively small air-launched RLV could improve space access and thereby enable new in-space transportation infrastructures that will deliver a significant increase in future space-based operations for the purposes of both exploration and resource exploitation. In short, it has shown that we do not necessarily need big launchers to enable big space operations!

In addition, it has also tried to show that such developments could be driven by commercial investments, though there is still much scope for governments to foster them in a synergistic manner by funding new capabilities (e.g. DARPA's XS-1) or procuring operational services (e.g. NASA's Commercial Resupply Services).

One major caveat of these results is that subsonic air-launch should be regarded as an enabling capability, since the majority of the technology/cost challenge resides within the RLV that performs the bulk of the work needed to place any payload into orbit. Nevertheless, it does relax the RLV design constraints significantly and so makes these challenges far more tractable, realistic and affordable.

As a final synthesis of all these ideas, an attempt has been made to consolidate them together by briefly sketching out some likely steps for achieving this new space paradigm. Table 11 presents these steps and includes a tentative timeline, covering the next decade, along with their likely impacts upon future in-space activities.

Clearly, many of these steps will slip, change or may never be realized. In fact, this new space paradigm may prove to be unachievable because of fundamental constraints that have yet to be discovered. So, although there is good reason for cautious optimism, it would be better to regard such developments as experiments within a process of Darwinian evolution rather than the milestones of some overarching space program, established by the directive of a government space agency.

Nevertheless, given the current number of new space ventures and their success to date, it seems reasonable to believe that some may manage to "boot-strap" themselves into orbit within the next decade and finally begin to open the space frontier in order to harness the infinite resources of outer-space for the benefit of all mankind.

Timeframe	*Future Steps*	*Impacts*
Proof of Concept (2012-2018)	COTS payload services to ISS (~2012)	MODEST: Increased microgravity experimentation
	Frequent reusable suborbital services for tourist passengers (~2016)	SIGNIFICANT: Rapid flight vehicle turn-around and passenger training
	COTS crew rotation to ISS (~2018)	MODEST: Improved human in-situ servicing and support
Concept Maturation (2018-2025)	Commercial space station & ELV support (~2020)	SIGNIFICANT: Increased human in-situ servicing and support
	Air-launched RLVs for ISS cargo and GEO satellite launch (~2020)	VERY SIGNIFICANT: Increased satellite missions and space infrastructure development
	Air-launched RLVs for passenger services to ISS and commercial stations (~2023)	VERY SIGNIFICANT: Increased human in-situ activities supporting complex space developments
	In-orbit propellant depots for crewed exploration missions (~2025)	VERY SIGNIFICANT: Enables deep space exploration missions and exploitation of space resources

Table 11. Steps towards a new space paradigm

REFERENCES

[1] Space Foundation publication "The Space Report 2012" URL: http://www.spacefoundation.org/programs/research-and-analysis/space-report/20-space-economy

[2] "Wal-Mart Annual Report, Eleven-Year Financial Summary", p.26/27 URL: http://www.walmartstores.com/sites/annual-report/2012/WalMart_AR.pdf

[3] "The Commercial Space Transportation Study, 1994" URL: http://www.hq.nasa.gov/webaccess/CommSpaceTrans

[4] "NASA ASCENT Study Final Report", Futron Corporation 2003 URL: http://www.futron.com/pdf/resource_center/reports/ASCENTFinalReport_V1.pdf

[5] NASA Get Away Special Wikipedia page URL: http://en.wikipedia.org/wiki/Getaway_Special

[6] "Salt, D.J. "Could a subsonic air-launched RLV enable a paradigm shift in space operations?", AIAA-2014-1897 SpaceOps 2014, Pasadena, California, 5-9 May 2014

[7] "Horizontal Launch: A Versatile Concept for Assured Space Access", NASA SP 2011-215994 URL: http://ntrs.nasa.gov/archive/nasa/casi.ntrs.nasa.gov/20120000791_2012000836.pdf

[8] Breugelmans F., et al. "Process and device for collecting air, and engine associated therewith" US Patent 6644016 B2, Issued November 11, 2003 URL: http://www.google.co.uk/patents/US6644016

[9] Giffin R.G., "Methods and Systems for Operating Oxydizer Systems" US Patent US8127527 B2, Issued Mar 6, 2012 URL: http://www.google.com/patents/US8127527

[10] Verstraete D., Bizzarri D., Hendrick P. "In-Flight Oxygen Collection for a Two-Stage Air-Launch Vehicle: Integration of Vehicle and Separation Cycle Design" Progress in Propulsion Physics 1 (2009) 551-568 (DOI: 10.1051/eucass/200901551) URL: http://www.eucass-proceedings.eu/articles/eucass/pdf/2009/01/eucass1p551.pdf

[11] Crocker A. M., et al. "Alchemist - An Enabling Technology for Low Cost Access to Space" NASA Phase I Small Business Innovation Research (SBIR) contract Award ID: 51951, Program Year: 2001 URL: http://www.sbir.gov/sbirsearch/detail/89845

[12] Crocker A. M., et al. "ACES: An Enabling Technology for Next Generation Space Transportations" AIP Conf. Proc. 699, 213 (2004) & AIAA-2003-4890, July, 2003 & IAC-03-S.5.03, Sept/Oct, 2003 URL: http://dx.doi.org/10.1063/1.1649577

[13] Dujarric C., "Possible Future European Launchers – A Process of Convergence", ESA Bulletin 97 – February 1999 URL: www.esa.int/esapub/bulletin/bullet97/dujarric.pdf

[14] Stadler R. "Results for a Fully Reusable TSTO-Launch Vehicle Concept", AIAA-98-1504

[15] Bayer M. "Description of the FESTIP VTHL-TSTO System Concept Studies" unpublished study report URL: http://thehuwaldtfamily.org/jtrl/research/Space/Launch/Vehicles/FESTIP-FESTIP-Two Stage To Orbit Launch Vehicle, Bayer.pdf

[16] DeLong D., Stuhlinger E. "Proposed Concept for a Spaceplane" AAS 86-463, AAS Proceedings Volume 64, Part 1 & Spaceflight (ISSN 0038-6340), vol. 29, Dec. 1987, p. 413-416

[17] Bienhoff D. "From Importing to Exporting: The Impact of ISRU on Space Logistics" AIAA 2011-7112, AIAA Space 2011 Conference & Exposition, Long Beach, California, 27-29 September 2011 URL: http://arc.aiaa.org/doi/abs/10.2514/6.2011-7112

[18] "2013 Commercial Space Transportation Forecasts", a report by the FAA's Associate Administrator for Commercial Space Transportation (AST) and the Commercial Space Transportation Advisory Committee (COMSTAC), May 2013 URL: http://www.faa.gov/about/office_org/headquarters_offices/ast/media/2013_GSO_NGSO_Forecast_Report_June.pdf

[19] "Salt, D.J., & Lindroos, M, "The Business Case for Small RLVs", TN200, Issue 3, 20th January, 1999 (deliverable item under ESA Contract No. 970801).

Reinventing Space Conference
BIS-RS-2014-49

Sprite, a Very Low-Cost Launch Vehicle for Small Satellites

Nicola Sarzi-Amade, Thomas P. Bauer, and
James R. Wertz
Microcosm, Inc.

Markus Rufer
Scorpius Space Launch Company

12th Reinventing Space Conference
18-20 November 2014
London, UK

S. Hatton (ed.), *Proceedings of the 12th Reinventing Space Conference*, DOI 10.1007/978-3-319-34024-1_13

BIS-RS-2014-49

Sprite, a Very Low-Cost Launch Vehicle for Small Satellites

Nicola Sarzi-Amade, Thomas P. Bauer, and James R. Wertz
Microcosm, Inc.
4940 W. 147th St., Hawthorne, CA 90250; (310) 219-2700
namade@microcosminc.com

Markus Rufer
Scorpius Space Launch Company
4940 W. 147th St., Hawthorne, CA 90250; (310) 219-2700
mrufer@scorpius.com

ABSTRACT

The Scorpius low-cost launch vehicle architecture greatly reduces the cost of space access due to its emphasis on designing specifically for low total life cycle cost. Due to its simplicity, a pressure-fed launch vehicle is low in cost compared with pump-fed and solid rockets. The pressure-fed approach in the Scorpius architecture is enabled by the development of all-composite propellant and pressurization tanks, which have about half the mass of metallic tanks. The low-cost Scorpius "Pressurmaxx" composite tanks comprise half the dry mass of the vehicle. In addition, a high-performance pressurization system using heated helium reduces the mass of the pressurization system by half. Ablative, LOX/Jet A engines have acceptable performance and are very low cost. The mass savings of the tanks and pressurization system together with the engines yield a 3-stage launch vehicle that can be much lower in cost than a high-performance (pump-fed) vehicle. Sprite, which delivers 480 kg to LEO, is the vehicle in the Scorpius® family of low-cost, scalable launch vehicles that has progressed the furthest in terms of development. Propellant tanks, the pressurization system, and engines of the size needed for Stages 1 and 2 of Sprite have been built and tested. A prototypical "pod" of Sprite has been flown suborbitally. This paper describes the Scorpius architecture, its scalability into a family of low-cost vehicles capable of payloads to Low Earth Orbit (LEO) from 100 kg through 9000 kg and larger, and the responsiveness of the vehicles. The Sprite configuration is presented, its performance and sample missions are shown, and a market analysis is provided.

KEYWORDS: Scorpius, Sprite, Launch Vehicle, Low-Cost, Responsive, All-Composite, Pressure-Fed, SmallSat

1. INTRODUCTION

Microcosm and its sister company, Scorpius Space Launch Company (SSLC), developed the Scorpius® family of low-cost launch vehicles with the goal of greatly reducing the cost of access to space. The vehicles were developed through SBIR Phase I, II, and III programs, funded mainly by the U.S. Air Force and NASA. The Scorpius® technology is scalable to different sizes; therefore all vehicles can be built using essentially the same technologies. The family of orbital launch vehicles is comprised of the following vehicles:

- Demi-Sprite: 160 kg (350 lbs) to LEO
- Sprite: 480 kg (1,060 lbs) to LEO
- Liberty: 1,920 kg (4,240 lbs) to LEO
- Exodus: 8,940 kg (19,700 lbs) to LEO
- Heavy Lift: 18.1 t (40,000 lbs) to LEO
- Space Freighter: 36.3 t (80,000 lbs) to LEO
- Super Heavy Lift: 90.7 t (200,000 lbs) to LEO

Some of the above vehicles are shown in Fig. 1. The figure also shows the SR-S, SR-Q, and SR-M suborbital vehicles. The orbital vehicles up to the Liberty size are appropriate for small LEO missions, whereas Exodus and Heavy Lift are appropriate for Large LEO missions or typical GEO/Inter-planetary missions. The Space Freighter and Super Heavy Lift vehicles can be used for very large LEO or large beyond-LEO missions[1]. All orbital vehicles use the same vehicle configuration and key technology elements:

- 3-stage expendable vehicle
- 6 outer pods constituting the first stage
- A nearly identical inner pod as the second stage
- 3rd stage and payload, on top

- LOX/Jet A, pressure-fed, ablatively-cooled, essentially identical first and second stage engines
- All-composite propellant tanks
- High Performance Pressurization System (HPPS)

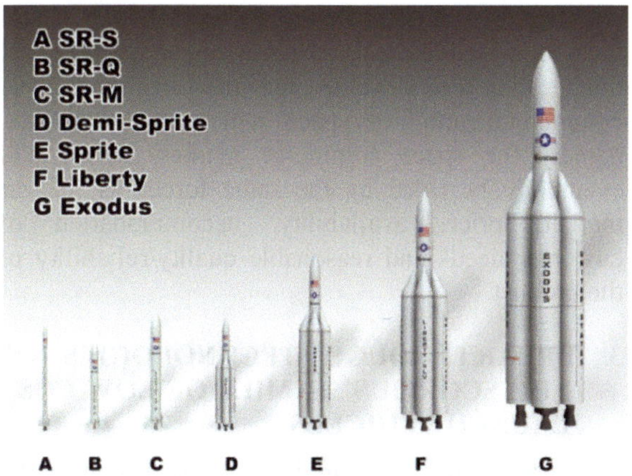

Figure 1. Scorpius® Launch Vehicle Family.

The 3-stage design was selected to minimize the cost to LEO. Even though, on one hand, the 3-stage approach has the disadvantage of increasing the overall parts count (e.g., more engines, more tanks) and of increasing the number of mid-air engine starts and separation events, it has several advantages. In particular, the 3-stage approach reduces the delta V required of each stage, decreases the sensitivity to mass, drag, and Isp in the lower stages, and increases the design margins, which drives down cost. For a given burn-out fraction, increasing the number of stages decreases the gross lift-off weight (GLOW) of the entire vehicle primarily because it means carrying less vehicle mass to a high velocity after it's no longer useful. This reduces the stage 1 engine size, which further reduces cost, and allows an Isp more tailored to altitude. To maintain the same GLOW as a 3-stage vehicle, a 2-stage vehicle would need to reduce the burnout mass fraction from 14% to 9%.

The Scorpius® launch vehicles are characterized by 6 identical outer pods (1st stage) and a central pod (2nd stage) that is virtually identical to the 1st stage pods. This means that, by having 7 virtually identical pods in each vehicle, if only 2 or 3 vehicles are produced annually, this is still 15 to 20 pods per year, which creates a small assembly line and the resulting economies of scale. The vehicles have very clean systems with almost no moving parts, and are characterized by very robust performance, being able to go to orbit with 1 engine out.

2. DESIGN FOR MANUFACTURABILITY

The design for manufacturability is a key aspect of the Scorpius® launch vehicle technology. The Scorpius® manufacturing process consists of setting up, maintaining, and using a low-cost launch vehicle manufacturing environment with build-to-inventory and launch-on-demand. Manufacturability, low cost, and assured availability are preferred over the traditional emphasis on performance optimization. In other words, Microcosm and SSLC adopt a throughput orientation versus a mission assurance orientation, they try to achieve a break from tradition, and utilize an industrial production mentality.

Backing off even a small fraction from the optimum and stepping into "robust design," "large margin" and "throughput orientation" territory has enormous benefits and a compounding effect on cost. Even a minimal level of modularization of sub-systems and standardization of interfaces and fasteners makes a significant difference. Adding flexibility to the material flow using multiple-use tools, movable production assets, wheeled setups, and cross trained personnel who work sliding scale and split shift schedules, reduces costly dependencies that are otherwise mandated by the demand for optimization. Reducing the requirements catalog to the essential functions of the product and designing multi-functionality into the production environment, as opposed to into the product, are essential requirements of a design-to-cost approach. Carry-over or adoption of manned flight or military quality standards are eliminated wherever possible. Microcosm and SSLC are breaking the chain of perpetual performance optimization.

In terms of technical performance, this vehicle architecture offers a very low parts count, virtually no secondary structure, high structural stiffness, large performance margins, low thermal effects sensitivity, high shock and vibration tolerance, and good operating agility. In terms of cost performance, in addition to low manufacturing cost, this design has a compounding impact on logistics and operational costs, since it facilitates the application of commercial standards for all non-flight procedures such as storage, check-in/check-out, crating, shipping/transportation, handling, corrosion prevention, and an industrial-type rapid production throughput.

If the commercialization of space is to be an achievable goal, we have to address cost as a key ingredient. The particular culture of the aerospace engineering profession plays a major role in terms of the cost of space missions and is therefore worth examining. For contrast, let's compare the performance optimization design mindset of an aerospace engineer to that of an automotive engineer, whose effort is driven by the need to achieve high Design Efficiency. That means simply, that his design considerations are guided by trades of total manufacturing time and total assembly time (impacting cost and schedule) contrasting the aerospace typical product function and performance optimization (impacting mission assurance). In the automotive world, a certain car model must fit a very narrow cost bracket in the market place, which mandates a highly disciplined design-to-cost approach. All other considerations pale in comparison — for if the car misses its selling price target, it will simply not sell and the development investment will go to waste. In contrast, the aerospace engineer is concerned with optimal vehicle/payload performance and maximum mission assurance, meaning that the design decisions are informed by the performance optimization goals of every part, as well as the mandate that the mission can never fail, with cost typically of significantly less importance. The following, very simple Design for Assembly (DFA) example illustrates the different mentalities (Fig. 2).

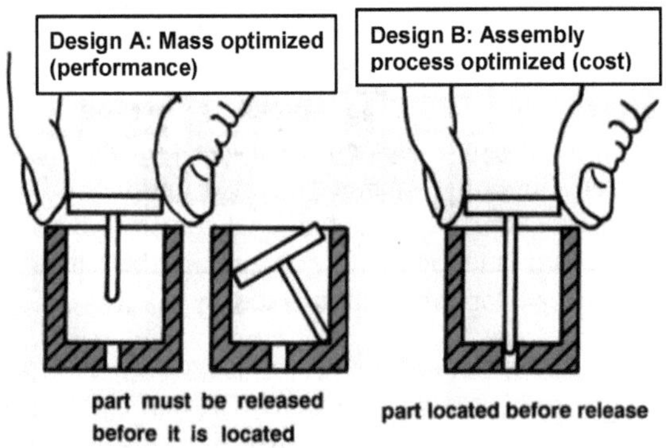

Figure 2. Aerospace and Automotive Part Design Comparison.

Example A: (aerospace) is a design guided by performance optimization, where the short pin helps to reduce mass at the expense of difficult insertion and a high operator skill requirement. Although not visible in this example, it can be safely assumed that this part would be made of some ultralight specialty alloy, probably machined out of a block of material, and that numerous inspections and material certifications would guarantee the flawless function that is demanded by the near 100% mission assurance requirement (i.e., typically > 99% even for unmanned missions). Considerations for such factors as ease of manufacturing, assembly, storage, cost, and availability do not rise to a level where they can compete with the mass optimization demands. Example B: (automotive) is a design for the same part, guided by cost goal demands, thus affecting aspects such as features that accommodate quick and error-free assembly at the cost of slightly higher mass. Although not visible in this example, it can be safely assumed that this part would be made of very common off-the-shelf materials and that it would include no additional treatments or processes that don't add direct value for the customer. The commercialization of space will create a market, namely the space commerce market, which will eventually be ruled by the same forces as the car market: price, availability, accommodation of customer needs and reasonable quality/reliability of the product.

3. BREAKTHROUGH TECHNOLOGIES OF THE SCORPIUS® FAMILY OF LOW-COST LAUNCH VEHICLES

The Scorpius® technologies, developed by Microcosm and its sister company, Scorpius Space Launch Company (SSLC), enable a low-cost architecture for the family of orbital launch vehicles. The three main technologies are the all-composite cryogenic tanks, the high-performance pressurization system, and the composite ablative engines. They not only are low-cost in themselves, but they also greatly reduce the vehicle's overall life cycle cost, including manufacturing, integration, and launch operations. These technologies make a high-performance pressure-fed propulsion system possible, therefore greatly reducing cost compared to pump-fed propulsion systems and solid rockets. In particular, pump-fed systems use turbopumps and regenerative cooling systems that make the vehicle as much as an order of magnitude more expensive than a pressure-fed system. Similarly, solid rockets use expensive and polluting propellant, are heavy and dangerous during ground operations, cannot be shut down or restarted, and require expensive steering mechanisms for controlled flight, making the pressure-fed system the option with the lowest cost; also, generally, compared with pressure-fed LOX/Jet A systems, solid motors for orbital launch have inferior performance. More details on the differences between pressure-fed and pump-fed propulsion systems are provided in Chakroborty and Bauer[2]. The key Scorpius® technologies are described more in detail below.

3.1 All-Composite Cryogenic Tanks

Microcosm and SSLC have developed an innovative line of all-composite, linerless, cryogenic tanks called Pressurmaxx®, which is characterized by a common carbon fiber material and SSLC's proprietary cryogenic resin formulation, termed Sapphire77. One of these tanks is shown in Fig. 3. The tanks can be built in a wide range of sizes and aspect ratios (Fig. 4), not only for the Scorpius® family of launch vehicles, but also for other applications such as spacecraft (manned and unmanned), aircraft, cryogenic fluids storage and transfer, high-pressure storage, unmanned aircraft systems (UAS), unmanned underwater vehicles (UUV), automotive, oxygen & air supplies, medivac vehicles, military ops, and Special ops diving. The tanks have been successfully tested with nitrogen at cryogenic temperatures down to –321 deg F, and at burst pressures up to 7,200 psi. This high performance is achieved with a shorter lead time than traditional composite-overwrapped pressure vessels (COPV) or metallic tanks, while maintaining very low cost. Another key characteristic of the tanks is their low weight, approximately half of that of metallic tanks of the same volume; a consequence of this advantage is that the tanks constitute only about half of the dry weight of any of the vehicles of the Scorpius® family.

Figure 3. Pressurmaxx® All-Composite Tank.

Figure 4. All-Composite Tanks of Different Sizes and Shapes.

The Pressurmaxx® composite tanks are characterized by unique technologies such as all-composite polar bosses, integration of external structural stringers (circumferential or longitudinal) and internal slosh baffles, and tooling production. Additive manufacturing techniques are employed for these integrated features (Fig. 5), which are not externally attached but built from "inside out." Thanks to these breakthrough technologies, the tanks are effectively unibody structures and can therefore become the primary structure of a launch vehicle or a spacecraft[3].

Figure 5. All-Composite Tank with Skirts and Longitudinal Stringers.

Tanks have been built from 0.5 cuft to 200 cuft volume for propellants, gases, pressurants, and cryogens up to 3,600 psi maximum expected operating pressure (MEOP), which translates into 7,200 psi burst pressure given that the tanks, currently, are built to a safety factor of 2.0. Figure 6 shows an example of a 200 cuft tank that was transported on a regular truck trailer, providing evidence of the great robustness and ease of handling of the tanks, which really is a revolution compared to current metal tanks used for space applications. Figure 7 shows an application of the PRESSURMAXX® all-composite tanks with the Armadillo Aerospace rocket vehicle, used in the X Prize Northrop Grumman Lunar Lander Challenge Level 2, for which two of our high-pressure helium tanks at 2,200 psi MEOP were used, allowing Armadillo to successfully complete the challenge.

Several qualification tests have been conducted on the tanks to date[3]:

- Chemical compatibility: compatibilities include petroleum-based fuels, e.g., kerosene; alcohol based fuels, e.g., ethanol; cryogens, e.g., liquid oxygen and nitrogen; various gases, e.g., methane, helium, oxygen, nitrogen; and propellants, e.g., turpentine, hydrazine, and AF-M315E green propellant.
- Pressure tests: pressurant tanks operating at 3,600 psi (7,200 psi burst rating) are in use, for which 50 fill/discharge cycles were performed.
- Temperature range: 25 temperature cycles and rapid chill-down testing have been conducted with nitrogen from +175 deg F to –321 deg F.
- Load / Impact / Vibration tests: a vibration test has been conducted on a spacecraft bus characterized by a unibody composite pressurized structure.
- Radiology tests: NASA White Sands Test Facility (WSTF) shearography, pressure, and leak tests have been conducted.

Figure 6. Local Transport of a 200-cuft., 500-psi LOX Tank.

Figure 7. Prize Winning Armadillo Lunar Lander GHe Tanks, 2,200 psi MEOP.

Additionally, Microcosm and SSLC have developed tanks incorporating a positive expulsion device (bladder), which is used for spacecraft in-space propulsion[3]. The positive expulsion device, or PED, is made from an EPDM (ethylene propylene diene monomer) rubber material that has already been qualified by both NASA and ESA and has flown to space on multiple missions. The development effort of the Microcosm/SSLC bladder tank technology was funded by NASA Glenn Research Center. This tank uses the linerless all-composite PRESSURMAXX® unibody technology already successfully demonstrated in various applications, and is designed for use in either blow-down or external accumulator mode. The bladder and tank body are Hydrazine and AF-M315E green propellant compatible. This technology does not require standard propellant management devices (PMD's), and is therefore simple and reliable.

3.2 High-Performance Pressurization System

The high-performance pressurization system (HPPS) developed by Microcosm and SSLC is based on Tridyne, which is a concept first developed in the 1950's by Rocketdyne (hence the name) in Canoga Park, CA. Microcosm and SSLC improved this concept, first through extensive analytical work and comprehensive test programs under IR&D, and then through various government contracts. These contracts were issued as part of the DARPA FALCON program, and they substantiated the viability of the system. The system configuration is illustrated in Chakroborty *et al.*[4]

Heating the helium through this process was shown experimentally to reduce the mass and volume of the required helium and the associated tankage by nearly 50% compared to a cold gas system, resulting in substantial payload gain. Figure 8 shows one of the tests: the tank on the right (white) is the regulated propellant, while the one on the left (black) is the actual Tridyne tank. There are virtually no moving parts in the system. Successful liquid oxygen (LOX) expulsion tests with a flight-like HPPS system validated the technology for a Sprite size vehicle. This technology qualification program also verified the scalability of the system to both smaller and larger sizes.[4]

Figure 8. High-Performance Pressurization System Test.

Thanks to Microcosm's HPPS, uniform and constant pressure is maintained in the launch vehicle's propellant tanks (LOX and Jet A). Therefore the performance is predictable all throughout the flight. The successful tests show that this system is expected to work in space without issues.

3.3 Composite Ablative Engines

The Scorpius® ablatively cooled engines are designed by both Microcosm and Scorpius Space Launch Company (SSLC) and are built in-house by SSLC. An image of the Scorpius® 20K lbf thrust engine is shown in Figure 9. These engines are characterized by an external layer (structural) of carbon fiber and Sapphire77® cryogenic resin, and an internal ablative layer. The simple design of the engine enables very low cost production. These engines have almost no moving parts and do not need expensive components like the turbo pump or the regenerative cooling system. The preferred propellant combination for the propulsion system is LOX/Jet A, both very low cost and compatible with the ablative layer. The SSLC-built engines are characterized by an ease of production and integration. Engines providing 5K lbf of thrust and 20K lbf of thrust have been built and tested. In particular, the 5K lbf engines have flown on two successful suborbital flights, the SR-S (1999) and SR-XM (2001), both from the White Sands Missile Range, NM. Examples of 20K lbf engines are shown in Figure 10 (these are the engines used by the Sprite vehicle). Microcosm has conducted firing tests on both the 5K lbf and the 20K lbf thrust engines (Fig. 11).

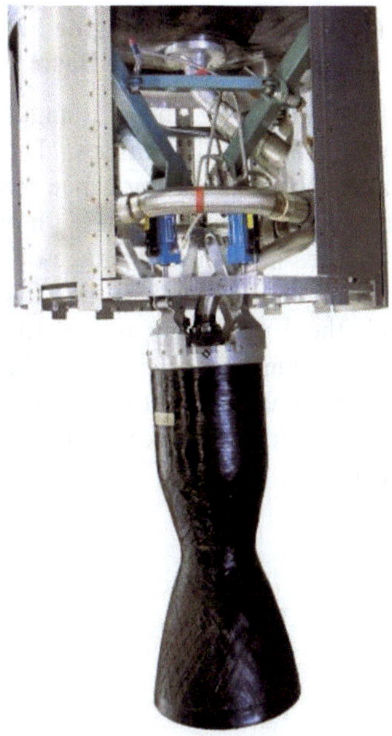

Figure 9. 20K lbf Thrust Engine Apparatus.

Figure 10. 20K lbf Thrust Engines.

Figure 11. 20K lbf Thrust Engine Firing Test (Edwards Air Force Base).

The specific impulse of the 1st stage engines is 286 sec. (vacuum) for all vehicles. The Scorpius® engines have moderate performance compared to other liquid propellant engines; however, the Scorpius® engines have the advantage of being much lower in cost. Additionally, the manufacturing process of these engines is relatively simple compared to pump-fed engines, because the Scorpius® engines have almost no moving parts (only valves and gimbals can move), making this system very low cost.

4. SPRITE CONFIGURATION AND PERFORMANCE

The configuration of the Sprite launch vehicle is shown in Fig. 12, and some of its key characteristics are presented in Table 1. Sprite's payload capacity to low Earth orbit (100 nautical miles altitude with launch due east) is 1,060 lb (480 kg). The launch price is less than $6.0M, in 2014 dollars, which is a very appealing aspect of the vehicle. Sprite uses pressure-fed engines that are small, light-weight, and simple. The vehicle has 3 stages, the first of which is made of 6 identical pods each with a 20,000 lbf of thrust engine, 1 fuel tank, 1 oxidizer tank, and 2 Tridyne tanks. The outer pods surround the core pod, which is the vehicle's 2nd stage. This pod is almost identical to the outer pods, as the only real difference is the slightly dissimilar engine configuration. On top of the 3rd stage sits the payload bay surrounded by a bi-conic fairing. Each pod and the 3rd stage are 42 inches in diameter, and the whole vehicle is 11.2 ft in diameter; additionally, the vehicle's height is 54.2 ft, and its gross lift-off weight (GLOW) is 80,500 lb (9,300 lb dry weight).

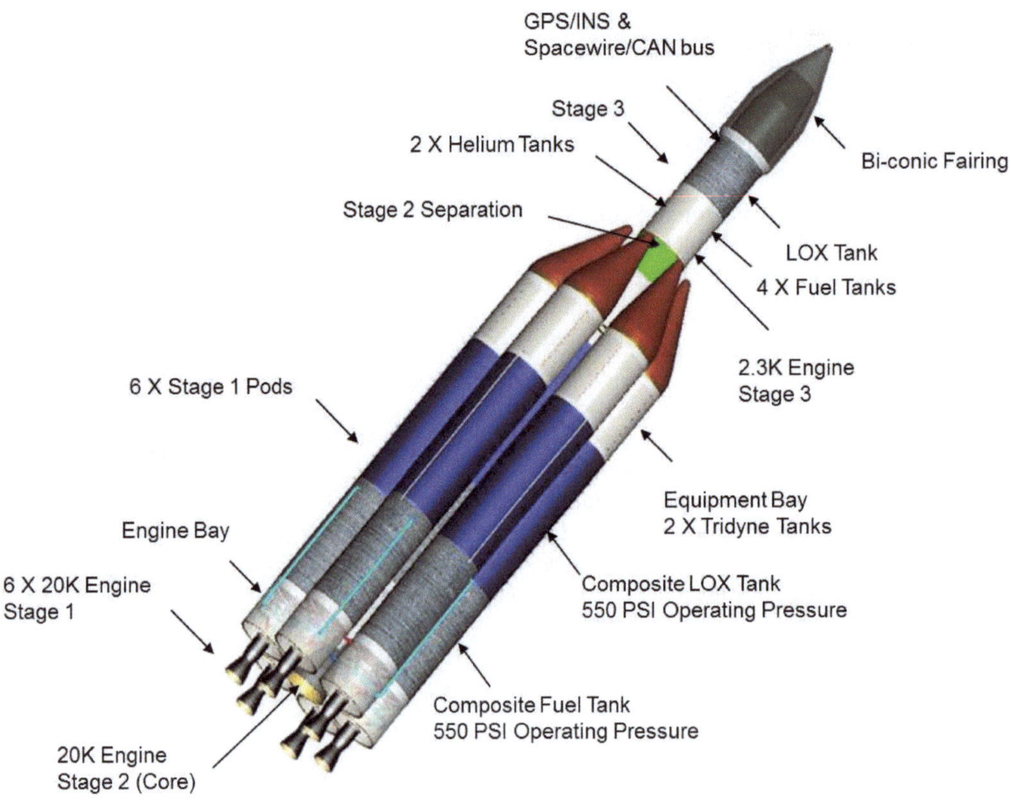

Figure 12. Sprite Configuration.

Table 1. Key Characteristics of the Sprite Launch Vehicle.

Characteristic	Sprite
LEO Payload (100 Nmi due East)	1,060 lb
Launch Price	< $6.0 M ($FY14)
Overall Height	54.2 ft
Pod Diameter	42.0 in
Vehicle Diameter	11.2 ft
GLOW	80,500 lb
Dry weight	9,300 lb
Propellant	LOX/Jet-A
Pressurization	Tridyne
Max Axial g's	5.9
Engine Configuration	
Stage 1	6 × 20K
Stage 2	1 × 20K
Stage 3	1 × 2.3K
Stage 1	
Number of pods	6
Thrust, vac (lbf)	120,000
Thrust, sl (lbf)	101,000
Gross Mass (lbm)	65,600
Stage 2	
Number of pods	1
Thrust (lbf)	22,300
Gross Mass (lbm)	10,900
Stage 3	
Thrust (lbf)	2,300
Gross Mass (lbm)	3,005

Sprite provides true launch-on-demand service from a flat pad with minimal infrastructure within 8 hours of arrival of the payload at the launch site, and it is capable of all-weather launch through 100-kt ground wind and 99.9% of winds aloft. This capability is possible thanks to Sprite's squat configuration and, therefore, low moments of inertia, which allow much better steering control, and also thanks to its very strong all-composite tanks which are also the load-bearing structure of the vehicle. Additionally, Sprite is scalable to much larger (or smaller) vehicles using the same technology and basic vehicle design. Finally, all the Scorpius® launch vehicles are very easy to launch, because they do not need a flame bucket, just a flame deflector, so they can launch from virtually anywhere. These key properties make Sprite and the other vehicles of the Scorpius® family extremely responsive and ready to meet any of the world's launch needs. Specific applications of the Sprite vehicle are presented in Sec. 6.

One of the scaled-down versions of Sprite, called Demi-Sprite, can put up to about 160 kg into LEO for a recurring launch cost of about $3.6 M. The main application consists of launching NanoEye or equivalent category spacecraft to LEO.

5. SPRITE'S STATE OF DEVELOPMENT

Sprite is the Scorpius® vehicle that has progressed the furthest in terms of development, both in terms of design and testing. The technology for Sprite has been revised since Chakroborty *et al.*[5] resulting in an increase in LEO payload performance from 318 to 480 kg (700 to 1060 lbs.) mostly thanks to the advancements in the tank technology. The metallic bosses of the first generation of composite tanks have been eliminated, resulting in truly all-composite propellant and pressurant tanks, which saves weight. Factor of safety of 2 has been established providing ample margin and assurance for the ranges. A high density ablative chamber has been incorporated to provide longer life. The avionics system has been updated taking advantage of ongoing developments in electronics that save weight, power, and size. Moreover, subsequent efforts have increased the confidence in the technology and approach through extensive analyses, simulation, wind tunnel testing, Tridyne expulsion testing, and 20K engine testing. As mentioned, considerable experience at building all-composite tanks in a variety of sizes for a range of applications with different pressures, temperatures, and fluid types has increased maturity in this most crucial of the Scorpius technologies. A GPS-based range operation has been adopted, which reduces range cost and flight hardware.

The SR-M suborbital launch vehicle is shown in Fig. 13. This vehicle is very similar to Sprite's 2nd stage, and has already been designed and built by Microcosm and SSLC, but has not yet flown. The SR-XM suborbital launch vehicle, which has successfully flown in 2001 (Fig. 14), represents a prior version of the SR-M launch vehicle, and therefore, of a pod of Sprite. The SR-XM was assembled, erected, fueled and ready to launch within 8 hours of arrival at the launch site (West Center 50, White Sands Missile Range).

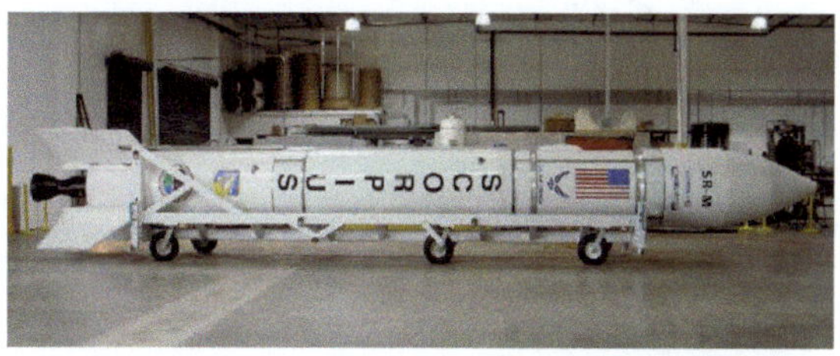

Figure 13. SR-M Suborbital Launch Vehicle.

Figure 14. Scorpius® SR-XM.

6. SPRITE VEHICLE — SAMPLE MISSIONS

The performance of the Sprite launch vehicle to various LEO orbits is depicted in Figure 15. Representative missions in LEO include launch of small satellites up to 480 kg for observation, remote sensing, science, and military. Another application is to launch to Sun-synchronous orbits (SSO) from dedicated locations such as the Vandenberg Air Force Base in California (for example for weather monitoring missions); additionally, Sprite can launch to transfer orbits for the International Space Station (ISS, circular orbit in a range of 330 km – 435 km altitude) with the purpose of delivering commodity cargo (e.g., water, food), to the station itself. Microcosm's NanoEye spacecraft, whose baseline configuration is in the nano/microsatellite category,[6] could potentially perform most of the above missions.

Sprite can also deliver payloads to orbits beyond LEO. For example, payloads can be delivered to Geostationary Transfer Orbit (GTO, 168 kg maximum payload); in particular, applications in Geostationary orbits (GEO) can include communication satellites, space situational awareness (e.g., space debris monitoring), or scientific observation missions. A longer-term application could be the launch of small satellites to GPS transfer orbits to allow future generations of GPS satellites to either replace or augment existing, much older satellites. Finally, with a Scorpius® upper stage, Sprite can launch small satellites to interplanetary orbits (i.e., very high energy orbits); in particular, small satellites like Hummingbird[7], in the 100 kg mass range, could be launched to escape orbits (and then the satellite can use its own propellant to maneuver to a desired interplanetary orbit). Additionally, smaller satellites, up to 54 kg, can be launched directly to a Mars transfer orbit. Potential missions for the Sprite launch vehicle are summarized in Table 2.

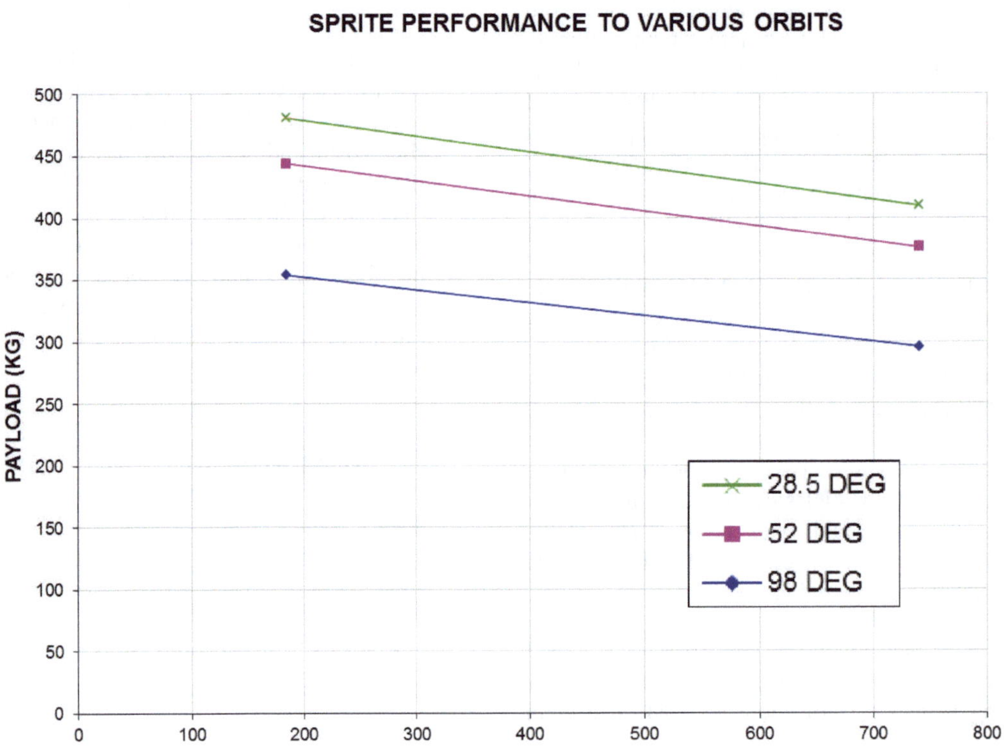

Figure 15. Sprite Performance to LEO.

Table 2. Potential Missions for Sprite.

Mission	Inclination (deg)	Altitude (km)	Payload (kg)	Note
NanoEye	= Target Latitude + 5	200 × 500	434	At 45 deg
Earth Observation	98	740	296	Sun-Synchro-nous
Experimental Satellite	35	400	443	
Comm Satellite	52	615 × 750	383	Satellite = 172 kg
ISS	52	330	426	
GPS Transfer Orbit	55	191 × 20,182	100	
GTO	28	185 × 35,746	168	w/Scorpius Stage 4
Mars Transfer	28	Escape	54	w/Scorpius Stage 4

7. MARKET ANALYSIS

Microcosm has reviewed several recent studies of the market need for low-cost access to space for small satellites. The main sources of information that were found by Microcosm are Snow *et al.*[8], Buchen and DePasquale[9], Bauer *et al.*[10], and Foust *et al.*[11]

SpaceWorks made an assessment of the 2013 global launch vehicle market[8] and an assessment of past and future nano/microstellite launch demand.[9] In particular, SpaceWorks projected the global launch demand in the nano/microsatellite market segment from 2014 to 2020 (note that SpaceWorks placed no value judgment on whether developers will successfully meet their announced launch date). The satellites' masses considered range from 1 kg to 50 kg (Fig. 16); this range can be served by several of the Scorpius® vehicles, in particular Sprite and Demi-Sprite. The Sprite vehicle can potentially deliver several nano/micro satellites to LEO with just one launch. A thorough study for payloads weighing up to 480 kg (i.e., small satellite range, 100–500 kg), which is Sprite's capability to LEO, has not yet been conducted, but Microcosm expects that the market trend will be very similar to that of nano/micro satellites (nominally, 1–100 kg).

The data source for this study is the SpaceWorks Satellite Launch Demand Database (LDDB), a continually updated database cataloging historical and future satellite missions; spacecraft masses included in this database range from less than 1 kg to over 10,000 kg, with over 3,800 historical and planned satellites identified. The nano/microsatellite projection was developed from a combination of two data sets: publicly announced projects and programs, and quantitative and qualitative adjustments to account for the expected sustainment of current projects and programs, as well as the continued emergence and growth of commercial companies.

The projections based on announced and future plans

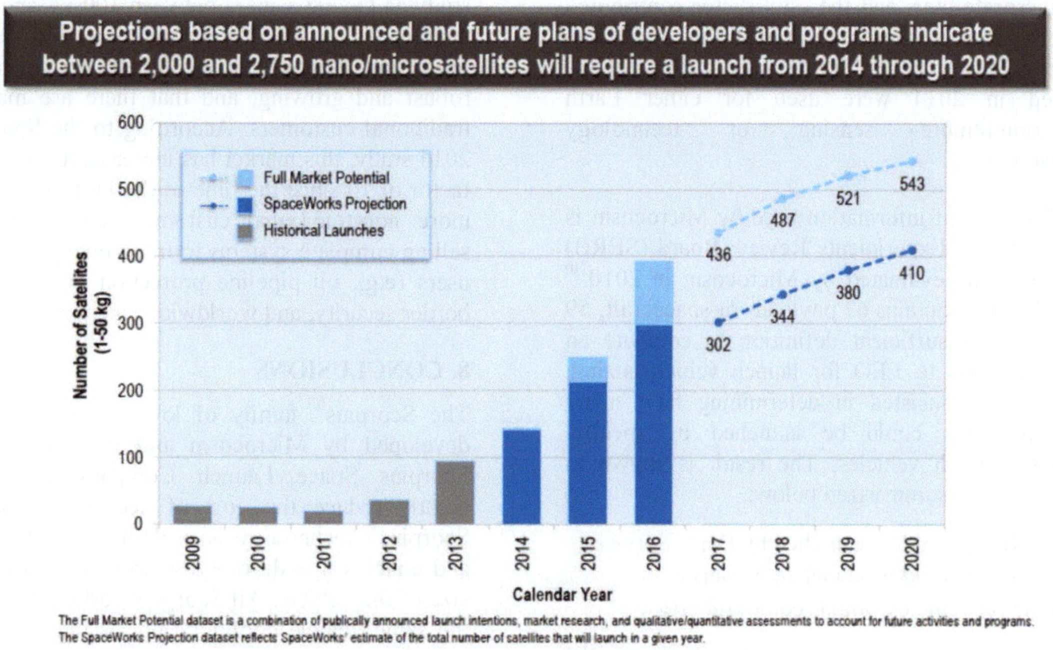

Figure 16. SpaceWorks Assessment of the Nano/Microsatellite Launch Demand. Reproduced from Buchen and DePasquale,[9] with Permision.

of developers and programs indicate that between 2,000 and 2,750 nano/microsatellites will require a launch during the period from 2014 through 2020 (compared to 92 in 2013 alone). According to SpaceWorks, the nano/microsatellite industry continues to thrive, with an estimate of roughly 140 satellites requiring launch during 2014. Additionally, the commercial sector contributed 64% of 2014 nano/microsatellites, and the civil sector contributed ~25%; future launches suggest that this trend will continue. Finally, 91% of the nano/microsatellites launched in 2014 were used for either Earth observation/remote sensing or technology demonstration.

The 2nd source of information used by Microcosm is the DoD Space Experiments Review Board (SERB) list, which was evaluated by Microcosm in 2010.[10] The SERB list contains 62 payloads or spacecraft, 59 of which with sufficient definition to compute an equivalent mass to LEO for launch vehicle sizing. The analysis consisted in determining how many SERB payloads could be launched by specific Scorpius® launch vehicles. The result is shown in Fig. 17 and also summarized below:

- 41 (70%) could be launched by Demi-Sprite
- 50 (85%) could be launched by Sprite
- 9 (15%) vehicles would require Liberty

As the figure shows, currently the knee of the demand curve falls generally near the Demi-Sprite launch capability.

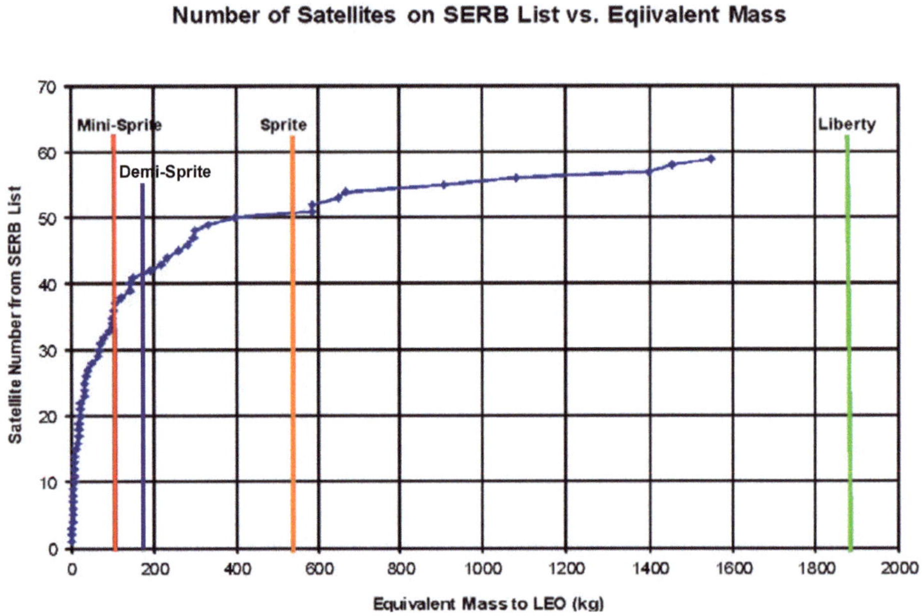

Figure 17. Distribution of Required Scorpius® Launch Vehicle Size for DoD SERB List.
(Mini-Sprite has now been replaced by Demi-Sprite in the launch manifest.)

The 3rd relevant source used by Microcosm for its market analysis is a Futron Study conducted in 2006 for AFRL, and presented at the 2008 USU SmallSat Conference.[11] The study identified over 30 markets in 4 principal areas: military (the largest market), civil/commercial remote sensing, civil/commercial communications, and other. The total addressable market for small satellites (which were defined in the study as having a mass between 100 kg and 200 kg) resulted to be 39 to 76 satellites per year. This projection showed that the SmallSat market is very robust and growing, and that there are many non-traditional customers. According to the SpaceWorks 2014 study, this market has increased by more than a factor of 10 since the time of the Futron study; many more non-traditional customers could come from selling complete systems to traditionally non-satellite users (e.g., oil pipeline protection in Mexico, U.S. border security, and worldwide emergency response).

8. CONCLUSIONS

The Scorpius® family of low-cost launch vehicles developed by Microcosm and its sister company, Scorpius Space Launch Company (SSLC), can greatly reduce the cost of access to space. The Scorpius® technology is scalable to different sizes and enables a wide range of missions based on the size of the vehicle. All Scorpius® orbital vehicles use the same vehicle configuration: 3-stage expendable, 6 outer pods constituting the first stage, a nearly identical inner pod constituting the second stage, and a smaller restartable third stage. (A fourth stage may be added for some missions.) All vehicles use the same key technology elements: a pressure-fed propulsion system based on LOX/Jet A, ablatively-cooled engines, all-composite cryogenic propellant tanks, and a high performance pressurization system based on Tridyne. By having 7 virtually identical pods in each vehicle, even if only a few vehicles are produced per year, a small assembly line can be created, which further reduces cost due to the economies of scale. The unique vehicle architecture offers a very low parts count, virtually no secondary structure, high structural stiffness, large performance margins, low thermal effects sensitivity, high shock and vibration tolerance, and a high controllability launch environment.

This design also has a compounding impact on logistics and operational costs. The smaller vehicles are easy to transport in a standard cargo container; they are also easy to move thanks to their robustness and compact dimensions. They do not need a flame bucket for launch but just a flame deflector, and therefore can launch from virtually anywhere. They are all characterized by a squat configuration, which lowers the moments of inertia and enables greater steering control. They can launch through 100 kt ground winds and 99.9% of winds aloft, thanks to their better controllability and strong structure.

The design for manufacturability is a key aspect of the Scorpius® launch vehicle technology. The Scorpius® manufacturing process consists in setting up, maintaining, and using a low-cost launch vehicle manufacturing environment, with build-to-inventory and launch-on-demand. Manufacturability, low cost, and assured availability are preferred over the traditional space industry's emphasis on performance optimization. A throughput orientation and an industrial production mentality are adopted, versus a traditional mission assurance orientation that is anchored in large systems and manned flight requirements. This approach does not trade away quality or reliability – it trades accommodations to manufacturability (cost) against performance optimization. Additionally, significant cost reductions are achieved by robust design, large margins, modularization, and standardization. Microcosm and SSLC are breaking the chain of perpetual performance optimization.

The Sprite small launch vehicle can deliver up to 480 kg to LEO for less than $6.0M. It is clear that a substantial market for small satellite launches exists, and will almost certainly grow significantly over time as small spacecraft become increasingly competent. The Sprite launch vehicle is expected to fulfill the need of potential customers to launch small satellites by providing access to various orbits and enabling numerous missions. The vehicle greatly reduces the cost of access to space and is very responsive thanks to its capability for launch-on-demand within 8 hours of payload arrival at the launch site. Sprite is expected to introduce a breakthrough, disruptive capability in the launch vehicle market.

9. REFERENCES

1. Bauer, T. 2014. "Lowering the Cost of Transportation beyond LEO with a Scorpius® Depot Architecture," *National Space Society's 33rd International Space Development Conference*, Los Angeles, CA, May 14–18.
2. Chakroborty, S. and T. Bauer. 2004. "Using Pressure-Fed Propulsion Technology to Lower Space Transportation Costs," AIAA 2004-3358, *40th AIAA/ASME/SAE/ASEE Joint Propulsion Conference and Exhibit*, Fort Lauderdale, FL, July 11–14.
3. Sarzi-Amade, N., M. Rufer, and T. Bauer. 2014. "Scorpius® All-Composite Launch Vehicle Technology," *National Space Society's 33rd International Space Development Conference*, Los Angeles, CA, May 14–18.
4. Chakroborty, S., M. Wollen, and L. Malany. 2006. "Development and Optimization of a Tridyne Pressurization System for Pressure Fed Launch Vehicles," AIAA 2006-4716, *42nd AIAA/ASME/SAE/ASEE Joint Propulsion Conference and Exhibit*, Sacramento, CA, July 9–12.
5. Chakroborty, S., J. Wertz, R. Conger, and J. Kulpa. 2003. "The Scorpius Expendable Launch Vehicle Family and Status of the Sprite Small Launch Vehicle," AIAA-LA Section/SSTC 2003-9005, *1st Responsive Space Conference*, Redondo Beach, CA, April 1–3.
6. Wertz, J., R. Van Allen, and T. Barcley. 2010. "NanoEye — Military Relevant Surveillance for Less than $5 Million Total Recurring Mission Cost," RS8-2010-1008, *8th Responsive Space Conference*, Los Angeles, CA, March 8–11.
7. Taylor, C., A. Shao, N. Sarzi-Amade, R. Van Allen, and J. Wertz. 2013. "Hummingbird: Versatile Interplanetary Mission Architecture," *Interplanetary Small Satellite Conference 2013*, Pasadena, CA, June 20–21.

8. Snow, A., E. Buchen, and J. Olds. 2014. "Global Launch Vehicle Market Assessment. A study of launch services for nano/microsatellites in 2013," SpaceWorks Enterprises, Inc., Atlanta, GA, July 17.

9. Buchen, E., D. DePasquale. 2014. "2014 Nano / Microsatellite Market Assessment," SpaceWorks Enterprises, Inc., Atlanta, GA, February.

10. Bauer, T., R. Conger, J. Wertz, and N. Sarzi-Amade. 2010. "Design, Performance, and Responsiveness of a Low-Cost Micro-Satellite Launch Vehicle," RS8-2010-5003, *8th Responsive Space Conference*, Los Angeles, CA, March 8–11.

11. Foust, J., D. Vaccaro, C. Frappier, and D. Kaiser. 2008. "If You Build It, Who Will Come? Identifying Markets for Low-Cost Small Satellites," SSC08-I-1, *Proceedings of the 22nd Annual AIAA/USU Conference on Small Satellites*, Logan, UT, August 11–14.

BIS-RS-2014-60

An Emerging Marketplace: Low Earth Orbit And The International Space Station

Mr. Sam Scimemi
Director, International Space Station
National Aeronautics and Space Administration
Washington, D.C., USA
sam.scimemi@nasa.gov

Low Earth orbit, as seen from the International Space Station in July 2014.

ABSTRACT

The United States has announced plans to continue supporting the International Space Station (ISS) through at least the year 2024. NASA, working with the other ISS International Partners, will continue to foster greater use of the ISS platform, for both research and commercial activities, while using the ISS as a base for expanding the commercial use of low Earth orbit (LEO). In the United States, NASA remains the primary supplier of capabilities and services in LEO, such as habitation systems, power, cooling, crew health equipment, upmass and sample return, research facilities, cold stowage, crew time, and data transmission. Access to LEO for ISS cargo has already been transitioned from a primary government activity to a commercially supplied capability through the development and operations of the Commercial Cargo Services providers. NASA is in the process of developing commercial crew transportation system capabilities also in support to the ISS. It is the goal of NASA to evolve these systems and capabilities through the ISS Program in such a way that they will support market driven commercial research, as well as NASA's long-term exploration plans. NASA will continue to make investments in these areas through at least 2024 to ensure continue access to LEO. This paper will examine the intersection of the growing commercial transportation and research markets, as well as the ways in which the transition from government to commercial activity in LEO might unfold.

S. Hatton (ed.), *Proceedings of the 12th Reinventing Space Conference*, DOI 10.1007/978-3-319-34024-1_14

THE INTERNATIONAL SPACE STATION: A GOVERNMENT OUTPOST READY FOR TRANSITION

The International Space Station (ISS) is humanity's outpost in low Earth orbit. Possessing more electrical power, crew time, data transfer capabilities, and experiment and research housing than any space vehicle in human history, the ISS is the anchor tenant of this most accessible region of space. Occupied continuously from November 2, 2000, through today, the ISS has hosted more than 200 people from 15 countries, and represents the largest and most complicated international engineering effort ever attempted. The pressurized crew modules, built in the United States, Japan, Europe, and Russia, launched onboard the U.S. Space Shuttle and Russia's Proton vehicle, were assembled on-orbit by spacewalking crews using a Canadian robot arm; most of the modules had never met on the ground, and every one connected successfully in space. The amount of crew time available for research aboard the ISS is already equal to every other human spaceflight program in history, worldwide, combined, and will more than double that amount over the next decade, with operations continuing through at least 2024. However, until now, the ISS has been a primarily government undertaking. This is beginning to change dramatically.

NASA's mission on the ISS encompasses four main areas:

- Research many science disciplines aboard this space laboratory
- Establish a global partnership for future exploration beyond Earth
- Serve as a technology development testbed for deep space exploration
- Grow a commercial marketplace in space

This paper will primarily address the fourth area: the ways in which the International Space Station is facilitating the growth of a robust commercial market in low Earth orbit for scientific research, technology development, and human and cargo transportation. The ISS will show that LEO is an emerging marketplace, ripe for commercial development and utilization.

RECENT ACTIVITIES IN THE COMMERCIALIZATION OF LOW EARTH ORBIT

While launch vehicles have traditionally been the realm of governments, or at least government-sponsored providers, actual users of Earth orbit have included many commercial players, from Telstar-1 in 1962, built by AT&T, through the commercial radio, satellite internet and phone service, and Earth observations satellites on-orbit today. NASA has, at various points, attempted to encourage commercialization; much of the justification for building the Space Shuttle was the promise of a cheap, easy way to quickly and reliably launch commercial satellites.

Onboard the ISS, commercial companies have been playing an increasingly important role in many different areas. NASA sponsored the Commercial Orbital Transportation System (COTS) program, which resulted in the development of two new launch vehicles and two new cargo capsules. Today,

SpaceX's Falcon 9 launch vehicle and Dragon capsule are three flights into a 12-flight Commercial Resupply Services (CRS) contract, delivering 20 metric tons of pressurized and unpressurized cargo to the ISS, as well as returning science samples to the ground; Orbital Science Corporation's Antares launch vehicle and Cygnus capsule are two flights into an eight flight contract, also delivering 20 metric tons of cargo to ISS and providing much-needed refuse disposal upon reentry. With the proven and ongoing success of these services, a CRS-2 contract is in work to continue essential cargo delivery to ISS for the duration of its life on-orbit.

NASA, using lessons learned from the COTS development effort, is also providing funding and expertise for a new round of vehicle development, this time to take crew to and from the ISS. Crew transportation, using Russia's historically reliable Soyuz vehicle, stands as the only mission-critical single-fault-tolerant link in the entire Space Station chain. For both mission and national needs, NASA is helping several different private companies develop human-rated vehicles under the Commercial Crew Program. While development is ongoing, first flights are expected in 2017.

Cargo and crew transportation, however, are only the most obvious and visible ways in which the ISS is contributing to the inevitable commercialization of low Earth orbit. With the ISS, NASA has built the runways, radios, and beacons of the 1920s airmail system. It has helped develop the private craft necessary to fly between Earth and space. Now, through the NASA National Laboratory initiative and the Center for the Advancement of Science in Space (CASIS), the market is open for new, commercial developments in nearly every area of research and technology, including communications, biotechnology, human health, habitats, material sciences, Earth and Space science and observations, and other uses that may not even be hinted at yet.

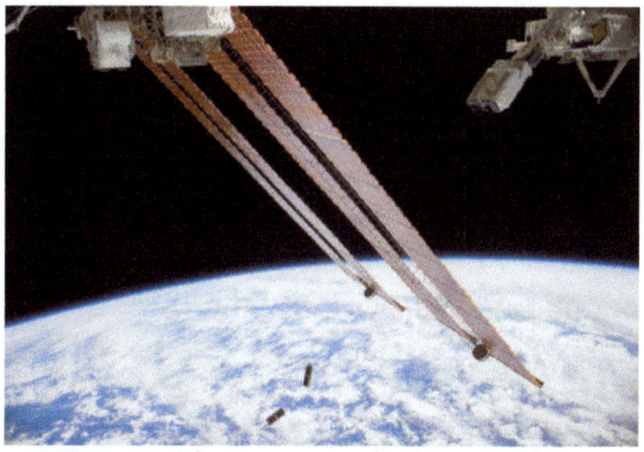

Figure 1 - The Japanese robotic arm, using a Nanoracks deployer, releases a pair of commercial CubeSats on 20 August 2014.

For example, through an arrangement with CASIS, the Nanoracks company of Webster, Texas, has developed a suite of facilities that are now onboard ISS. Other interested parties, from biotech development firms to elementary schools, can contract with Nanoracks and CASIS to use these facilities for whatever purposes they require, from vaccine development to Cubesat launches. Because transportation to ISS is still in a relatively early stage of maturation, NASA provides launch services and on-orbit accommodations for all U.S. National Lab and CASIS payloads, eliminating this not-insignificant risk for smaller developmental organizations.

Also under development and slated for launch in 2015 is the Bigelow Expandable Activity Module (BEAM). Developed by Bigelow Aerospace, the

BEAM will be a demonstration of the feasibility of using lower-cost, lower-weight inflatable modules for human habitation in low Earth orbit. By utilizing the ISS, Bigelow is able to save on development costs of independent power, data, and environmental control systems, while getting experience with actual humans moving around and working in an inflatable structure- something that has never before been demonstrated in space. If BEAM is successfully demonstrated on ISS, it will open the door to large-scale, lower-cost inflatable habitats that can be deployed in low Earth orbit as an eventual commercial successor to the ISS.

Figure 2 - An artist's conception of BEAM attached to ISS (Photo: Bigelow Aerospace)

Nanoracks and Bigelow are only two examples of the many companies currently pursuing research using the capabilities of the ISS. In order to fully understand the potential of the commercial marketplace that is developing in space, we must examine the constraints- financial, technical, and political- to opening up this final frontier.

CREATING AN ECONOMIC DEVELOPMENT ZONE IN LEO

Commercial activity in low Earth orbit, for industries besides communications and remote sensing, has always faced a chicken-and-the-egg problem; supply of transportation and resources cannot exist without demand for these capabilities, but demand for services in LEO cannot develop without a consistent supply chain. Economically viable supply and demand is just one of the barriers or constraints to developing a commercial market in LEO.

NASA, along with its international partners, hopes to help show that LEO is a viable economic development zone, with unique resources that are available to anyone with the ability to exploit them. Similar to emerging markets around the world, LEO is a geographic area with its own benefits and challenges; as with any market, those that can maximize the benefits while mitigating the challenges will profit the most. As the scope and breadth of the activities at the ISS show, this profit can take many forms- economic, research results, technology development, new operational models, and so on. However, like traditional emerging markets, there are barriers to commerce and development in LEO. Unlike traditional emerging markets, however, they are not simply matters of culture and regulatory differences (though these also factor as well).

Despite NASA's efforts to serve as a pathfinder in utilization of LEO, significant barriers still exist outside NASA's control. These include economically viable transportation for cargo and crew; intellectual property rights derived from government activity; and investment and tax incentives to encourage private industry to risk their own capital in LEO. By committing to operating the ISS through at least 2024,

NASA has at a minimum another decade to continue to work with stakeholders through the technical, financial, and policy barriers that inhibit the development of a commercial market in LEO.

Another significant barrier is that of demand for LEO, which is still primarily NASA-driven. Going forward, NASA can guarantee at least 10 more years of operations in low Earth orbit, with the crew and cargo transportation this will require (approximately five cargo and two crew flights per year, at a minimum). NASA can also define areas of operations that are open to commercially-provided services for which NASA can be a customer; for example, the Sabatier carbon dioxide reduction system that is on-orbit was originally a commercial model where NASA purchased the water it produced rather than the hardware that generates it. NASA is in the process of determining what capabilities and/or services that are required for NASA's mission that are also applicable to the transition and development of a LEO commercial market. It is hoped that a commercial demand for services like this will flow out of the proof-of-concept demand that NASA is providing. Until such time, industry is likely to remain dependent on government to provide demand for services in LEO.

Through examples like Nanoracks and BEAM, it is becoming apparent that there is demand for space station-type capabilities in LEO. While many communications and Earth-observation satellites can justify the expense of a free-flying satellite bus, many smaller, experimental or developmental payloads cannot afford that type of investment. However, by using the power, data, crew time, attitude and control, and transportation resources provided by the ISS, experiments and technology development activities can be undertaken that would never before have been economical on their own. As profitable commercial endeavors become routine, it can be expected that other commercial providers will enter the market to provide ISS-like resources in LEO, offering services to allow continued development or production that can only be accomplished in this environment. For some activities, reliable and economically feasible crew transportation will also be a key factor; by 2024, ISS will have proven this capability as well.

For the NASA side of this emerging market, there are steps the government can take to continue the positive path that industry is on. NASA, with input from industry, can begin creating a strategic plan for the gradual transition of being a supplier of many services in LEO- data transmission, environmental control and life support, research facilities- to being a consumer of and customer for these services, and can forecast its own need for continued LEO services post-ISS. As a pathfinder, NASA is able to take the types of risks for new types of operational models that industry cannot. Working with industry, NASA can help identify the areas of ISS operations that can be privatized, in order to provide operational experience for the commercial successors to ISS.

However, NASA still needs help from industry to create the kind of effective strategic plan necessary for this marketplace to flourish. New avenues to create demand need to be identified, especially in areas where government does not or cannot operate, such as space tourism. Methods for protecting intellectual property (IP) created in an international environment by non-employees of the IP creators need to be strengthened. Government and industry need to work together to find ways to create investment incentives for microgravity research and applications. Fortunately, much of this will be accomplished in due course as the full research potential of the ISS becomes apparent. "Mission success" for ISS will be defined as the day when a private space station is launched because there is no room to do more research onboard ISS.

CONCLUSIONS

Spaceflight is an entirely different endeavor than computer development, commercial air transportation, or standard economic emerging markets. It is unrealistic to draw straight-line comparisons between development timeframes for these or any other industries. However, the historical model of government development followed by a transition to commercial utilization and further development is sound. NASA and the space industry are quickly reaching this tipping point, and are beginning to show that low Earth orbit is an emerging commercial marketplace, similar to any other. The next few years will be a critical time for commercial transportation to low Earth orbit. The next decade will also be a critical time for NASA, along with industry partners, to show that research and development in low Earth orbit need not be confined only to the launch and communication industries. Biomedical firms, educational institutions, and technology development efforts need help in realizing the full potential of the amazing facilities and capabilities offered by the ISS. Full utilization of all sectors of low Earth orbit will be required to ensure a vibrant and viable marketplace. Working together, we can achieve new heights.

The constellation Orion rises in this view from the International Space Station, taken in June 2014.

BIS-RS-2014-36

The Austral Launch Vehicle: 2014 Progress in Reducing Space Transportation Cost through Reusability, Modularity and Simplicity

Adriaan Schutte
Heliaq Advanced Engineering
Manchester, United Kingdom;
adriaan@heliaq.com

ESTACA Space Odyssey (ESO) Team
ESTACA
Paris, France;
nicolas.clemencin@estaca.eu

Peter Thoreau
University of Queensland
Brisbane, Australia;
peter@thoreau.com.au

ABSTRACT

The re-use of launch vehicles (RLVs) is accepted as the most promising method for significantly reducing space access cost, increasing responsiveness and increasing reliability. Unfortunately, flight experience has proven otherwise. This leads to the current situation where all operational launch vehicles are expendable, and the majority of LVs under commercial development are also expendable systems.

The Austral Launch Vehicle (ALV) project is an international effort to develop a cost-optimized partially Re-usable Launch Vehicle (RLV). Now in its fourth year, the ALV project originated as an investigation into the additional requirements that will ensure that re-usability can provide real launch cost reduction. These requirements were identified as modularity, flexibility and simplicity. Modularity is required to increase the vehicle flight rate (through larger payload range and more module flights per launch) and to reduce development cost (through reduction of the number and size of newly developed elements). Flexibility is required to ensure a wide market can be captured, while simplicity leads to critical reductions in development and operational costs as well as increased reliability.

The ALV family of RLVs are being developed based on these principles. In the ALV architecture, only the first stage modular boosters are re-used since they represent the bulk of the launch vehicle mass and cost, while being the simplest to recover. The ALV first stage boosters use a deployable wing and aero engine to return to the launch site after re-entry. Re-use of upper stages does not appear economically feasible, except for small satellite launches where the structural mass fraction of expendable stages becomes excessive. For this reason the ALV project is partnered with the SPARTAN scramjet powered, reusable second stage project of the University of Queensland. The SPARTAN uses the ALV boosters as first stage.

The ALV project gathers students and professionals' knowledge from several companies, associations and universities in the UK, Australia, France and South Africa. The project is organized in four phases: Phase 0 studies are almost complete, and Phases 1 and 2 consist of the development of test vehicles of increasing complexity as precursors to the full-scale development in Phase 3. The Phase 1 vehicle currently being designed, the ALV-1, is a small scale, low cost, analogue test vehicle to prove key concepts of ALV architecture. It will also be extensively used to develop and test the avionics for the larger vehicles, and is designed to be simple and extremely low cost. Work on the ALV-1 is progressing well, with prototyping set to commence in 2014 and a flight test planned for the end of 2015.

KEYWORDS: [Reusable Modular Launch Vehicle Flyback Booster]

S. Hatton (ed.), *Proceedings of the 12th Reinventing Space Conference*, DOI 10.1007/978-3-319-34024-1_15

1. INTRODUCTION

The re-use of launch vehicles is the only practical way to reduce space transportation cost. But despite this potential for significant reduction in the cost of space access when compared to expendable vehicles, this has not materialized. Experience with reusable vehicles (e.g. the Space Shuttle) has shown that reusability alone is not sufficient to guarantee that a RLV will have lower overall costs than an expendable vehicle. Evidently some additional design drivers are required to ensure real cost reductions.

This project originated from a study that aimed to identify these additional design drivers. Since the goal was the reduction of mission costs, the study was based on commercial considerations rather than technical. Both sides of the cost equation were evaluated using market analysis and programme cost estimates. The result of this study, which was presented in detail in a previous paper[1], was the identification of three key design drivers that are required for commercial viability. These drivers, namely Modularity, Flexibility and Simplicity, were then applied to a RLV design. The resulting concept is now called the Austral Launch Vehicle (ALV).

Since the original study in 2011, the ALV project has made significant progress. This paper presents an overview of the main design drivers, the selected configuration and its advantages, ALV project execution details as well as recent progress on the ALV-1 test vehicle.

2. OVERVIEW OF ALV DESIGN DRIVERS

This section presents a brief overview of the three key design drivers for the ALV project.

Fundamental Commercial Constraints

The original study identified the two fundamental commercial problems that any new RLV design has to address, namely low flight rate combined with high development costs. These two issues conspire to negate most of the advantage that RLVs offer over expendable vehicles. As a consequence these issues have to be addressed at the earliest phases of project if the design is to be commercially viable.

However, basing a vehicle design on commercial considerations presents a significant problem, since it is not possible to do an economic evaluation of a vehicle that does not yet exist. For the ALV project this problem was overcome by translating the two fundamental commercial constraints (low flight rate and high development cost) into three key design drivers on which the new design was to be based. These design drivers had to be sufficiently generic to allow a wide variety of vehicle configurations to be considered, but also needed to be sufficiently concise to ensure the design addressed the fundamental commercial constraints.

Design Driver 1: Modularity

Modularity is required to increase the flight rate by allowing a multitude of vehicles with wide payload range to be integrated using standardized modules. The development cost of the vehicle is therefore spread over a larger number of flights, and due to the smaller size of each module the development cost is also reduced.

For the vehicle in the original study, the booster flight rate was increased an order of magnitude through the use of modularity. Given the uncertainty in future launch rates, modularity is seen as an absolute necessity for a new rocket-powered RLV design.

Design Driver 2: Simplicity

Even for a modular vehicle the high development costs of reusable vehicles may prove uncompetitive with expendable systems. Only by reducing development costs by a substantial amount will competitiveness be assured. The graph below, taken from the original study, shows how a new RLV design does not meaningfully reduce total programme cost if statistical (US Government) development costs are used (red line).

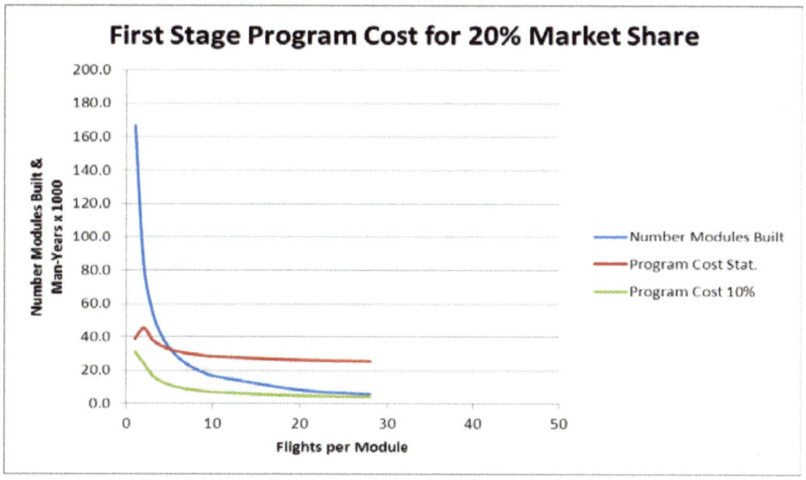

Figure 1: Total Programme Cost Estimates

However, if development costs can be reduced to 10% of the Government costs (as some private space companies have claimed) then the RLV may realize significant savings (as shown by the green line).

Such a cost saving is only possible if the design is simplified to the maximum extent possible. For the ALV project this is implemented by:

- Use of operational concepts that are well proven
- Use of commercially available components, preferable those from the aviation industry
- Simplified operations
- Reduction of the part counts
- Removal of the requirement for high performance systems, parts and materials

An added advantage of simplicity is the gains in reliability when compared to more complex systems.

Design Driver 3: Flexibility

To fully capture the additional launch opportunities afforded by a modular design with a large payload range, the design has to be sufficiently flexible to allow any orbit and mission type. Any reduction in flexibility will inevitably lead to lost launch opportunities, reducing the vehicle flight rate. To be fully flexible, the vehicle must not have any restrictions on:

- Launch direction
- Launch time (including time of day, season, weather conditions within reason)
- Payload orbital elements

3. ALV SYSTEM ARCHITECTURE

Vehicle Overview

The ALV-2 is a modular, three stage, liquid propellant, small satellite launch vehicle of which only the first stage is reusable.

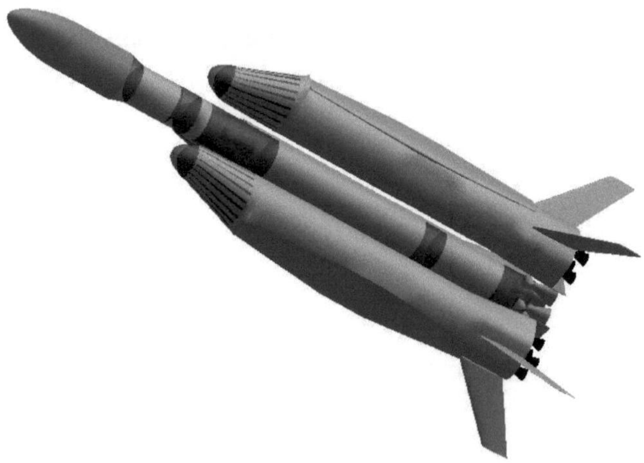

Figure 2: Typical ALV-2 Integrated Launch Vehicle with 3 Boosters

The ALV integrated vehicle consists of from 1 to 6 flyback boosters, a core second stage and a third stage. It should be noted that the core second stage uses either the same technology (including rocket engines and tank construction methods) as the third stage (for 1-3 boosters) or the first stage (for 4 and 6 boosters). The various combinations (i.e. 1, 2, 3, 4 and 6 boosters and matching upper stages) can place from 100kg to over 1,000 kg into typical Sun-Synchronous Orbits.

Operation

The ALV integrated vehicle is launched vertically from a launch pad (refer to Figure 3). After a short vertical section the gravity turn is commenced and the vehicle ascends at approximately zero angle of attack until first stage booster separation. The second stage is ignited some seconds before first stage separation and operates at full thrust during separation. After separation the upper stages place the payload in the desired orbit, similar to an expendable system (note fairing separation is not shown in the figure, this occurs shortly after first stage separation). The second and third stages are not recovered but are destroyed during subsequent re-entry.

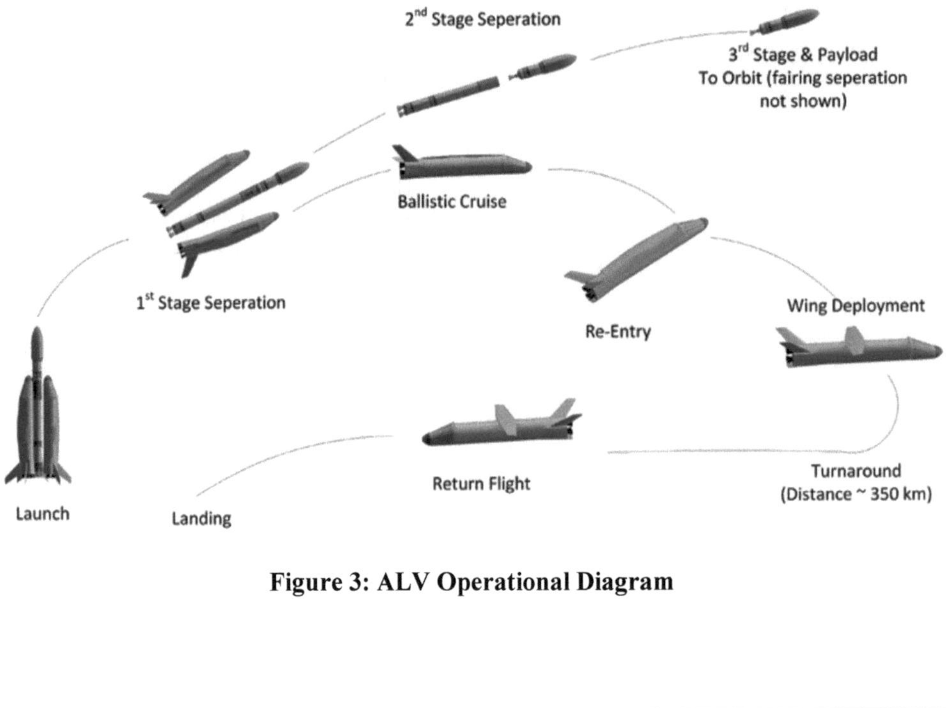

Figure 3: ALV Operational Diagram

After separation, the first stage booster modules maneuver to impart a mutual separation velocity before main engine cut-off. The boosters then coast unpowered during an exo-atmospheric ballistic flight until re-entry (note only one booster is shown for simplicity). During re-entry the vehicle is held at a high angle of attack. Pitch stability provided by the ruddervators as well as a rear body flap for pitch trim. The high angle of attack is maintained until low supersonic speed, when the vehicle enters a shallow dive to cross the transonic region.

Once sub-sonic, the angle of attack is again increased until the vehicle reaches low sub-sonic velocity. At this point the vehicle pitches down and the wing is deployed. Best glide speed is assumed and the engine started. With the engine operational the vehicle performs a turn and follows pre-programmed route of 3D waypoints to the landing approach holding point (a distance of approx. 350 km). From here a pilot takes control and performs a remotely piloted landing of each module in turn.

Minimum refurbishment is required between flights, and once a booster is refurbished it is re-integrated into another vehicle. As described in section 4, the booster rocket engines are only use and average of 5 times on a first stage module after which they are mounted on a second stage and expended. This low number of engine re-uses significantly reduces the development and manufacturing cost of the rocket engines, while still unlocking most of the savings afforded by reusability.

For small payloads, boosters can be integrated with the SPARTAN scramjet accelerator. The operational sequence is similar to that described above, with the exception that the booster does not perform an extended exo-atmospheric ballistic flight but rather decelerates gradually after separation due to the relatively high dynamic pressure. The flyback distance is therefore significantly shorter.

The following sections describe important aspects of the ALV-2 vehicle architecture and highlight the advantages of the selected configuration.

Use of Liquid Propellant Engines

A study of the comparative costs of liquid and solid rocket engines revealed that the operational cost of a reusable solid rocket engine is high, primarily driven by the cost of propellant. Conversely the manufacturing cost of the casing is relatively low, due to the simple nature of the structure. This high ratio of operational cost to manufacturing cost does not justify reusability; which is supported by operational experience from the Space Shuttle and Ariane 5 boosters. With this in mind liquid rocket engines are used for the ALV-2. The use of liquid propellant engines has several other advantages, including very low structural mass resulting in a light vehicle during the return flight.

Re-use of First Stage Only

A core aspect of the ALV architecture is that only the first stage is reused. A cost-benefit analysis showed that only reuse of the first stage is commercially attractive. The fundamental reason for this is the relatively small mass and cost contribution of the upper stages compared to the significantly increased complexity of re-entering vehicles at higher velocity. For example, the first stage mass of a typical three stage vehicle can be 75% or more of the total vehicle mass, making the re-use of the lower stages significantly more commercially important than for upper stages. Furthermore, it is significantly less complex to re-use lower stages than upper stages; a typical first stage separation velocity is around 1,800 m/s, whereas for a third stage it can approach 8,000 m/s. Since the energy that has to be dissipated during re-entry increases as the square of speed, a reusable third stage needs to dissipate up to 16x more energy during re-entry. This requires complex and expensive Thermal Protection Systems (TPSs), which are not justifiable considering the small mass and value of the third stage.

The above evaluation holds true for larger rocket second stages, but for very small satellites (<= 500 kg) fixed costs dominate the expendable stage costs. This results in the cost per kilogram of small satellites being much higher than for larger satellites. For very small satellites it is therefore imperative that the number and size of expendable stages be kept to a minimum. The scramjet powered SPARTAN reusable second stage (which is partnered with the ALV project) shows potential to reduce the cost of small satellite launches. The SPARTAN concept uses a fully reusable second stage in conjunction with a very simple, low cost third stage to reduce mission cost. SPARTAN also has distinct advantages in terms of mission flexibility, reduced propellant (fuel only) cost and ease of reuse.

Figure 4: UQ SPARTAN Scramjet Accelerator and 3rd Stage

Three Stage Configuration

Three stages were selected as the optimal configuration after several vehicle architectures with two to four stages were designed and analyzed. It was found that two stage vehicles require complex engines with high specific impulse and that a TPS is required, as the first stage re-enters at a significantly higher velocity. Both of these requirements significantly increase the complexity of the project and negate any advantages of a reduced number of stages. The four stage configuration is typically only used with solid propellant engines where the incremental development cost of each stage is small and there is little weight penalty for using smaller stages. For liquid propellant vehicles this is not the case and the development cost and overall vehicle mass is significantly increased with multiple small stages. Furthermore, with an increase in size the speed contribution and therefore mass of the first stage decreases, resulting in a lower fraction of the overall vehicle mass being re-used. Three stages provide the optimal combination of simplicity and first stage velocity contribution.

Modularity of First and Other Stages

Modularity is a key feature of the ALV-2 design, allowing multiple configurations with a wide payload range to be integrated at minimal additional cost. This is most visibly manifested in the multiple identical flyback boosters that form the first stage. Modularity is also implemented in several other ways on the ALV-2, for example various sizes of second stages are manufactured exclusively from either first or third stage components, including tanks and rocket engines.

The importance of modularity cannot be overstated. It is the only practical way to increase flight rate in a constrained launch market. The primary mechanism for increasing flight rate are (1) increasing the payload range significantly resulting in higher launch rate and (2) increasing the module flight rate by using multiple modules for each launch. An increased flight rate results in the development cost being amortized over more flights, resulting in a significant reduction of the cost share per flight.

Modularity not only increases the flight rate, but also significantly reduces the development cost of the modules. This is clearly illustrated by the conceptual cost model used for this project (Transcost[2]). The primary input to all cost equations is a parameter related to vehicle mass or engine size. In all cases the development cost of a vehicle or engine is significantly reduced when the mass / size of the vehicle or engine is reduced. In a modular vehicle each module within a stage is significantly smaller (and therefore cheaper to develop) than a unitary stage of the same capability would have been.

Additionally, the technical risk of using smaller individuals modules compared to a large unitary vehicle is lower. Reasons for the decrease in technical risk for smaller modules include:

- Reduced likelihood of combustion instability in small engines
- Significantly reduced cost of exhaustively testing small rocket engines

- Decreased likelihood of structural anomalies including aero elasticity, flutter, oscillations, etc.
- Specialist manufacturing facilities are not required for small structures and engines
- Significantly reduced logistics complexity and cost

Lastly, the higher flight rate of multiple modules results in increased opportunities to monitor the performance of the system and improve the design. This results in the design reaching maturity much quicker than for a large system with a low flight rate.

Use of Flyback Boosters

Reusable boosters can potentially be recovered in a number of ways, including parachute recovery, vertical rocket powered descent, deployment of rotors for a helicopter or gyrocopter style landing, etc. Trade-off studies were performed on all known feasible methods, and the flyback booster concept was shown to be the optimal recovery method. Some drawbacks of the other methods that resulted in their disqualification were:

- Short return flight range resulting in either an ocean landing or a strongly reduced launch azimuth range (most other options)
- High likelihood of damage to tank structures (parachutes, especially without vehicle aerodynamic surfaces)
- Significant technological challenges and complexity and excessive parasitic propellant load (rocket powered descent)

4. ALV-2 FLYBACK BOOSTER CONFIGURATION

This section focuses the ALV-2 flyback boosters that form the first stage. This stage will be discussed in detail, since it is a new design and the focus of the ALV project. The second and third stage will not be explicitly discussed, since they are considered relatively standard.

Propellant Selection and Rocket Engine Design

The ALV-2 boosters use Ethanol and Liquid Oxygen (LOX) as propellants. This combination produces thrust at sufficient efficiency (i.e. specific impulse), while the relatively low temperature in the combustion chamber simplifies engine design. Furthermore the propellants are non-toxic, environmentally benign (in case of an incident), non-polluting and extremely low cost.

The rocket engine uses four chambers fed by a single turbopump. This configuration has several advantages, including avoiding combustion instability due to small chamber size, significantly reducing the cost of developing, manufacturing and testing the rocket chambers as well as the ability to create several second stage engine configurations of varying thrust levels by using fewer chambers and changing the impellers in the turbopump. Sharing the turbopump further reduces the weight of the module.

The original programme cost study clearly showed that there is no commercial justification for flying a reusable booster more than approximately 25 times (refer to Figure 1: Total Programme Cost Estimates). Furthermore, it showed the bulk of the benefit can be gained by re-using the booster only 5-6 times. For the ALV-2 the shared use of rocket engine components on the first and second stages provides a significant opportunity for cost savings. The rocket chambers, turbopumps and other high wear components will be flown on the ALV-1 first stage boosters for approximately 5 flights, before being removed and flown on the expendable second stage. This significantly relaxes the re-usability requirements on the rocket components (a major cost saving) at virtually no additional cost (since second stage engines would have been manufactured in any case).

Note that the actual number of times an engine is flown is dependent on the distribution of the customer payloads and therefore number of first stage modules used in each launch. But since the third stage engines are used on the second stage for integrated vehicles with 1-3 boosters, and the first stage engines are only used as second stage engine for 4 or 6 booster integrated vehicles, the average number of times a first stage is flown can be assumed to be 5 flights. Note that the third stage engines are never re-used.

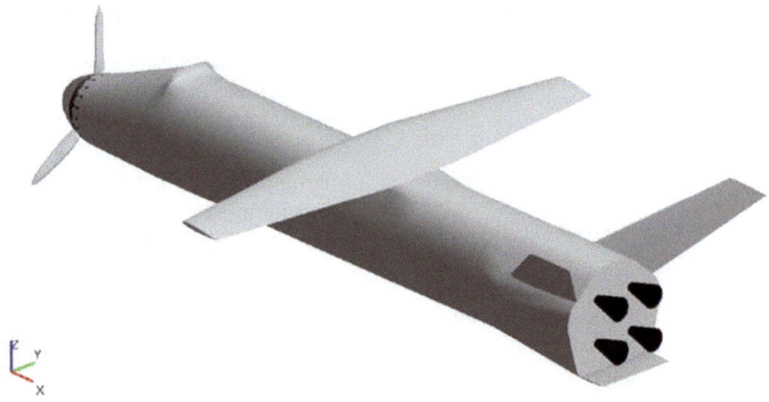

Figure 5: ALV-1 Booster in Aircraft Mode

Influence of Modularity on Wing and Tail Configurations

The modular nature of the first stage places a key restriction on the design of the flyback boosters, namely the avoidance of physical interference between adjacent modules around the core second stage.

Most flyback booster concepts can be classified into two groups, those using delta wings and those using deployable wings. The advantage of delta wings is that they are stationary and do not require a deployment mechanism. But, in a modular vehicle the delta wings will physically interfere with each other when integrated around a core second stage (if more than two boosters are present), requiring them to be folded for launch. This completely removes the simplicity advantage of the delta wings, and makes pivoting wings the preferred option due to the significantly reduced drag during ascent and weight saving of the structurally efficient wing. Of the various types of pivoting wing, the single pivot wing that pivots approximately at the centre is the simplest, strongest and lightest option and thus selected for the ALV-2 boosters. The ALV-2 wing can only be deployed during flight, and has to be stowed on the ground using external force.

Similar to the main wing, the requirement of modularity introduces interference issues for low or mid mounted horizontal stabilizers and elevators. These configurations will require folding control surfaces, which increases the complexity. The only configurations that will not experience interference are the T-Tail and V-Tail. The ALV-2 was originally designed with a T-Tail, but this proved to be aerodynamically ineffective due to body shadowing during re-entry and required heavy vertical tail structures. Recent iterations were re-designed with a V-tail and rear mounted body flap, similar to the system used on the X-37B spaceplane.

Aero Engine

Trajectory analysis showed that the ALV-1 first stage modules will re-enter the atmosphere at approximately 350 km ground distance from the launch site. This large distance will require an aero engine for the return flight and landing. It should be noted that several gliding flyback boosters have been proposed, but the separation speed of these boosters cannot be much higher than Mach 3 in order to glide back to the launch site. The low velocity contribution of such a first stage increases the size of the expendable upper stages, defeating the advantage of a reusable first stage.

Further advantages of using an aero engine includes the ability to make more than one attempt at landing and the capability to loiter in order to allow other modules to land. The aero engine can also be used to land at a runway not co-located with the launch site, increasing the flexibility of the system. Lastly, the aero engine allows a significant amount of vehicle testing to be performed in aircraft mode, an important development consideration.

The aero engine has to be located at the front of the booster to position the Centre of Gravity (CG) at the correct location. Several large boosters have been proposed with turbojet engines located in the nose, using exhaust ducting to route the hot exhaust away from the propellant tanks. This concept has however not been proven and adds weight to the system. In all likelihood the engine air intake will also require a moveable shield during ascent due to the hypersonic air stream velocity.

Furthermore, the ALV-2 boosters have been designed to cruise and land at relatively low speeds (Mach 0.3), increasing safety and reducing drag. At these low speeds jet engines are relatively inefficient. Lastly, turbine engines are expensive and will form a major share of the overall booster's cost, overshadowing the initial motivation for reusing the booster. The piston engine and propeller combination is at least an order of magnitude cheaper than an equivalent turbine engine.

With these and other considerations in mind, a front mounted piston engine and folding propeller system is used. Mounting a piston of the front of an aircraft is common practice, and folding propeller systems have been used on motorgliders for many decades. The ALV-2 system uses a nose cap that is moved forward before the propeller blades are swung open by centripetal force when the engine is started.

Fuel System and Landing Gear

The ALV-2 fuel tank for the aero engine is located in the intertank section near the CG. From here fuel is pumped to the nose mounted engine along the front tank and through a firewall.

The ALV-2 uses a tricycle landing gear arrangement for ease of landing. The main landing gear is stowed in the intertank section while the nose landing gear is stowed in the nose cone. Similar to the wing, the gear can only be deployed in flight, and retraction has to occur on the ground using external force.

5. ALV PROJECT EXECUTION

Project Roadmap

The ALV project is divided into four phases, where the overall goal of phases 1 to 3 is to design, manufacture and successfully test a vehicle of increasing size and complexity. This approach is common in the aerospace industry for the development of new types of vehicles that have not flown previously. There are a multitude of advantages to this approach, including the minimization of overall risk and the incorporation of gains in technical and operational knowledge into the design early in the project.

The initial phase (Phase 0) consisted of several studies and investigations and is now complete. The remaining three phases each have a clearly defined set of objectives and a vehicle (ALV-1, ALV-4 and ALV-2) to be developed.

Phase 1: ALV-1

The ALV-1 is a small, low cost test vehicle that aims to prove some of the main concepts of the ALV architecture, and give developers early access to a flying test bed. The main objectives of the ALV-1 vehicle are:

- Prove the following concepts in the sub-sonic speed range, including
  - Vertical launch and horizontal landing
  - Transitioning from rocket mode to aircraft mode (i.e. wing deployment)
  - Opening of the nose cap, unfolding of the propeller and engine start
  - Autonomous return flight and approach
  - Remotely piloted landing
- Act as a test platform for the avionics package
- Provide development and operational experience to the team

The ALV-1 is designed to be a close analogue of the larger vehicles. However, due to the small scale and low cost of the ALV-1, many sub-systems and components will not be scalable to the larger vehicles (for example the rocket propulsion sub-system). The project is not spending development effort on these "dead-end" sub-systems, and will buy Commercial-Off-The-Shelf (COTS) equipment to perform the required functions.

Work on the ALV-1 is currently nearing the end of the preliminary design phase with the full structural and configuration layout completed and operational procedures (e.g. sequence of events) documented. The first flight of the ALV-1 is planned for the end of 2015. The ALV-1 is described in the following section of this paper.

Phase 2: ALV-4

The ALV-4 will be a significant increase in capability, size and complexity over the ALV-1. This phase has the following goals:

- Prove key ALV-1 operational concepts:
  - Vertical launch and control using rocket thrust vectoring

    - First stage separation (approx. Mach 5 and 30 km altitude)
    - Second stage ignition and guided flight
    - First stage exo-atmospheric ballistic flight
    - First stage controlled re-entry at full velocity (approx. Mach 5)
    - Sonic transition
    - Sub-sonic return flight and landing (as per ALV-1)
- Launch of the SPARTAN-1 scramjet accelerator (approx. Mach 5 and 25 km altitude) and return flight
- Prove the design of key sub-systems:
    - Liquid rocket engines
    - Chamber steering mechanisms
    - Propellant tanks and pressurization systems
    - Heat shielding

The ALV-4 is in conceptual design stage and is awaiting the finalization of the SPARTAN-1 conceptual design before proceeding. In the meantime several technical studies are being performed to identify suitable technology options to allow the vehicle to be built and tested on a restricted budget.

The first stage flyback boosters and the second stage will utilize identical technologies, thereby reducing development cost. The intent is not to construct a third stage for the ALV-4 in order to limit project cost and technical effort, although this decision might be revisited in future. A simple third stage will be developed for SPARTAN-1.

Phase 3: ALV-2

The ALV-2 will be the full scale small satellite launch vehicle. The ALV-2 will be capable of placing satellites from 100 kg to 1,000 kg in typical sun-synchronous orbits. The ALV-2 design is nearing the end of conceptual design stage. The major body of outstanding work is a detailed aerodynamic study of the ALV-2 geometry. This work will be the subject of a PhD at the University of Queensland from 2015 onwards.

Funding

The ALV project has not required any external funding to date. This has been made possible by the academic nature of the project and the hard work of more than 20 contributors and volunteers over the last 4 years. As an academic endeavor the ALV project has also benefitted these students in gaining real world aerospace design experience.

It is expected that ALV-1 development and testing will be completed without any requirement for external funding. It is also expected that the bulk of the ALV-4 development work will proceed without external funding. However, manufacture and testing of the ALV-4 will most likely require some level of funding, depending on the final size of the vehicle.

The ALV-2 will be a large scale engineering project that will not be suitable as an academic exercise. Although some design work on the ALV-2 might proceed regardless, full-scale development will only be possible through significant private and / or government investment.

Contributors and Locations

The ALV project is an international academic collaboration spread across four continents. The project, managed by Heliaq Advanced Engineering, uses industry accepted Project Management and Systems Engineering procedures. Avionics development work is carried out by Heliaq in the UK.

Design work on the aero propulsion system and all components firewall forward (nose cone, nose cap and also the nose landing gear) is performed in Australia by current and former students from the University of Queensland (UQ). The ALV project is also tightly integrated with UQs hypersonic research SPARTAN-1 scramjet powered accelerator project, and UQ staff contribute significantly to both projects.

Design work on the fuselage, wing and ruddervators is performed in France by the ESTACA Space Odyssey (ESO) student group. ESO has a rich history in the launch of sounding rockets, and the ALV-1 project is an evolution of this work.

A vehicle manufacture and testing workshop is being built by Heliaq in South Africa. This workshop relies heavily on computer controlled (CNC) equipment for parts manufacture. Ground infrastructure for the ALV-1 and ALV-4 is also being prepared in South Africa.

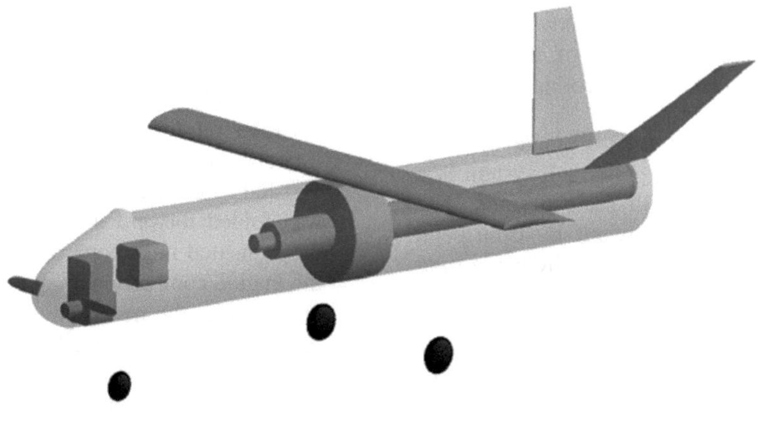

Figure 6: ALV-1 Test Vehicle in Aircraft Mode (Wing & Landing Gear Deployed)

6. ALV-1 VEHICLE DESCRIPTION AND DEVELOPMENT PROGRESS

This section describes the design and operation of the ALV-1 test vehicle. ALV-1 development has progressed well over the last year, to the point where prototyping will commence before the end of 2014.

Please note the Figure 6 is only a conceptual design sketch. Detailed CAD models are shown in the accompanying presentation.

ALV-1 Concept of Operation

The ALV-1 will be vertically launched using rocket propulsion with its wing, propeller and landing gear stowed. The launch elevation angle will be less than 90 degrees to ensure optimal conditions for mode transition (from rocket to aircraft mode) at the trajectory apogee. Control during ascent will be limited to small corrections near the apogee to ensure the correct attitude and velocity for transition.

The transition to aircraft mode involves the wing opening, engine starting and propeller deployment. From here the ALV-1 will follow pre-programmed waypoints to a holding pattern from where the approach for landing will commence. The approach and landing will be flown, using remote control, by a pilot on the ground.

ALV-1 Specifications

The following approximate specifications are still subject to change up to the Critical Design Review:

- Physical:
  - Overall Length: 2000 mm
  - Fuselage Diameter: 275 mm
  - Gross Weight: 40 kg
  - Landing Weight: 25 kg
- Performance:
  - Max. Velocity: Mach 0.9
  - Apogee Altitude: 2500 m
  - Endurance (Aircraft Mode): > 2 hours

ALV-1 Rocket Propulsion and Launch

The ALV-1 design does not attempt to develop or use any rocket components that can potentially be used on the larger vehicles. Since the ALV-1 rocket propulsion system is not required to resemble that of the larger vehicles, a Commercial-Off-The-Shelf (COTS) solution was selected. Advantages of this approach over a new design include: significantly reduced effort, increased reliability and very low cost.

No commercial liquid rocket motors are available in a sufficiently small size to suit the ALV-1. Furthermore, hybrid rocket motors of the correct size have significantly reduced performance compared to solid rocket motors, and are not as widely available. As such a simple commercial solid rocket motor intended for high powered amateur rocketry is used for the ALV-1 ascent. No attempt is made to provide thrust vectoring during ascent, as this mechanism will not be transferable to the larger vehicles.

ALV-1 Aero Propulsion

In contrast with the rocket propulsion, the ALV-1's aero propulsion system is being designed to closely approximate that of the larger vehicles. The aero propulsion system is located at the front of the vehicle to offset the weight of the heavy rear mounted rocket motor. The aero propulsion system makes a significant contribution in moving the CG forward to the approximate centre of the vehicle, which is a prerequisite for using a pivoting wing.

The aero propulsion system uses a single cylinder piston engine. The ALV-1 propeller sub-system is a close approximation of the full scale designs. The propeller blades are swung outward by centripetal force after engine start and held open by blade drag and centripetal force.

Since the aero engine is required to be started in flight, a starter / generator system is installed. The starter / generator is connected to the same power transfer system that connects the engine and propeller shafts, resulting in little additional weight. Power flow is reversed once the engine is started to generate power for the on-board control systems, ensuring flight endurance of several hours.

ALV-1 Firewall, Nose Cone and Cap

The aero propulsion system is contained in the truncated nose cone, forward of the firewall. The firewall, nose cone, nose cap and aero propulsion system is being designed as an integrated package that can be built and tested independently from the rest of the vehicle. All structural components are manufactured from machined or sheet metal aluminium alloy.

The engine is mounted to the firewall at the front, with the main avionics stack mounted to the rear facing the fuselage compartment. All penetrations through the firewall are by means of sealed connectors and fittings. The nose landing gear is mounted on the lower rear of the firewall.

The nose cone is a structural assembly that supports the propeller shaft and the power transmission system. The nose cone also has and air intake and internal baffles for engine cooling, with the hot air exhausted at the top of the nose cone from the wingtip protection protuberance. Note that the engine is only started once the wing is deployed and the protuberance opening is clear. The nose cap is moved forward with a powered actuator to expose the propeller blades before engine start.

ALV-1 Fuel System

The ALV-1 uses unleaded fuel for the piston engine, and the fuel system has been designed to comply with light aircraft requirements. The fuel is stored in a fuel tank located near the vehicle's CG. The fuel tank has been designed to operate safely in both vertical (rocket mode) and horizontal (aircraft mode) attitudes.

ALV-1 Landing Gear

The ALV-1 is equipped with tricycle landing gear that can only be extended during flight. No brakes are installed on the ALV-1. The main gear is folded forward into the fuselage so that gear opening is assisted by drag and gravity. Originally, the nose gear was folded forward into the engine compartment in the truncated nose cone. Due to space constraints in the engine compartment, and the landing gear design progressing, the nose gear was moved and mounted on the rear of the firewall. The nose gear is steerable, and extends with a sprung system strong enough to overcome aerodynamic drag and then locks in place. During flight landing gear doors cover the bays ensuring that the aerodynamics are not compromised.

ALV-1 Wing

The ALV-1 wing is a single composite assembly that rotates about a pivot point on the fuselage. The wing is held closed during ascent by a latch at the front wingtip. At apogee the latch is released and the wing is opened by a pre-charged mechanism. Once fully open the wing is latched by a locking mechanism at the pivot point. The wing can only be deployed during flight, and it has to be manually stowed on the ground using externally applied force. The wing is equipped with full length flaperons for roll control and for the increasing lift coefficient during landing.

ALV-1 Ruddervators

Pitch and Yaw control of the ALV-1 is accomplished through two all-moving ruddervators on the top of the empennage arranged as a V-Tail. These control surfaces are used for vehicle attitude control in both rocket and aircraft modes.

The larger ALVs are also equipped with a body flap at the rear of the empennage, below the rocket motors. This is used solely for pitch trim during the high angle of attack hypersonic re-entry, similar to the X-37B and Space Shuttle. However, since the ALV-1 is a sub-sonic vehicle, the rear body flap has been omitted to save weight.

ALV-1 Fuselage

The ALV-1 fuselage is externally geometrically similar to the larger ALVs. It is manufactured from riveted aluminium alloy sheet and machined components, using traditional aircraft construction methods.

ALV Avionics

The ALV-1 uses the full avionics stack of the larger ALVs. In fact, provision of an analogue test vehicle for development and testing of the full ALV avionics stack is one of the primary drivers for building the ALV-1.

The multiple flight modes (rocket, re-entry vehicle and aircraft), wide operational envelope (up to Mach 5 and 90 km altitude) and large gross vehicle mass range (from < 50kg for ALV-1 to >15 ton for ALV-2) of the ALV family of vehicles presents a significant challenge to the design of the avionics stack. In essence the avionics stack must be capable of controlling almost any type of vehicle of any size (within reason) over a very wide operational range. There is thus a requirement for a universal control system architecture that is equally applicable to vehicles types ranging from a light UAV to large Launch Vehicle.

To address this challenge, the Heliaq Aerospace Vehicle Bus (HAVBUS) control system architecture has been developed. HAVBUS is an architecture that allows modular and scalable control system to be configured that can be physically concentrated (for small vehicles) or distributed around a large vehicle's airframe.

The HAVBUS architecture uses one or more lightweight stacks to aggregate modules that can be located in a single physical location. Stacks are then interconnected using redundant vehicle and payload data buses, as well as power, to form a single distributed control system. The functioning and configuration does not change with physical location, allowing arbitrary control systems to be configured.

A HAVBUS interface specification has been produced and may be released for public use in the future.

7. CONCLUSIONS

The ALV project originated from a study to identify the key drivers for a cost effective Reusable Launch Vehicle. Using these drivers (modularity, simplicity and flexibility), a family of cost optimal partially reusable launch vehicles were designed. The ALV architecture has numerous advantages that promise significantly reduced costs compared to expendable vehicles. The ALV flyback boosters can be integrated with the scramjet powered SPARTAN second stage, resulting in a very small satellite launch vehicle with two re-usable stages and cost effective operations.

The ALV project has progressed significantly over the last year. Moving on from the initial studies, the conceptual design of the ALV-2 is nearing completion. Moreover, work on the small scale ALV-1 test vehicle is progressing well and the design is maturing rapidly. The ALV-1 will prove key aspects of the ALV architecture, including transition to aircraft mode and avionics testing, while providing development and operational experience to the team. Prototyping of ALV-1 is due to commence in 2014, and a first flight is planned for the end of 2015.

REFERENCES

1. Schutte, A.N. and Tolyarenko, N. September 2013. "The Austral Launch Vehicle: Reducing Space Transportation Cost Through Reusability, Modularity and Simplicity", *Proceedings of the 64th International Astronautical Congress, Beijing, China.*
2. Koelle, D. 2011. "Handbook of cost engineering for space transportation systems with TRANSCOST 8.1: Statistical-analytical Model for Cost Estimation and Economical Optimization of Launch Vehicles", *Ed. 8.1 Rev 3a, Germany.*

Quantifying the Cost Reduction Potential for Earth Observation Satellites

Anthony Shao
Microcosm/University of Southern California

Elizabeth A. Koltz
University of Southern California

James R. Wertz
Microcosm/University of Southern California

12th Reinventing Space Conference
18-20 November 2014
London, UK

S. Hatton (ed.), *Proceedings of the 12th Reinventing Space Conference*, DOI 10.1007/978-3-319-34024-1_16

BIS-RS-2014-40

Quantifying the Cost Reduction Potential for Earth Observation Satellites

Anthony Shao and James R. Wertz
Microcosm/USC, 4940 W. 147th Street, Hawthorne CA 90250; 1-310-219-2700
ashao@microcosminc.com; jwertz@microcosminc.com

Elizabeth A. Koltz
University of Southern California (USC), 854B Downey Way, Los Angeles, CA, 90089; 1-310-968-0452
chapman@usc.edu

ABSTRACT

In the present budget environment, there is a strong need to dramatically drive down the cost of space missions. There is the perception that SmallSats are inherently much lower cost than more traditional larger satellites and can play a central role in reducing overall space mission cost, but this effect has been difficult to quantify. Without quantifiable evidence of their value, SmallSats are under-utilized as a method for reducing space mission cost. The purpose of this study is to quantify the relationship between cost and performance for space systems, by creating a *Performance-Based Cost Model (PBCM)*. Today, most acquisition performance analyses focus on cost overruns, or how much the system costs relative to what it is expected to cost. Instead, PBCM allows us to focus on more important questions, such as, how much performance we can achieve for a given cost, or what the cost is for a given level of performance. In this paper, we present the relationship between cost vs. orbit altitude for a fixed resolution and coverage requirement, cost vs. resolution, and cost vs. coverage. Traditional cost models for space systems are typically weight-based, primarily because mass allocation is determined early in mission design and has historically correlated well with actual hardware cost. To provide the underlying cost data for this study, we apply 3 cost models widely used throughout the aerospace cost modeling community: the Unmanned Space Vehicle Cost Model (USCM), the Aerospace Corp. Small Satellite Cost Model (SSCM), and the NASA Instrument Cost Model (NICM).

Our first application of the PBCM is for Earth observation systems. Past Earth observation systems have used traditional space technology to achieve the best possible performance, but have been very expensive. In addition, low-cost, responsive dedicated launch has not been available for SmallSats. Space system mass is proportional to the cube of the linear dimensions—equivalent to saying that most spacecraft have about the same density. This means that by flying at lower altitudes, satellites can reduce their payload size and therefore the entire mass of the satellite, thus reducing the cost of the system dramatically. We conclude that for an Earth observation system, an increase in performance, reduction in cost, or both, is possible by using multiple SmallSats at lower altitudes when compared to traditional systems. Specifically,

- By using modern microelectronics and light-weight materials such as composite structures, future SmallSats observation systems, **operating at a lower altitude than traditional systems**, have the potential for:
  - Comparable or better performance (resolution and coverage)
  - Much lower overall mission cost (by a factor of 2 to 10)
  - Lower risk (both implementation and operations)
  - Shorter schedules
- Relevant secondary advantages for the low-altitude SmallSats include:
  - Lower up-front development cost
  - More sustainable business model
  - More flexible and resilient
  - More responsive to both new technologies and changing needs
  - Mitigates the problem of orbital debris

The principal demerits of the approach are the lack of low-cost launch vehicles, the need for a new way of doing business, and changing the way we think about the use of space assets. This paper provides the basis for this assessment, estimates for the level of cost reduction, and reports on additional results since the 2013 Reinventing Space Conference and AIAA Space 2014 Conference.

KEYWORDS: Reinventing Space, Cost Reduction, Observation Satellites, Low-altitude, SmallSats, Cost Estimation, Cost Modeling

I. Background

At the start of the space program in the 1960s, spacecraft were inevitably massive and large given the technology constraints at the time. As a result, spacecraft were very heavy and flew in a higher altitude regime due to the dense atmosphere at lower altitudes. Above approximately 500 km altitude, it is relatively easy for a satellite to stay in a circular orbit above Earth for many years to several hundred or even thousands of years [Wertz, Everett, Puschell, 2011]. This capability allowed engineers to design spacecraft for long on-orbit lifetimes, typically in the range of 5 – 15 years. Because spacecraft had to last this long, many processes and requirements were put in place to ensure that the spacecraft, its subsystems and parts were above certain reliabilities (i.e., > 99.99%). Parts redundancy and testing was a method utilized to increase reliability. However, this further increased the cost of the spacecraft, and thus the overall cost of the mission. In turn, schedules were elongated due to all the processes, testing, and design reviews. The ever-increasing cost of space missions leads to longer schedules and fewer missions. This leads to a demand for higher reliability, which, in turn, leads to higher cost, longer schedules, and fewer missions. This current mentality is represented by Fig. 1. We believe that space systems today have the following major problems: (1) they cost too much, (2) they take too long to build and launch, and (3) they are not as responsive or robust as they should be. It is often assumed that in order to reduce cost, you must chose to reduce performance or reliability, for example. This research on SmallSats will show how the claim "faster, better, cheaper—pick any two" is flawed.

Currently, there is a clear and present budget problem that must be addressed. Arati Prabhakar, DARPA's Director, was quoted in Space News [2014a] saying there is "something going on inside the national security community in space that's actually quite troubling, that has to do with how slow and costly it is for us today to do anything we need to do on orbit for national security purposes." The USAF has announced a series of studies to determine the future of its big satellite programs. General Shelton was quoted in Space News [2014b] stating, "Do we want to continue with the military dedicated constellation? Can we turn either a portion or all of this over to a commercial provider and contract for a service?" To add context to these remarks, the commercial providers Gen. Shelton refers to have offered the same technologies but at less cost. Mark Valerio, VP of Lockheed Martin's Military Space business, quoted in Space News [2014b] saying, "We're looking at innovative options for hosting payloads, and we are suggesting ways to reduce costs while maintaining our technology edge to address evolving threats." Google has also demonstrated that the demand for low cost satellite imagery is high by announcing plans to buy Skybox, a company that makes Earth imaging microsatellites [Space News, 2014c].

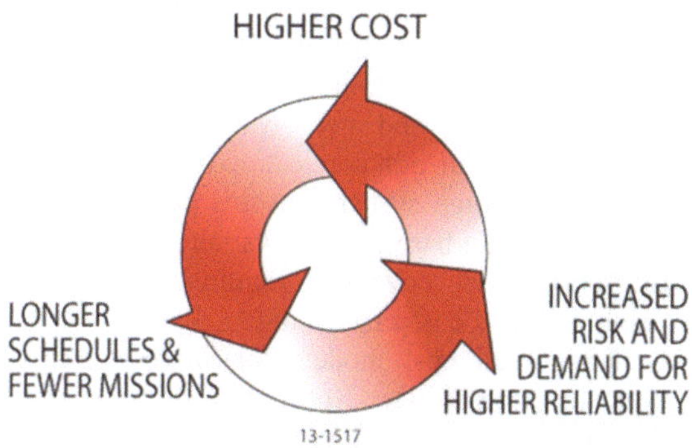

Figure 1. The Space Spiral [Wertz, Everett, Puschell, 2011].

In the present budget environment, likely extending into the future, there is a strong need to drive down the cost of space missions. The main goal of this paper is to quantify the relationship between cost and performance, or measures of effectiveness (MoEs) and determine ways to reduce space mission cost for Earth Observation systems. This cost and performance relationship can ultimately allow us to pursue potentially useful mission design alternatives, such as systems that are lower cost, have better performance, or both. Questions that would be useful to ask when designing a system are:

- What is the cost per level of performance (e.g., cost/resolution, cost/coverage rate, cost/photo)?
- What is the best performance that can be achieved for a fixed cost?
- What is the lowest cost option for a mission with fixed requirements?

Performance-Based Cost Modeling (PBCM) is a mission engineering approach to enable programs to be able to ask these questions early in the design phase in order to drive down cost from the outset. In this paper, we will explore how various factors such as satellite size and orbit altitude affect the cost of space mission. In Sec. II, we introduce the PBCM approach, discuss the technique used to perform the study, and show how we can quantify the relationship between cost and performance. The results are then presented in Sec. III.

II. Performance-Based Cost Modeling (PBCM)

Today, most acquisition performance analysis focuses on cost overruns, or how much the system cost relative to what it is expected to cost. PBCM allows us to instead focus on the more important questions of how much performance we can achieve for a given cost, or what the cost is for a given level of performance. The goal of PBCM is not to create a new cost model, but to use existing and widely used cost models to find new ways to reduce space mission cost. Our first application of the PBCM is for Earth observation systems. In this paper, we present the relationship between cost vs. altitude (for a fixed resolution and coverage requirement), cost vs. resolution, and cost vs. coverage.

Traditional cost models for space systems are typically weight-based, primarily because mass is determined or assigned early in mission design and has historically correlated well with actual hardware cost. To provide the underlying cost data for this study, we use three cost models widely used throughout the aerospace cost modeling community [Apgar, 2011]:

- Unmanned Space Vehicle Cost Model (USCM8) [Tecolote Research, 2002]
- SmallSats Cost Model (SSCM) [Aerospace Corp., 1996]
- NASA Instrument Cost Model (NICM) [Habib-agahi, 2010]

Our goal is to determine cost as a function of performance for an Earth Observing (EO) system. To do this, we predict the life-cycle costs by using the models listed above (USCM8, SSCM, and NICM) and define the performance as measured by two parameters: (i) the resolution at nadir, and (ii) the area coverage rate. For a baseline mission, we will assume the following performance:

- Imaging in the visible
- Resolution = 0.5 meter (at nadir)
- Area Access Rate = 14,200 km^2/sec
- Mission Duration = 8 years

Cost can then be measured by the cost per year to achieve this level of performance.

The coverage rate of 14,200 km^2/sec corresponds to the area access rate (AAR) of a system in a circular orbit at 800 km with a minimum working elevation angle of 30 deg. In order for a satellite at this altitude to meet the 0.5 m resolution requirement, a system with diffraction-limited optics will have a 0.88 m aperture telescope. We will define this system with an 8-year design life as our baseline. If the satellite life at a particular altitude is, for example, 4 years, then we will need twice as many satellites to cover the full 8-year mission duration. Similarly, if the coverage at a given altitude is one third of the baseline value, then we will need triple the number of satellites to provide the baseline coverage.

In order to achieve the same resolution with diffraction-limited optics, we vary the aperture size in direct proportion to the altitude. Thus, at 400 km, we use an aperture of 0.44 m to achieve the same 0.5 m resolution. We assume that mass is proportional to the cube of the linear dimensions, which translates to assuming that the spacecraft dimensions scale linearly with the aperture and that the density of the various spacecraft are approximately the same (validated by Reeves [1999]). Our baseline spacecraft dry mass at 800 km is then estimated to be 1,559 kg, corresponding to a typical observing satellite at that altitude. (The actual value has very little effect on the results when comparing costs, since it is the ratio of the masses that matters.)

At lower altitudes, we assume a shorter satellite design life. To make the model simple, we assume a design life proportional to the altitude, such that the design life is 8 years at 800 km, 4 years at 400 km, and 2 years at 200 km. Therefore, we will need more satellites at lower altitudes due to the shorter design life and the reduced coverage. Because the design life is shorter, we can assume less redundancy, and therefore lower mass at lower altitudes. We essentially reduced the mass per satellite as a function of altitude and also required 10% more satellites to cover potential launch failures.

Finally, there are financial issues associated with the satellite lifetime and the number of satellites required for the mission. We have defined an upfront cost equal to the non-recurring development cost plus the first production unit, often called the *theoretical first unit* (*TFU*). The remainder of the spacecraft are built assuming a 90% learning curve, which is conservative for space systems [NASA, 2008a]. An advantage to building multiple satellites is that they don't all have to be built prior to the first launch. The production of the satellites can be spread out over time and, therefore, paid for over time. For this effect, we have initially used an 8% interest rate and an impact of amortization of a 19% reduction in cost for units built after the first one, based on the results of a 90% learning curve analysis over the 8 year mission duration [Shao and Koltz, 2013]. The list of numerical assumptions is shown in Table 1.

In summary, the steps for PBCM are as follows:

1. Identify the numerical performance requirements
2. Size the payload required to meet the desired resolution
3. Size the spacecraft bus to support the payload

4. Determine the spacecraft wet mass
5. Determine the number of satellites required for coverage and lifetime requirements
6. Input mass estimates into weight-based cost model Cost Estimating Relationships (CERs) to predict costs
7. Determine launch cost
8. Determine recurring and non-recurring engineering (NRE) costs
9. Estimate total mission cost

A more detailed description of each of these steps can be found in Shao et al. [2013] and Koltz et al. [2013].

We have not taken into account any variations in ground system performance on cost. To first order, we do not anticipate any major changes due to the ground system. Recall that the initial assumption was that the resolution and coverage rate were both held constant as the altitude changed. This implies that the data rate will also remain constant. Changing the altitude and the satellite lifetime will have three principal secondary effects:

- At lower altitudes the time in view of a single ground station will be less, but there will be more satellites viewing more ground stations
- At lower altitudes the same power-aperture on the spacecraft will result in higher data rates
- With shorter lifetimes and newer technology, the ability to store data on board the spacecraft becomes much higher (and will be effectively unlimited in the future)

The net effect is that we do not anticipate a substantive impact on the cost or performance results of the study due to the ground station, although it could result in a somewhat further reduction in cost for the lower altitude system.

III. Results for Earth Observing Systems

We selected three mission altitudes of 200 km, 400 km, and 800 km and applied the technique and assumptions described Sec. II. We also have provided three real observation system examples for reference, which include NanoEye [Wertz, Van Allen, and Barcley, 2010], Quickbird [Digital Globe, 2013; Spaceflight Now, 2000], and GeoEye-2 [GeoEye, 2013; Space News, 2012]. We start off by determining the payload aperture diameters using diffraction-limited optics and we see that the aperture is linearly proportional to the mission altitude (i.e., 0.22 m at 200 km, 0.44 m at 400 km, and 0.88 m at 800 km). As can be seen in Table 2, the payload power and datarate scale proportionally to the mission altitude as well. **For a fixed resolution, the spacecraft mass required at 200 km is 17 kg, but is almost 2 orders of magnitude larger (1,559 kg) at 800 km.** This is a very significant difference in mass and will generate a substantial difference in mission cost, as will be seen in Table 3a and 3b.

Table 1. List of numerical input assumptions.

Assumptions	Value
Resolution (m)	0.5
Area Access Rate (AAR) at 800 km Altitude (km^2/s)	14,217
Mission Duration (yrs)	8
Wavelength to Observe (nm)	550
Spacecraft/Payload Average Density (kg/m^3)	79
Propellant Density (kg/m^3)	1000
Dry Mass/Aperture[3]	2287
Payload % of Total S/C Dry Mass	31%
Spacecraft Power/Spacecraft Dry Mass (W/kg)	1.3
Payload Power Percentage of Spacecraft Power (W)	46%
Spacecraft Datarate at 800 km Altitude (kbps)	800,000
Drag Coefficient	2
Solar State (Min, Mean, Max)	Mean
Minimum Working Elevation Angle (deg)	30
Percentage of Launches that Fail	10%
Min. No. Sats for No System Redundancy	2
Spacecraft Propellant I_{sp}	235
Learning Curve	90%
Interest Rate	8%
Cumulative Savings Effect of Amortization	19%

The area access rate (AAR) is less at lower altitudes, and therefore will require additional satellites to satisfy the coverage rate requirement of 14,200 km^2/sec. To support the same coverage rate a single satellite at 800 km, requires 2.9 satellites at 200 km and 1.6 satellites at 400 km. Then, based on the design life of each spacecraft and accounting for launch failures, we determine the number of satellites required for the entire 8-year mission. For the baseline mission providing 0.5 m resolution in the visible at 14,200 km^2/sec, for 8 years, our 3 options are:

1. 1 traditional large satellite (1,559 kg) flown at 800 km
2. 3.6 moderate-size satellites (156 kg each) flown at 400 km
3. 12.9 SmallSats (17 kg each) flown at 200 km

The projected cost values, in constant year dollars, for several cost items using USCM8 and NICM are displayed in Table 3a, and for comparison using SSCM in Table 3b. The key cost values here are:

- The total upfront cost (line 2)
- The remaining recurring cost with learning curve (line 6)
- The total adjusted system cost after amortization (line 12)

Table 2. Physical Parameters of 3 Select Mission Altitudes and 3 Example Observation Systems.

	Physical Parameters	Model Predictions			Examples		
					NanoEye	Quickbird	GeoEye-2
1	Orbital Altitude (km)	200	400	800	215	482	681
2	Resolution (m)	0.5	0.5	0.5		0.65	0.32
3	Payload Aperture Diameter (m)	0.22	0.44	0.88	0.23	0.60	1.10
4	Spacecraft Dry Mass (kg)	24.4	194.8	1,558.6	23.0	995.0	2,086.0
5	Non-Redundancy Mass Reduction	30.0%	20.0%	0.0%			
6	Corrected Spacecraft Dry Mass (kg)	17.0	155.9	1,558.6			
7	Spacecraft Wet Mass (kg)	292.5	181.2	1,559.4	76.4	1,028	2,540
8	Payload Power (W)	10.2	93.2	932.0			
9	Payload Datarate (kbps)	273,345	489,309	800,000			
10	Spacecraft Area Access Rate (km^2/sec)	4,858	8,696	14,217	5,177	10,034	12,819
11	Satellite Orbital Period (min)	88.5	92.6	100.9	88.8	94.2	98.4
12	Spacecraft Design Lifetime (yrs)	2	4	8	2.15	4.82	6.81
13	No. of Sats Needed for Same Coverage at Any Given Time	2.9	1.6	1.0	2.7	1.4	1.1
14	Number of Satellites Required for Entire Mission	11.7	3.3	1.0	10.2	2.4	1.3
15	Number of Redundant Satellites	1.2	0.3	0.0	1.0	0.2	0.0
16	No. of Satellites to Build w/ System Redundancy*	12.9	3.6	1.0	11.2	2.6	1.3
17	Total Launch Mass (kg)	3,767	652	1,559	859	2,659	3,309

* Note that fractions of satellites have been allowed in this model for purposes of comparison simplicity and a smoother display of results

Table 3a. Cost Predictions for the 3 Selected Altitudes using USCM8 [Tecolote Research, 2002] and NICM [Apgar, 2011], and 3 Example Observation Systems.

	Cost Estimates - USCM8 and NICM (from SME)	Model Predictions			Examples		
					NanoEye	Quickbird	GeoEye-2
1	Orbital Altitude (km)	200	400	800	215	482	681
2	Total Upfront Cost (FY13$M)	$47.45	$178.85	$991.29	$15.5	$87.5	$835.0
3	Total NRE Cost (FY13$M)	$14.89	$100.79	$708.75	$10.0		
4	TFU or T1 Cost (FY13$M)	$24.95	$73.31	$244.88	$2.0	$60.0	$784.4
5	Total RE Production Cost w/ Learning Curve (FY13$M)	$217.84	$217.07	$244.88	$22.5	$134.3	$981.7
6	Remaining RE Production Cost w/ Learning Curve (FY13$M)	$192.90	$143.76	$0.00	$20.5	$74.3	$197.3
7	Average RE Unit Cost per Spacecraft (FY13$M)	$16.92	$60.35	$244.88	$2.0	$51.9	$784.4
8	Nth (Last) Unit Cost (FY13$M)	$14.62	$55.65	N/A	$2.0	$50.6	N/A
9	Equivalent Present Value of Amortized Cost (FY13$M)	$203.94	$123.97	$0.00	$36.0	$87.8	$170.1
10	Total System Cost Before Amortizing (FY13$M)	$299.27	$331.93	$991.29	$36.0	$161.8	$1,032.3
11	Total System Cost to be Amortized (FY13$M)	$251.82	$153.07	$0.00	$20.5	$74.3	$197.3
12	Total Adjusted System Cost After Amorizing (FY13$M)	$251.39	$302.82	$991.29	$51.6	$175.3	$1,005.1

Table 3a. Cost Predictions for the 3 Selected Altitudes using SSCM [Aerospace Corporation, 1996], and 3 Example Observation Systems.

	Cost Estimates - SSCM (1996) (from SME)	Model Predictions			Examples		
					NanoEye	Quickbird	GeoEye-2
1	Orbital Altitude (km)	200	400	800	215	482	681
2	Total Upfront Cost (FY13$M)	$12.27	$48.05	$790.88	$15.5	$87.5	$835.0
3	NRE Cost (FY13$M)	$2.30	$26.59	$569.37	$10.0		
4	TFU or T1 (FY13$M)	$2.35	$16.71	$183.84	$2.0	$60.0	$784.4
5	Total RE Production Cost w/ Learning Curve (FY13$M)	$20.49	$49.48	$183.84	$22.5	$134.3	$981.7
6	Remaining RE Production Cost w/ Learning Curve (FY13$M)	$18.14	$32.77	$0.00	$20.5	$74.3	$197.3
7	Average RE Unit Cost per Spacecraft (FY13$M)	$1.59	$13.76	$183.84	$2.0	$51.9	$784.4
8	Nth (Last) Unit Cost (FY13$M)	$1.37	$12.68	N/A	$2.0	$50.6	N/A
9	Equivalent Present Value of Amortized Cost (FY13$M)	$62.41	$34.08	$0.00	$36.0	$87.8	$170.1
10	Total System Cost Before Amortizing (FY13$M)	$89.33	$90.13	$790.88	$36.0	$161.8	$1,032.3
11	Total System Cost to be Amortized (FY13$M)	$77.07	$42.08	$0.00	$20.5	$74.3	$197.3
12	Total Adjusted System Cost After Amorizing (FY13$M)	$74.68	$82.13	$790.88	$51.6	$175.3	$1,005.1

The total upfront cost for both the 200 and 400 km mission are much less than the upfront cost for the 800 km mission. However, both missions at the lower altitude have additional costs associated with the mission (i.e. the remaining production cost). **Even without adjusting the cost due to advantages of amortization, the total system cost (Table 3, line 10) shows that at lower altitudes the life-cycle costs are much less, even with many more satellites to build.** (Again, the life-cycle cost does not include operations cost. Section V describes how adding operations cost will not impact the relative results of the study.) Results from Table 3b have notably different values because USCM8 is developed by parametric cost modeling of traditional large satellite systems, and the SSCM is derived from parametric cost modeling of SmallSats [Apgar, 2011].

Our model estimates the required mass to operate at each altitude for a given resolution and coverage rate, and then inserts them into separate costs models (USCM8 & NICM, and SSCM). We run the model twice over a range of altitudes: first with projections from USCM and NICM, and again with projections from SSCM, and then plot them on the same graph for comparison. This means the missions are compared within the same class each time. In a sense, what we are doing is comparing apple to apples and oranges to oranges at the same time.

Performance vs. Cost

A. Cost vs. Coverage

Figure 2 shows the relationship between cost and coverage for two mission altitudes at a fixed resolution of 0.5 m. In order to have twice the coverage at a given altitude, it takes twice as many satellites, which increases the cost by approximately 1.8 times. (Recall that we introduced a 90% learning curve in this model to account for the production of multiple units.) **Flying high increases cost because it is more expensive to achieve a given resolution.**

B. Cost vs. Resolution

Figure 3 shows a sample relationship between cost and resolution for two mission altitudes. For a given mission altitude, if a higher resolution is desired, you must build a larger satellite and, therefore, spend more money. **At any altitude, twice the resolution increases the spacecraft mass by 8 times and increases the cost by up to 4.5 times.**

C. Cost vs. Altitude for Fixed Resolution and Coverage

The relationship between total mission life-cycle cost and altitude for a fixed resolution and fixed coverage requirement is shown in Fig. 4 over a range of altitudes in LEO. In the figure, the blue lines represent predictions using USCM8 and NICM, and the red lines represent predictions using SSCM. The solid lines represent the cost predictions using spacecraft bus mass values that fall within the range specified by the cost models. Extrapolated predictions based on values that are outside these specified mass ranges are indicated by the dotted lines in Fig. 4, which according to Aerospace Corporation [Mahr and Richardson, 2002] is less certain but not an unreasonable estimate. Beardon [1996] gives an in-depth analysis of this using the planetary spacecraft NEAR (Near Earth Asteroid Rendezvous), which went beyond the SSCM database range in several cases, and provided decent correlation between the model results and the actual spacecraft costs.

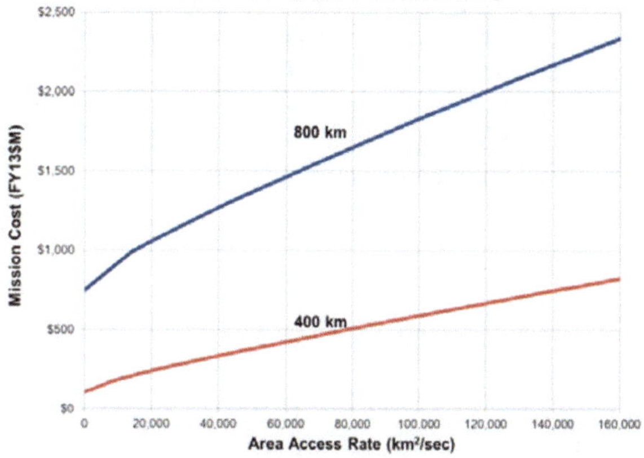

Figure 2. Cost vs. Coverage for a 0.5 m Resolution Requirement at 400 km and 800 km.

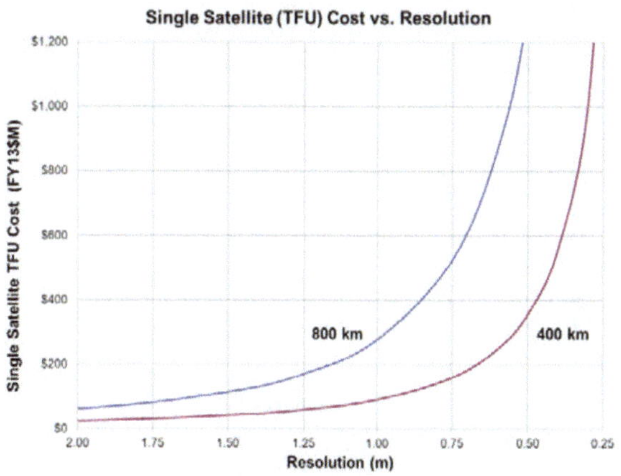

Figure 3. Single Satellite Theoretical First Unit Cost vs. Resolution at 400 km and 800 km.

As can be seen, the results from the two sets of models correlate very well with each other. (Note that the shapes of the two curves are essentially the same, suggesting that the extrapolation is reasonable.) The

Standard Error of the Estimate (SEE) of 34% has been added to the plot as vertical dashed bars.

IV. Impact of Altitude and Size on Cost and Performance

By reducing the altitude, you can reduce the size of the spacecraft, and this has a significant impact on cost. This can be seen in Fig. 4. The results clearly show that using smaller satellites at lower altitudes can provide much lower cost missions for an observation system while achieving the same performance requirements in terms of both resolution and coverage. We have also included three real observation systems as examples for comparison against this model. There have been many assumptions made to produce the results of this PBCM. However, changing the values of these assumptions, does not change the shape of the curves in Fig. 4. That is, **the relationship between mission cost and altitude remained the same over a very wide range of assumed inputs** because the shape of these curves depends only on physics and the empirical mass-based cost models.

Our most substantive conclusion is that **by significantly reducing the altitude of an Earth observation system, we can achieve the same performance in terms of resolution and coverage, but at dramatically lower cost**. Why is that the case? Basically, if we reduce the altitude by a factor of 2, we will also reduce the sensor aperture and linear dimensions of the spacecraft by a factor of 2. This reduces the volume and mass of the spacecraft by a factor of 8, which, according to the traditional mass-based cost models, reduces the cost by a factor up to 4.5. There will likely be the need of more spacecraft at the lower altitude because of reduced coverage per satellite and possibly a shorter design life, or greater atmospheric drag, but even with more spacecraft, it will be a much lower cost and more robust system that is less sensitive to spacecraft or launch failures. In addition, schedules are shorter, spending is spread out over time, and the problem of orbital debris essentially goes away below roughly 500 km [Wertz, et al. 2012]. This path has the potential to be an important option for Earth observing systems, particularly in times of critical budget problems.

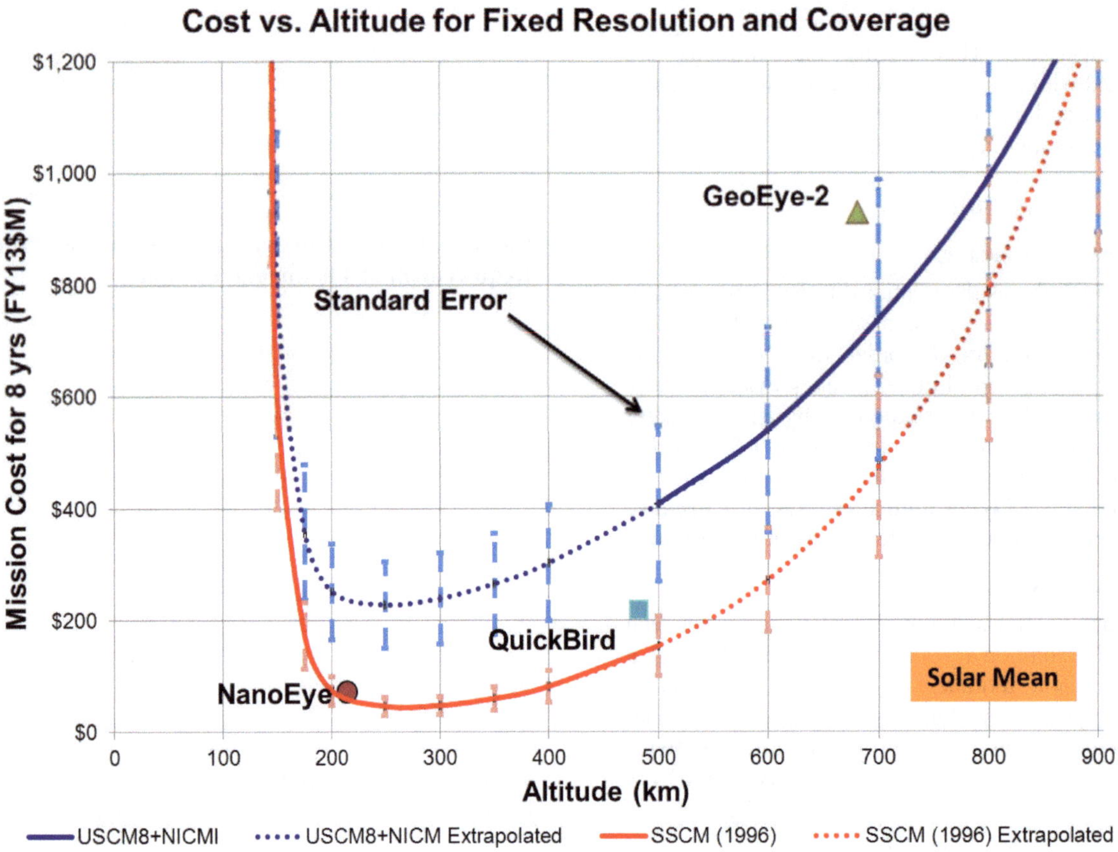

Figure 4. Cost vs. Altitude for a Fixed Resolution (0.5 m) and Coverage Rate (14,200 km^2/sec). The total mission cost includes launch, but excludes cost associated with operations.

The primary disadvantage of low altitude is that there is higher drag, which can result in a short mission lifetime. However, compared to traditional large satellites, there are many advantages to SmallSats at lower altitudes. Below is a list of advantages of low-altitude SmallSats over traditional satellites:

- Shorter development schedules
- Lower implementation and operations risk
- More flexible and resilient
- More responsive to new technologies and changing needs
- More sustainable business models
- Greater attitude agility due to smaller moments of inertia

The specific advantages of low altitude systems as identified by Eves [2013] are:

1. If the resolution of the required system is already adequate, a reduction in orbit height potentially allows a smaller, lower cost, and lighter sensor to be used.
2. The lower the satellite orbit, the greater the mass of hardware and/or payload that can be placed into orbit.
3. For a given (passive or active) imaging sensor, the resolution or performance improves proportionally as you lower the altitude.
4. A shorter path length makes it easier to establish an adequate communications link budget to a terminal on the ground.
5. For a given aperture size, the effective surveillance footprint size of the mission actually increases as the orbit altitude decreases, so the timeliness of revisit is better.
6. Flying lower permits the collection of unique data sets that would not otherwise be possible (e.g., the gravity map resolution of the GRACE mission [Tapley, et al. 2004]).
7. There is no need to perform de-orbit maneuvers since atmospheric drag can bring the satellite down "for free."
8. The problem of long-term orbital debris environment is mitigated since spacecraft below approximately 500 km will decay within a few days to several months.

V. Operations Cost

The relationship between cost and performance in Fig. 4 does not include operations costs. So far as we are aware, none of the publicly available space cost models that include operations cost break that cost down into elements that reflect the size, cost, or complexity of the spacecraft that is being operated. Nonetheless, it is reasonable to assume that we won't use the same Ops Concept for a $200 million spacecraft with a 10-year intended life as we would with a $2 million spacecraft with a 2-year intended life. There is also empirical evidence that this difference is real, as discussed below. Creating an operations cost model that is a function of the spacecraft size or complexity will likely be a challenging task. Adding operations cost to the current model will likely either move the model vertically, without changing the shape of the curve or possible tilt it a bit further in the in the direction of favoring SmallSats. We do not expect the change to be substantial in either case, but would welcome any data that others may have that reflects the impact of spacecraft size and complexity on operations cost.

Typically, operations cost depends on the following factors [Apgar, 2014]:

1. The number, complexity, and location of control and other ground stations and whether the control stations are dedicated to a single program (e.g., GPS) or allocated to multiple programs (e.g., JPL robotic missions)
2. The number of operators and hours per day required, the requirement for data recovery or additional data processing, and the level of automation (See Chap. 28 of Wertz, Everett, and Puschell [2011])
3. The amount of on-going R&D required (e.g., the need to upgrade operating software)
4. The amount of contactor support during the early years of the mission

In addition, small satellites naturally have lower operating cost. NEAR, Clementine, SAMPEX, ALEXIS, UoSat-05 are all examples of low cost small satellite programs with low operations costs. Operations costs for these specific missions were approximately 5-10% of their total mission cost, and their associated data can be found in Wertz and Larson [1996]. Therefore, multiple SmallSats flying at lower altitudes can have comparable operations cost to a single traditional satellite mission. Chapter 6 of Wertz and Larson [1996] gives detailed methods and concepts for reducing the cost of mission operations.

VI. Schedules, Reliability, and Risk

SmallSat missions provide much shorter schedules, comparable reliability, and significantly less risk than traditional large satellite missions. SmallSat schedules are much shorter than for traditional satellites. For instance, according to the Performance of Defense

Acquisition System Annual Report [DoD, 2013a], traditional major defense programs take 8.8 years in development (Milestone B) and well over 10 years from Milestone A to implementation. Reliability of SmallSats (including single-string SmallSats) is essentially similar to that of traditional large satellites according to a Goddard study [NASA, 2008b] of over 1,500 spacecraft launched between 1995 and 2007.

Risk is defined as the probability of a negative event times the impact or consequences of that event. Non-recurring cost for SmallSats is 1 to 2 orders of magnitude less than for traditional satellites [NASA, 2008b]. Therefore, implementation risk is low due to low non-recurring cost and short schedules. The consequences of failing to implement a SmallSat system will not endanger the larger, more traditional system. Operational risk of SmallSats is also much lower than traditional systems due to shorter operational life and the availability of spares (on orbit or on the ground) or back-up. An immediate result of having shorter schedules, reduced risk, and increased reliability is that SmallSats support the DoD objective of disaggregation [DoD, 2013b].

SmallSat missions are developed in less than 3.5 years while more traditional, large satellites an average of 10 years to develop. SmallSats have comparable reliability to larger satellite programs, despite often having single-string configuration and using COTS products. SmallSats poses significantly less risk (both implementation and operational) than traditional large satellite missions because failure rate is comparable to that of large satellites and consequence of failure is reduced due to low development cost. In addition, a paper by Hurley and Purdy at NRL "Designing and Managing for a Reliability of Zero" [2010], points out that most of today's space systems are designed for a reliability of zero, in the sense that for every day that the system is not operational or the data available to the end user, it has a reliability of zero. If the data isn't there, it doesn't matter to the warfighter who was killed or the scientist who's data was lost whether it wasn't there because of a parts failure or because the program was delayed or canceled due to more reviews or a lack of funding.

VII. Conclusions for Earth Observation Systems

The United States has more missions that need to be done than there is time and money available to do them. If the U.S. continues with the traditional way of doing business, there is the potential of physical gaps between missions that need continuity, such as weather and climate data and surveillance. Additionally, the U.S. does not have and has never had launch on demand, other than for ICBMs. Without responsive dedicated launch vehicles, it is impossible for the U.S. to respond to emergencies. This is a capability that Russia/Soviet Union has had for the past 3 decades. SmallSats can never replace traditional large satellites, but it is reasonable to believe there should be some sort of mix of both large and SmallSats in order to fill in mission gaps and increase the number of missions without added cost.

SmallSats are under-utilized as a method to dramatically reduce space mission cost. Without quantifiable evidence of their value, SmallSats will continue to be overlooked and under-recognized for their potential. By space mission cost, we mean the total mission cost from design and fabrication of a spacecraft, through launch and operations for the entire duration of the mission. Traditional (large) satellites have been used since the start of the space program, in the 1960s. These programs have done a tremendous job in terms of engineering and meeting the goals of NASA, DoD, and the United States. However, the U.S. has gotten to the point where there are many more missions that we need to or would like to accomplish, than there is funding available for them. If there are methods to dramatically reduce space mission cost, then it is clearly a benefit to implement them, or at least consider them.

Past Earth observation systems have used traditional space technology to achieve the best possible performance, but have been very expensive. In addition, low-cost, responsive dedicated launch has not been available for SmallSats.

Due to advancements in technology and modern microelectronics, SmallSats at lower altitudes now have the potential for much lower overall mission cost, comparable or better performance, lower implementation and operations risk, and shorter schedules.

SmallSat observation systems need greater field of view (FoV) agility than larger, higher altitude systems. The needed agility is inversely proportional to altitude, but moments of inertia are also much smaller. Responsive, low-cost, small launch systems are needed for operational missions. All of this requires changing the way we do business in space and how we think about using space systems. This culture change is probably the most challenging thing, and the USC/Microcosm *Reinventing Space Project* [2014] is directed at continuing to find ways to make progress in this direction.

VIII. Future Work

While the PBCM provides sufficient detail to draw significant conclusions about the use of LEO satellites for Earth observation, there are several areas that should be researched in the future to broaden and strengthen the model. The current cost model is based solely on circular LEO Earth Observation satellites, but other constellation configurations such as LEO Elliptical Orbits and other types of missions such as communications or interplanetary science missions can be studied. Also, there are significant characteristics of LEO missions that require adaptation from more traditional, large missions, such as using autonomous orbit control, propulsion systems, checkout time and calibration process, and the responsive capabilities of SmallSats.

IX. Acknowledgements

We would like to acknowledge the California Space Grant Consortium for providing seed funding to the Microcosm/USC Reinventing Space Project. We would like to thank Azad Madni from the Astronautical Engineering Department at USC for his academic advisement and guidance on this research. We would also like to acknowledge the assistance of International Cost Estimating and Analysis Association (ICEAA) which was kind enough to let us present our results at their March 19, 2014 Workshop [Koltz, Shao, and Wertz, 2014]. Participants at the meeting asked a number of penetrating and useful questions that substantially improved the results. These questions and answers are available on the Reinventing Space Project website [2014].

X. References

1. Aerospace Corporation. 1996. *SmallSats Cost Model (SSCM)*. El Segundo, CA: The Aerospace Corporation.
2. Apgar, Henry. 2011. "Cost Estimating." Wertz, J., Everett, D., and Puschell, J. *Space Mission Engineering: The New SMAD*, (SME) Chap. 11. Microcosm Press: Hawthorne, CA.
3. Bearden, David, A., et al. 1996. "Comparison of NEAR Costs with a Small-Spacecraft Cost Model." AIAA/USU Conference on Small Satellites, September 16-19.
4. Department of Defense. 2013a. "Performance of the Defense Acquisition System," 2013 Annual Report. Washington, DC: Office of the Under Secretary of Defense, Acquisition, Technology and Logistics, June 28.
5. Department of Defense. 2013b. "Resiliency and Disaggregated Space Architectures." Air Force Space Command, White Paper, AFD-130821-034.
6. Digital Globe. 2013. Quickbird Datasheet. DS-QB 07/13.
7. Eves, Stuart. 2013. "Very Low Orbits." 2013 Reinventing Space Conference. International Panel – Presentation, El Segundo, CA.
8. GeoEye. 2012. GeoEye-2 Fact Sheet, Rev 03/12.
9. Habib-agahi, Hamid. 2010. *NASA Instrument Cost Model (NICM) Version IV*. Pasadena, CA: NASA JPL.
10. Hurley, Mike and Bill Purdy. 2010. "Designing and Managing for a Reliability of Zero." ESA 4S Symposium, Funchal, Portugal, Paper No. 1885505, May 31–June 4.
11. Mahr, Eric, and Greg Richardson. 2002. "Development of the Small Satellite Cost Model (SSCM) Edition 2002." IEEEAC paper #1159, Updated December 19, 2002.
12. NASA. 2008a. *2008 Cost Estimating Handbook (CEH)*. National Aeronautics and Space Administration, NASA HQ, Cost Analysis Division, Washington, DC.
13. NASA. 2008b. NASA Small Explorers Office SmallSat Study, Final Report. JHU/APL Space Department, June.
14. Reeves, Emery I. 1999. "Spacecraft Design and Sizing." Wertz, J. R. and W. J. Larson. 1999. *Space Mission Analysis and Design. 3rd ed.*, Chap. 10. Hawthorne, CA and New York: Microcosm Press and Springer.
15. Spaceflight Now. 2000. "Commercial Eye-in-the-sky Appears Lost in Launch Failure." Nov. 21.
16. Space News. 2012. "NGA Letters Cast Cloud Over GeoEye's Enhanced View Funding." June 23.
17. Space News. 2014a. "DARPA Chief Says Space Programs Are Too Slow and Costly." January 15.
18. Space News. 2014b. "Thule Air Base USAF Examining Alternative for All of Its Big Satellite Programs." February 17, page 14.
19. Space News. 2014c. "Google to Buy SkyBox for $500 Million." June 10.
20. Shao, Anthony, Elizabeth A. Koltz, and James R. Wertz. 2013. "Performance Based Cost Modeling: Quantifying the Cost Reduction Potential of Small Observation Satellties." AIAA Reinventing Space Conference, AIAA-RS-2013-1003, Los Angeles, CA, Oct. 14–17.
21. Shao, Anthony, and Elizabeth A. Koltz. 2013. Microcosm Technical Report, PBCM Detailed Analysis. Microcosm: Hawthorne, CA.
22. Tapley, B.D., S. Bettadpur, M. Watkins, C. Reiger. 2004, "The Gravity Recovery and

Climate Experiment: Mission Overview and Early Results." American Geophysical Union.

23. Tecolote Research. 2002. *Unmanned Space Vehicle Cost Model (USCM)*. El Segundo, CA: US Air Force, Space and Missile Systems Center.
24. *The Reinventing Space Project.* 2014. Website: http://www.smad.com/ReinventingSpace.html
25. Wertz, James R., David F. Everett, and Jeffery J. Puschell. 2011. Space Missions Engineering: The New SMAD. Microcosm Press: Hawthorne, CA.
26. Wertz, James R., Nicola Sarzi-Amade, Anthony Shao, Christianna Taylor, and Richard E. Van Allen. 2012. "Moderately Elliptical Very Low Orbits (MEVLOs) as a Long-Term Solution to Orbital Debris." 26th Annual AIAA/USU Conference on SmallSats, SSC12-IV-6, Utah State University, Logan, UT, Aug. 13–16.
27. Wertz, James R., Richard R. Van Allen, and Tina Barcley. 2010. "NanoEye—Military Relevant Surveillance for Less Than $5 Million Total Recurring Mission Cost." 8th Responsive Space Conference, RS8-2010-1008, Los Angeles, CA, Mar. 8–11.
28. Wertz, James R., and Wiley J. Larson. 1996. *Reducing Space Mission Cost*. El Segundo, CA: Microcosm Press and Springer.

AIR LAUNCHED SPACE TRANSPORTATION SYSTEMS

Mr. Oleg A. Sokolov
Commercial Space Technologies Ltd.,
Kosmonavta Volkova 5, building 1, premise 19, 127299 Moscow, Russia
+7499 150 1741 (Russia)
+44(0)208 840 1082 (UK)
cst@commercialspace.co.uk

ABSTRACT

The air launch concept of using a carrier aircraft to launch a normal type of launch vehicle in the air, promises several benefits including, if the carrier is sufficiently large, lower specific launch costs. These benefits are defined in this paper by exposing the influence of various factors, including the classes of payloads to be launched and the sorts of carrier aircraft to be used. A comparison of a normal expendable surface-launched launch vehicle and an air-launch system (chosen for their similar launching capabilities and levels of technical perfection), the Ukrainian 'Cyclone-4' and the Russian 'Air Launch' system, is carried out from the point of view of their economic efficiencies.

KEYWORDS: launch systems, reduction of launch price, commercial success, future developments

INTRODUCTION

The 'air launch' launch system concept, in which the launch of a common type launch vehicle is carried out from a carrier aircraft in air, is well known in the world of astronautics. Although only one example, the U.S. 'Pegasus' (Fig. 1), is in regular operation (but at an uneconomic launch price) a great number of corresponding projects have been developed. This interest for the concept is generated by the serious advantages that would be provided by it.

Figure 1. The 'Pegasus' system that is the only operational example of the air launch concept but is not featured with a high economic efficiency

Credit: Adverts of Orbital Science Corporation.

The list of these advantages includes the absence of a necessity to build and use a complicated and expensive launch facility with all the supporting infrastructure, a common airfield of corresponding class can be used instead. An air-launch system can also provide a large range of launch azimuths and can carry out its launches to include safe drop zones. Lastly, the system's carrier aircraft can be considered as a reusable first stage. The use of reusable components enhances economic efficiency as these components are not repeatedly manufactured for every mission. Furthermore, common aircraft types can be used as carriers without serious upgrading and, therefore, the price of development for the carrier aircraft would be, if not excluded entirely, then at least decreased significantly.

As shown in numerous studies, for example, [1], these advantages could be found to be either realistic or assumed, depending on various factors, including the sorts of system applications and purposes. However, most of these factors would be beyond the interest of a commercial customer, for which the system would be contracted: these customers would be interested in their satellites getting launched into their required orbits at the lowest possible launch price.

Considering that a suitable launch system of any sort is chosen for every specific mission under the condition of guaranteed implementation of technical requirements for this case of launching, let us consider that the first part of the customers' interest should be met in any case and let us see how air-launch systems would meet the

S. Hatton (ed.), *Proceedings of the 12th Reinventing Space Conference*, DOI 10.1007/978-3-319-34024-1_17

second part (i.e. an achievement of higher economic efficiency than by common expendable launch vehicles), in which cases of application and under what conditions.

APPROACH TO THE CHOICE OF CANDIDATES FOR THE COMPARISON

The stereotypical 'rocket' section of any air-launch system (i.e. in which a carrier aircraft lifts an expendable launch vehicle but not a reusable space vehicle) is actually a common multi-stage launcher, with either a liquid-propellant or solid-propellant option that differs from surface-launched analogues only by a less powerful first stage. Although a carrier aircraft is not a real substitution for a launch vehicle's first stage (the gains of velocity and of altitude provided by these aircraft are significantly less), even a partially efficient 'quasi-first stage' can provide certain energetic benefits. At the same time, this 'weakened' launcher is somewhat cheaper to produce and, more importantly, does not need a surface bound launch facility. This feature promises the greatest contribution towards the increased economic efficiency of air-launch systems.

However, a carrier aircraft itself is not a gratuitous product: it requires significant non-recurring expenditures for its development and manufacture, or for its purchase or leasing. Following the adaptation for its new application, recurring expenditures are then required for its maintenance, airfield servicing, etc. In certain studies, e.g. [1], it was even supposed that the total costs relevant to the carrier aircraft's introduction would be comparable with the costs of construction and operation for a fixed surface launch site. Therefore, in order to expose an economic benefit from the supposed substitution of a reference common launch vehicle with an analogous air-launch system (in terms of launching capability), it is necessary to compare the costs for building a surface launch site and operating it with the relevant items of a carrier aircraft. If this comparison does not show an evident preference for the air-launch system, then the results of calculations on the size of the economic benefit from a diminution of the launcher's first stage for the air-launched option would be included.

A search of candidates for this comparison can be carried out in all classes of space launchers, but the range for this search can be narrowed if one takes into account two important circumstances. The first of them is relevant to the choice of carrier aircraft. Heavy class launch vehicles, which are mostly intended for launching communication satellites into geostationary orbit (GEO), which is the most profitable segment of the world's current market of launch services, have a mass of no less than several hundred tons and even air-launch systems with a diminished first stage would be too heavy for any currently available heavy airplane used as a carrier aircraft (the heaviest cargo airplane in the world, the Ukrainian An-225 'Mria' (Fig. 2), with only one existing flight example, has a cargo capacity of 250 tons). A suitable carrier aircraft would have to be specially developed and built for this supposed heavy class air-launch system but, as shown in [2], the task of creating an airplane with a cargo capability of more than 300 tons is too complex (although feasible) for the current aviation industry. As a result, this unique airplane would evidently cost more than the construction of a launch site for a common heavy launch vehicle. The largest cargo capability, which can be provided by currently operated super-heavy cargo airplanes, for example, the Russian-Ukrainian An-124 'Ruslan' and the U.S. C-5 'Galaxy', is at the level of 100 tons. Taking into account that an air-launched launch vehicle of this mass has to correspond in payload capability to a surface-launched launch vehicle with a launch mass around 140-150 tons, it is possible to consider that the most powerful air-launch systems, created with current technologies and the adoption of available cargo airplanes for use as a carrier aircraft, would be comparable to intermediate class launchers. Although these launchers would be significantly inferior to heavy class launchers in their GEO payload capabilities, they could nevertheless launch 'small geostationary satellites' with masses of 0.5-1.0 tonnes into GEO (especially considering that launching from near-equatorial areas could be easily achieved by air-launch systems). Therefore, these launchers can be players in the most profitable segment of the current market in launcher services.

Figure 2. The heaviest-in-the-world cargo airplane Ukrainian An-225 'Mria'.

Credit: Aerospace Review

It is necessary to underline that the use of common cargo airplanes for the role of carrier aircraft would be most suitable from an economic point of view if their adaption for this role does not disturb the airplane's original purpose in the air transportation of cargo. This would provide the opportunity to use the carrier aircraft for regular air freighting between space launches, thus gaining additional revenue that would enhance the air-launch system's economic efficiency. Neither a military nor passenger aircraft could provide this opportunity if they were used as the carrier aircraft.

The second circumstance, after the correct choice of carrier aircraft, is that, as found in a process of recent developments, the application of the air-launch concept for the creation of a super-small class launch system (with payload capabilities of up to 500 kg) would not provide serious economic benefits in comparison with surface-launched super-small, common type, launch vehicles. The demand for services from this class of launcher is very high but only if the specific launch price is comparable with that of piggy-back launches for super-small satellites. Meanwhile, whilst the typical super-small launchers, which are currently being developed, cannot promise this price decrease, their potential competitors with air-launch concepts promise even worse values. 'Virgin Galactic's' 'LauncherOne' system (Fig. 3), with a payload capability of 100-225 kg, should have a launch price of about US$ 10 mln. i.e. at the same level of a small-class 'Falcon-1', but with a payload capability several times less, even the Russian 'Ishim' air-launch system project proposed to launch 160 kg of payload into a low-Earth orbit (LEO) for US$ 3-4 mln., corresponding to a specific launch price of about US$ 20-25 per kilogram of payload. That itself was even higher than the average specific price for a launch of the small class Russian 'Start-1' (700 kg of payload for US$ 10-12 mln.).

Figure 3. An artistic image of the 'LauncherOne' air-launch system.

Credit: Adverts of 'Virgin Galactic'

As supposed in [3], this increment in the launch costs for super-small air launch systems is caused by two main factors. Firstly, by the necessity to use either high speed military aircraft, or specially developed aircraft as carriers (which as mentioned earlier, decreases the economic efficiency). Secondly, the development of small but sufficiently powerful launch vehicles for launching from these carrier aircraft would require complex task solving leading to enhanced expenditures. This statement still requires confirmation from special studies, but it is sufficient for the rejection of this class of candidate in the comparative analysis amongst air-launch systems.

So, potential candidates would have to represent either small or intermediate class launchers. It is necessary to note that these candidates should be sourced from amongst existing launchers, as well as those proposed in various sufficiently developed projects. However, it seems difficult to find within one of these classes a pair of suitable candidates, for which their main launching characteristics, both for a surface-launched candidate and for an air-launched candidate, are sufficiently similar. Fortunately, a pair of launchers can be found in the intermediate class. One being the 'Cyclone-4' surface-launched launch vehicle (Fig. 4), which was developed and is also being built in Ukraine for the joint Ukrainian-Brazilian operation from the Alcantara (near-equatorial) Brazilian spaceport. The second candidate is the Russian 'Air Launch' project (Fig. 5), an air-launch system which has been developed up to a sufficient level for this comparison. These candidates chosen for comparison have already been described in detail in [4, 5].

Figure 4. An artistic image of the 'Cyclone-4' launch vehicle on its launch pad

Credit: Adverts of 'Yuzhnoye' NPO

Figure 5. An artistic image of the 'Air Lainch' system in flight at the moment of launch vehicle ejection

Credit: Adverts of Air Launch Aerospace Corporation

THE COMPARISON OF THE CHOSEN EQUIVALENT CANDIDATES

Although both candidates are not fully completed launchers, they are at a high level of development and their economic factors have been calculated with some of the world's leading experts. At the same time, both candidates, 'Cyclone-4' and 'Polet' launch vehicles, were designed with a maximum adoption of existing technologies in order to maintain a high level of technical perfection and to save on cost by using already established production techniques. Equally, the same approach that was aimed at avoiding extra expenditures, was also applied for the development of the 'Cyclone-4's' surface launch facility, as well as for the choice of carrier aircraft in the 'Air Launch' system.

As mentioned above, both candidates are classed as intermediate launch vehicles, defined by a payload capability in the range of 3 to 5 tons to LEO. Both of these candidates are intended to launch from near-equatorial locations on the globe. Their LEO capabilities are somewhat different (5.1 tons for 'Cyclone-4' and 4.0 tons for 'Air Launch', at an inclination of 0°), but the payload capabilities for launching into GTO at an inclination of 0° almost coincide: 1600 kg for 'Cyclone-4' and 1650 kg for 'Air Launch'. The last indices are most important for the comparison of equal technical capabilities as it is the most profitable and competitive segment of the current global market for launch services, thus success in this competition would define certain economic indices, namely launch price.

Despite both candidates representing projects that are not yet realized (this circumstance underlines once more the equal conditions for their comparison), the expected figures for each launch vehicle were repeatedly announced. Thus, representatives of the 'Cyclone-4's future operator, the 'Alcantara Cyclone Space' (ACS) company, disclosed in 2011 the price of US$ 70 mln., although an earlier figure of 40 mln. had been mentioned; meanwhile a figure of US$ 40 mln. was shown in the Business Plan for the 'Air Launch' project in 2010 [6].

For the purposes of comparing the candidates' economic efficiencies, these figures should be checked indirectly by assessing the required investments and expenditures and, as shown above, these assessments must be initially done for the launch site of 'Cyclone-4' and the carrier aircraft of 'Air Launch'.

It is necessary to once more note that both of the primary components of the candidates' launching structures were developed using the principle of maximum cost saving: thus, although the launch site for 'Cyclone-4' was developed on the basis of a similar structure for 'Cyclone-3', that features a very high level of automation, this level was decreased for the 'Cyclone-4' as it will not require such a high rate of launches as its predecessor. The mass of the 'Air Launch' system's 'Polet' launch vehicle was chosen in accordance with the maximum cargo capacity of the An-124 carrier aircraft i.e. this aircraft was not oversized for this application.

The expenditures for the 'Cyclone-4's' launch site at Alcantara (Fig. 6) had to include the cost of building the site's facilities and supporting infrastructure, plus the cost of Ground Support Equipment (GSE). Meanwhile the building works in Alcantara had to be funded by the Brazilian party, the total figure of this funding was clearly defined in [7, 8], equaling US$ 243.6 mln. from which it can be assumed, approximately 20% would have to be spent on organization and logistics, therefore the direct investment from the Brazilian party would be US$ 200 mln. The cost of the GSE, although with certain difficulties, can be defined, since it is being delivered by the Ukrainian party, with funding originating from within their share of the financial participation, which is equal to the Brazilian party's share. However, this figure for the GSE cost can be defined using a deductive method.

Figure 6. Artistic image of the launch site for the 'Cyclone-4'

Credit: Adverts of Alcantara Cyclone Space

As stated in [8], the Ukrainian share of the total investments should be equal to the Brazilian share, with US$ 200 mln. spent on the development of the launch vehicle (more exactly, on a slight upgrading of the two lower stages of the basic 'Cyclone-3' with the addition of a new third stage) and on the manufacture of four initial examples, as well as for the development and manufacture of the GSE. Based on certain available information, it is possible to assume that the development of the 'Cyclone-4' launch vehicle would cost about US$ 70 mln., whilst the cost of manufacturing four launchers would be US$ 80 mln.; meaning that the GSE should be about US$ 50 mln.

So, the total cost of the 'Cyclone-4' launch site should be US$ 250 mln. As noted in [7], these investments are planned to be reimbursed over a 20 year period of the launch vehicle's operation, with the annual rate of launches equaling 5-7.

For 'Air Launch', the total cost of the system's operation, including on-ground pre-launch preparations and the flight of the carrier aircraft up to the ejection of the 'Polet' launch vehicle, has to be assembled mostly from the cost of the carrier aircraft's purchase or lease alongside costs for its upgrading and for the adaptation of an airfield (including building special facilities for the pre-launch preparation of the launch vehicles and their payloads). The cost of the existing An-124 airplane (Fig. 7) can be assessed at over US$ 40 mln., whilst the cost for its upgrading and use as a carrier aircraft were mentioned in /5/ as US$ 7.5 mln. It is therefore possible to consider that this aircraft can be used in the 'Air Launch' system for US$ 50 mln.

Figure 7. The An-124 'Ruslan' super-heavy cargo airplane in flight.

Credit: Adverts of 'Polet' Aviation Company

The adaptation of the existing 'Frans Kaisiepo' 1-st class airfield on the Biak Island of Indonesia, as a base for the "Air Launch' system (this adaptation requires special facilities for the preparation and fuelling of the launch vehicles and spacecraft, Fig. 8), will require significantly more expenditure. The total sum was estimated at as much as US$ 120 mln. [5, 9]. By adding this figure to the cost of the carrier aircraft, the figure, which is roughly analogous to the cost of launch site for the 'Cyclone-4', is US$ 170 mln.

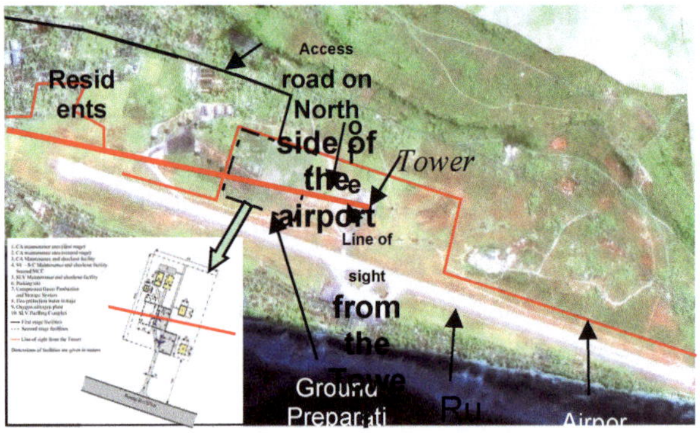

Figure 8. An aerial photograph of the 'Frans Kaisiepo' airfield with an indication of the facilities to be built.

Credit: Adverts of Air Launch Aerospace Corporation

As one can see, this index for the 'Air Launch' system is significantly less than for the 'Cyclone-4'. With regards to reimbursing these investments, the same conditions as the 'Cyclone-4' are foreseen for 'Air Launch' i.e. over a period of 20 years with an annual rate of 5 7 launches.

It is possible to supplement that these indices both for 'Cyclone-4' and for 'Air Launch' do not include the regular expenses for the permanent maintenance of the launch site and the inter-flight maintenance of the carrier aircraft. However, 'Air Launch' is in a better position even in this regard, as the necessary upgrades for the An-124 airplane sought to maintain its capability as an air transport for common cargo and, therefore, it could be used for gaining extra income in intervals between missions; this money could in turn be spent on the airplane's maintenance.

It is also possible to compare the costs of the launch vehicles to be used, the 'Cyclone-4', against the 'Polet' launch vehicle of the 'Air Launch' system. Both use liquid-propellant, although they use different components, but, at almost equal payload capabilities, the launch mass of the 'Cyclone-4' (201 tonnes) is nearly twice higher than that of the 'Polet' (102 tonnes) -this difference comes about not only from launching the latter in air but also from using a more efficient propellant. In any case, this difference is reflected in the total cost of the launcher: whilst the 'Cyclone-4's' manufacturing cost is US$ 20 mln., the planned production cost of the 'Polet' (including its upper stage) should be US$ 14-16 mln. [6]. (These costs are included into launch prices only as components to be added to other ones such as cost of the launch preparation servicing, the share of non-recurring expenditures, etc.)

With these indices and their comparison, it is understandable why the above mentioned figures of launch prices, equaling US$ 70 mln. for launches of the 'Cyclone-4' and US$ 40 mln. for 'Air Launch', are correct. In turn, this correlation evidently shows a preference for air launch based launch systems in an economic regard.

This analysis compared two launchers that were proposed as already developed projects with different degrees of completion readiness. Of course, it would be better to compare two or more recently completed launchers of the chosen concepts but, as mentioned earlier, only a sole air-launch system, the U.S. 'Pegasus' (that shows low economic achievement) is currently operating and with no true surface-launch counterpart, whilst the chosen candidates (despite being in the form of incomplete projects) provide the opportunity to compare launchers that should be equal in their chosen parameters.

A COMPARISON WITH OTHER AIR-LAUNCH SYSTEMS

It is necessary to note that the distinctive feature of both candidates outlined above is an optimum use of all capabilities that would be provided by the adopted components (for example, designing the air-launch system's launch vehicle at a mass that corresponds to the maximum lift capability of carrier aircraft). This approach should be used by designers of future systems of the air launch concept; otherwise the true economic benefits that this concept could offer could not be realised. An example of an incomplete application of this approach can be seen in the 'Space Clipper' project (Fig. 9) that was proposed in 2011 (in a deeply revised version of the 1990s' project) by the Ukrainian 'Yuzhnoye' Design Bureau [10].

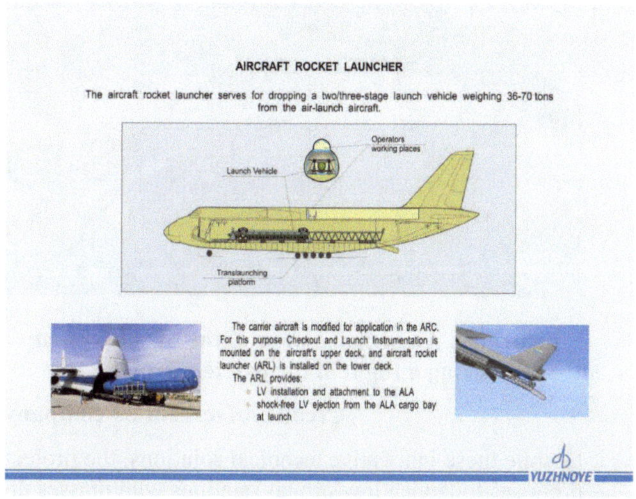

Figure 9. The 'Space Clipper' project.

Credit: Adverts of 'Yuzhnoye' DB

In accordance with this project, the same An-124 airplane (as in the 'Air Launch' project) should be used as a carrier aircraft but the launch vehicle, which has to be launched from this carrier, should have the launch mass either of 37 tons or of 67 tons (the project foresaw two options of the launcher) i.e., significantly less than the carrier aircraft's cargo capability. Without going in details (this was done in [11]), it is possible to underline that this discrepancy led to the result that the launch price for the system with the launch vehicle's heaviest option, which should be capable to launch up to 1000 kg of payload into the LEO with the altitude of 500 km and the inclination of 0°, should be as much as US$ 30 mln. [11]. However, the more optimized 'Air Launch' would launch a payload up to 3.5 tons into the same orbit at the same launch price (see in [5, 6])! An economic advantage of this system is evident.

At the same time, it is possible to mark that this approach, i.e. an application of common airplanes as carrier aircraft with using their complete cargo capabilities, does not bring satisfactory results in an application to air-launch systems of small and, moreover, super-small classes. Thus, a typical example is the Swiss SOAR project, in which not only a complete use of the A-300 carrier aircraft's cargo capability is foreseen but also the launcher's first stage should be implemented in the form of a reusable sub-orbital shuttle (this system, which is shown in Fig. 10, was described in details in [12, 13]).

Figure 10. An artistic image of the SOAR system during a flight of the carrier airplane.

Credit: Adverts of S3 company

Despite these innovative technical solutions, the project promises to launch low-orbital satellites with masses up to 250 kg at the launch price of US$ 10.5 mln. (for the rate of 2013). Even with a correction for the current inflation, this means that the economic efficiency of the proposed system will be worse than for common surface-launched launch vehicles, for instance, the U.S. 'Falcon-1' or Russian 'Start-1'.

One can think that an introduction of these innovations namely would enhance the total cost of the system realization which will reflect inevitably on the system's launch cost and price. However, the earlier developed Russian project of the 'Ishim' air-launch system of super-small class (it had to launch payloads with masses up to 160 kg into LEOs [14]), in which an upgraded option of the standard MiG-31 heavy fighter and a common type of two-stage solid-propellant launch vehicle should be used (Fig. 11), would provide a launch price at the level of US$ 4-5 mln. [14] which corresponds to the SOAR system's economic efficiency.

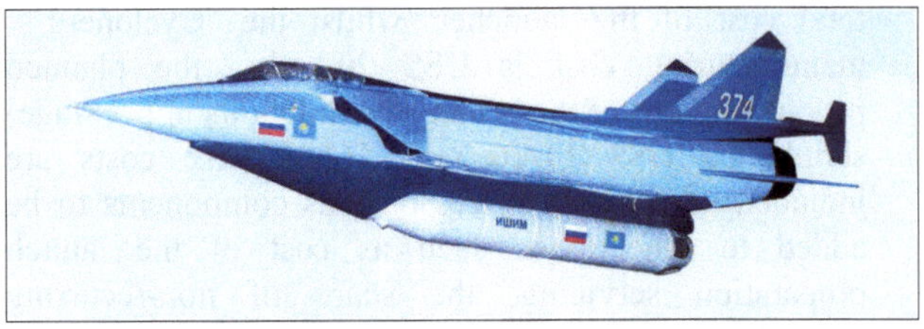

Figure 11. A model of the 'Ishim' launch system – the MiG-31I carrier aircraft

with a suspended launch vehicle

Credit: Novosti Kosmonavtiki

It is also necessary to note that the economically efficient application of the air-launch concept for super-small launchers requires separate special studies which could be carried out in the near future.

CONCLUSIONS

1. The implemented comparison of two launchers with almost equal payload capabilities, one of which is a common type of surface launched launch vehicle whilst the second one is an air launch system, shows a preference of the air launch concept-based launch systems in an economic regard.

2. An optimum use of all capabilities that would be provided by the adopted components of the proposed air launch system is an indispensable requirement for achieving an economic efficiency of any air launch system.

3. Nevertheless, the problem of the achievement of an economically efficient application of the air launch concept for super-small launch systems still requires separate special studies for its solving.

References

Space Transportation Systems Using the Air Launch Concept – Flight to the Future or a Dead-End, CST report, 2002.

A Choice of Carrier Aircraft for Air-Launch Systems, CST report, 2013.

Super-small Launch Systems for Super-small Satellites, CST report, 2010.

4. Launch vehicles from FSU Countries for the Brazilian Alcantara, CST report, 2009.

5. The 'Air Launch' Project, Its Current Status and Prospects for Its Realization, CST report, 2010.

6. Business Plan for the 'Air Launch' Project, Noerr Company, 2910.

7. http://www.brazildispatch.com/2012/06/.

8. 'Cyclone-4' Project Overview, presentation of ACS, 2011.

9. Novosti Kosmonavtiki, Vol. 16, # 3 (278), 2006.

10. Airspace Rocket Complex 'Space Clipper', presentation of 'Yuzhnoye' DB, 2011.

11. A Survey of the Current Inventory of Small Launch Vehicles in Russia and the FSU and Its Development to 2020, CST report, 2013.

12. www.space.com/21629-swiss-private-rocket-plane-evolution.html, 2014.

13. A Preliminary Assessment of Swiss Project of the SOAR Air-launch System, CST report, 2014.

14. Survey of FSU Countries' Launch Vehicles Suitable for Launches of Small Satellites as Prime Payloads during the Period of 2009-2015, CST report, 2009.

BIS-RS-2014-15

S.O.F.I.A: Solar Orbiters For Imaging Asteroids

Dr Markos Trichas
Airbus Defence & Space
Gunnels Wood Road, SG1 2 AS, Stevenage, UK; +44 (0)143877 4196
markos.trichas@astrium.eads.net

Alexander Elliott, Ian Harris, Graham Johnson, Stephen Kemble, Emanuele Monchieri, Joanna O'Rourke, Stephen Pulker, Julia Ryan, Matthew Stuttard, Matthias Winter
Airbus Defence & Space
Gunnels Wood Road, SG1 2 AS, Stevenage, UK

Dr Damon Landau
Jet Propulsion Laboratory
4800 Oak Grove Dr, Pasadena, CA, USA

Dr Martin Elvis
Harvard-Smithsonian Center for Astrophysics
60 Garden Street, Cambridge, MA, USA

ABSTRACT

The primary objective of this Airbus Defence and Space funded, Pre-Phase-0 study was to make an early assessment of the feasibility of a combined NEO/space weather monitoring system based on interplanetary Cubesats. During this study the original concept of surveying and in-situ visits of NEOs had to be de-scoped to just a survey system as the original assumptions on the performance specifications of the proposed JPL propulsion system were overestimated. The study mainly focused on assessing the currently available and in development technology that could be used on an interplanetary Cubesat. We have reviewed such technologies for critical subsystems such as AOCS, communications and propulsion with the aim of defining if such a mission is feasible within the current hardware availability, or if not, define the key design drivers for further review / development. In the absence of robust information for various parameters on technologies and subsystems, we have made assumptions based on the defined specifications for each of these systems. A cost analysis could not be performed as there was no available information on any of the studied technologies. All the missing information will be assessed in a further study funded again by Airbus Defence and Space.

KEYWORDS: [Cubesats, NEO, Space Weather]

I. CONCEPT

With plans to deflect, visit and even mine asteroids beginning to become a reality, it may well be that the Near-Earth Asteroids (NEOs) detection/characterization phase becomes the bottleneck to developing asteroid resources. However, in order to plan any mining, deflection or human visitation activities to NEOs, firstly we need to find the missing population of NEOs and then establish a wide range of selection criteria such as: mass, non-spherical morphology, surface boulder size distribution, mass moments, small companions, impulsive outgassing. Assuming that the probability to meet each of these criteria is as high as 75%, the combined probability to find a candidate to pass the selection is less than 20%. This means that in order to select 5 potential candidates, at least 25 targets need to be investigated. Assuming a robust schedule of launch windows every 6 months and taking into account current design strategies (e.g. OSIRIS-REX), in order to achieve the above goal, more than 15 years and $1 billion need to be spent. Solar Orbiters for Imaging Asteroids (SOFIA) aims in achieving the above goal in less than 10 years, in only a fraction of the cost.

S. Hatton (ed.), *Proceedings of the 12th Reinventing Space Conference*, DOI 10.1007/978-3-319-34024-1_18

The idea, based on the SWARM concept (D. Landau), is to put 360 Cubesats at a 0.9 AU heliocentric orbit, with a spacing of 0.015 AU. This formation solves the following problems: 1) synodic period problem 2) survey all of 1 AU orbit and 3) since it will observe NEOs at phase 0, optics requirements are minimal. Using the Microfluidic Electrospray Propulsion (MEP) developed in JPL, each of the Cubesats will require maximum 2 years for a trip to station. They will then spend 1-2 years as NEO Sentinels and on the third year they will start their 1-2 years round trip to their assigned NEOs. Another 3-6 months will be spent in orbit around the NEO gathering data. As soon as they are back in formation, each of the Cubesats will start relaying its data to the nearest Cubesat, until the nearest Cubesat to Earth relays everything back to the ground.

S.O.F.I.A. not only would be able to produce the largest ever database of NEOs but also to fully characterize hundreds of them. The latter can't be achieved from the ground as described above and it is absolutely crucial in planning any future NEO mission. For comparison, only 4 NEOs have been fully characterized so far. The mission relies in mass production of Cubesats hence not subject to single unit failures both in orbit but also on the ground. SOFIA/SWARM concept is shown in Figure 1.

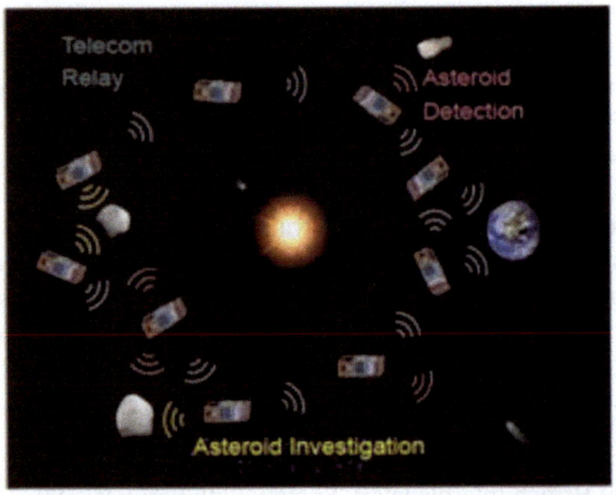

Figure 1: Schematic showing the basic idea behind the SOFIA/SWARM concept (D. Landau).

In the following sections of this paper we will provide an insight of the major outcomes of the feasibility study on this concept.

II. MISSION ANALYSIS

Requirements

In close collaboration with the science team (Landau & Elvis) we have identified a set of key mission requirements:

- Observe NEOs at distance <0.1 AU over all solar longitudes
- Observe the NEOs at phase-0 (Sun facing)
- Form a communications ring for data relay back to Earth with spacing ~0.015AU, i.e. 360 satellites in a 0.9AU radius ring.

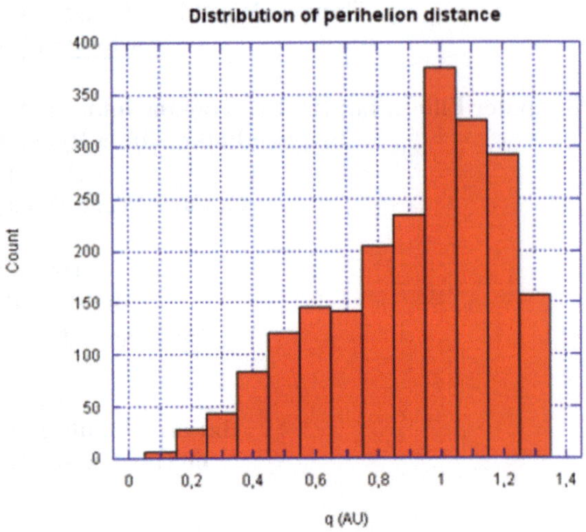

Figure 2: Perihelion distributions for NASA's 10 year Spaceguard survey sample of NEOs.

Figure 2 shows the perihelion distribution of all NEOs detected in NASA's 10 year Spaceguard survey. From this figure it becomes apparent that NEO perihelions lie in the range of 0.2 to 1.3AU. Our primary interest lies in NEOs with perihelion <= 1AU (ie Earth crossing). Therefore the best candidate orbit for this constellation is a circular (no drift or longitude libration) orbit with a radius between 0.8-0.9AU.

Potentially Hazardous Objects

Earth collisions can potentially happen for any NEO with perihelion < 1AU and inclination vector such that the ecliptic crossing is at ~1AU. For an arbitrary, unknown set of NEO's, risk of collision over an unlimited time period is related to the number of Earth orbit crossings within the resonance period, eg resonance = n NEO revs to m Earth revs – crosses Earth orbit at n different longitudes relative to Earth.

Such resonances can be for NEO's with semi-major axis (SMA) close to 1AU, eg 20:21 resonance, ie 20 crossings within the repeat period. But resonances are also possible with greater SMA, eg 20:41. Again has 20 crossings in the repeat period. The repeat period is now 41 years in comparison to 21 years which means a similar probability of Earth collision but probability per unit time is smaller.

Mission Baseline

The initial mission objective was:

- Transfer period of no more than 2 years
- 2 years observing NEOs
- Leave constellation and perform an in-situ visit of selected NEOs for approximately 2 years.

On-board propulsion should be a high specific impulse, micro-thrust system. A typical unit capability is 100 microN with Isp 5000secs (initial JPL cited impulse capability) which should be equivalent to 1000's m/s for few kg spacecraft. The initial baseline proposed transfer method was:

- Low energy escape via Lunar Gravity Assist (LGA) achieving 1500 m/s
- SMA of 0.9 AU with perihelion at 0.8 AU
- Use electric propulsion to speed up, slowdown in the drift orbit
- Use electric propulsion to circularise at required radius.

Following the availability of the detailed JPL MEP spreadsheet the mission baseline scenario had to be de-scoped and remove the in-situ visit as not-feasible. The latter statement is true for the studied technologies in development. Potential upgraded capabilities of the MEP system that JPL develops could offer this capability as well.

Starting with 1500 m/s (assuming LGA method for escape) and if we assume maximum available acceleration of 10^{-4}N/3kg then the target to achieve an equally spaced ring in solar longitude is achieved within 2 years. Example of relative longitude of 180 degrees after two years with DV 2km/sec and 0.9AU final semi-major axis is given in Figure 3 where the evolution is shown. Achieving relative longitude greater than 180 degrees and also achieving the required final orbit radius is more challenging than the previous case. Options are:

- Direct injection to further increase speed which has limited effectiveness
- Extension of transfer to 4 years
- Increase of thrust to 300μN
- Use a service module to deploy spacecraft in target orbits

Transfer Initialization

The following sequence can be used to initialise the transfer:

- Inject the constellation to an Earth elliptical orbit like GTO or sub-lunar orbit using a service module.
- A range of different Earth escape velocities should be targeted to achieve the required longitude spread to avoid excessive demands on the satellite propulsion. This can be achieved by different Lunar Gravity Assists (LGAs), each targeting a different velocity.
- Ideally each satellite would perform a different Lunar gravity assist to target its optimal escape velocity vector. In this case each satellite would need to execute and control the lunar gravity assist and would need a propulsion system compatible with that. This would include the apogee raising from sub-Lunar orbit.

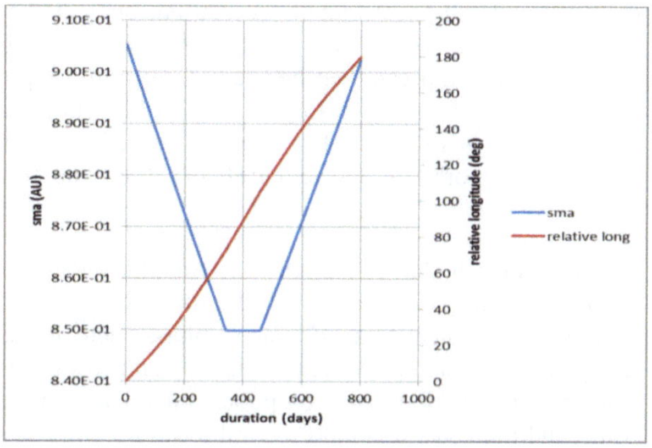

Figure 3: Evolution of semi-major axis and Earth relative longitude over 800 day transfer using updated mission baseline analysis.

Relatively high thrust is needed to perform the apogee raising/correction manoeuvres (ie acceleration of the order of 0.01 m/s^2. Alternatively groups of satellites can be collected onto a mini-dispenser equipped with a small propulsion unit. This dispenser unit would execute and control the lunar gravity assist. This would include the apogee raising from sub-Lunar orbit. The mini-dispensers are then separated from the main dispenser before each makes its LGA. Each of the spacecraft is then separated from the mini-dispenser post LGA and then achieves a range of target Earth relative longitudes by application of different thrusting strategies. All dispensers can start in the same initial orbit (ie on the same main dispenser), eg elliptical orbit such as GTO or sub-Lunar. Each group/mini-dispenser then instigates its own LGA by making a perigee manoeuvre to target the Lunar flyby.

- If starting from GTO, total DV to reach the moon and achieve LGA is 700 m/s

- If starting from 300000km, total DV to reach the moon and achieve LGA is 40 m/s
- High thrust is needed for this Lunar gravity assist and Earth escape mission phase.

Launch options

Low cost launch options are preferred. A dedicated launch to sub-Lunar orbit is possible, although potentially a more costly option than a shared launch to an intermediate orbit or injection with a small launcher to a lower orbit. Therefore two indirect scenarios can be considered:

Use of Rockot type launcher to LEO: In this case the injection orbit is similar to that used by Lisa Pathfinder (LPF): slightly elliptical with perigee altitude circa 200km and apogee circa 900km. A propulsion stage (eg the LPF propulsion module) can then be used to perform apogee raising to reach the sub-Lunar orbit. LPF allocates approx. 3100 m/s to reach L1. The DV needed to reach the sub-Lunar orbit is only circa 80 m/s less than this value.

Use of Ariane-5 shared launch: The Auxilliary Payload adaptor (ASAP 5) can potentially be used. This system was proposed when Ariane-5 was commissioned. It has rarely been used and its current availability and cost needs clarification. ASAP 5 has two options: micro-sats on the 'ring' and mini-sats in a more central location. The mass available for a mini-sat is approx. 300kg. Up to 4 mini-sats can be used on the adaptor. The injection orbit would be a shared launch to GTO. This can potentially be used as a starting point for an apogee raising sequence to reach a Lunar gravity assist. However the compliance required with the nominal midnight launch window may result in an unfavourable apse location for the transfer after launch at certain times of year. Additional transfer DeltaV is needed in these cases. The nominal DeltaV for apogee raising from GTO to a 300000km sub-lunar orbit is circa 700 m/s. This can rise to over 1000 m/s at unfavourable launch dates. Therefore a propulsion module is needed to deliver the required DeltaV.

NEO In-situ visit

Initial concept proposed a transfer to NEOs in order to perform an in-situ examination of a large sample of NEOs. Then each of the spacecraft would transfer back to the constellation in a round trip estimated to last two years. Transfers are limited by rate at which DV could be delivered which is calculated approximately to 2000 m/s per year with continuous thrust. To achieve a successful transfer to a NEO the following criteria need to be satisfied:

- DV is low enough (restricts range of apohelions, perihelions, inclinations, nodes, perihelion longitudes that can be accessed).
- The phasing of the NEO in its orbit must be suitable
- Given a large scale observation, the closest member of the constellation can be selected to minimise the transfer duration and required DV. This solves perihelion longitude and node issues.

Mission Design Baseline

Following all the analysis above there are two alternative options for this mission. Both of them use a single launch to an elliptical orbit, eg to sub-Lunar, or to lower orbit, followed by apogee raising with a dedicated propulsion stage to reach sub-Lunar orbit.

1) Separate mini-dispensers from the main dispenser are put in this orbit. Each mini-dispenser will then execute a manoeuvre to target with its required LGA: manoeuvre with 40 m/s plus 30 m/s for corrections. High thrust is needed (ie chemical propulsion). After the LGA the satellites are dispensed from the mini-dispensers.

2) All satellites separate from the main dispenser in sub-Lunar orbit. Each satellite would perform a different Lunar gravity assist to target its optimal escape velocity vector. In this case each satellite would need to execute and control the lunar gravity assist and would need a propulsion system compatible with that. This would include the apogee raising from sub-Lunar orbit. Manoeuvre capability of at least 40 m/s plus 30 m/s for all corrections is required.

Each satellite will then follow a transfer of up to three years until the required constellation is achieved. Satellite DV for transfer will be up to 5km/sec using a low thrust SEP system.

III. PROPULSION

The SOFIA mission analysis in Section II showed that the following propulsion requirements (Figure 4) are needed which assumes that the spacecraft start off in a sub-Lunar orbit. In order to achieve this, a separate propulsion module (for example, the Lisa PathFinder propulsion module) could be used to transport all the SOFIA satellites together from the launcher to the sub-Lunar orbit.

LGA

The LGA manoeuvre requires high thrust. For this, the proposed JPL Microfluidic Electrospray Propulsion (MEP) thrusters would not be suitable, as their thrust is in the order of micro-Newtons. A survey of existing technology was performed and a viable option found in the VACCO Chemically Etched Micro Systems (ChEMSTM) propulsion modules, specifically the Hybrid ADN Delta-V / RCS system, seen below in Figure 5.

	ΔV Requirement
Low energy escape via Lunar Gravity Assist (LGA)	80 m/s
Transfer to observation orbit	5000 m/s
Rendezvous with NEO	4000 m/s

Figure 4: DV requirements

Figure 5: VACCO propulsion system

The VACCO module is equipped with one 100mN ECAPS ADN thruster to provide the delta-V, and four 10mN cold gas ACS thrusters for attitude control. The ACS thrusters will be used to stabilise the spacecraft during the delta-V manoeuvres, but not for the operational AOCS manoeuvres (for which the JPL MEP thrusters will be used). As the system is a self-contained module, it can be easily integrated into the SOFIA spacecraft with reduced AIT effort in comparison to a bespoke system. A trade-off was performed for the 80m/s LGA manoeuvre between the VACCO system, the JPL MEP and an integrated Butane system from JPL, built for the Microinspector spacecraft.

From our trade-off analysis, we concluded that the VACCO module can achieve the delta-V required in just over 2 hours. This assumes the thruster is firing constantly, which may not realistically be the case; however the order of magnitude is acceptable. The fuel mass required for this manoeuvre is 400g which fits within the module's total fuel capacity of 528g.

Transfer to Operational Orbit

The transfer to the operational orbit requires a very high delta-V of 5000 m/s. This would be achieved using the JPL MEP thrusters (Figure 6) due to their high predicted ISP. JPL claims that they will be able to achieve an ISP of 5000 seconds, so this is the figure used in the calculations.

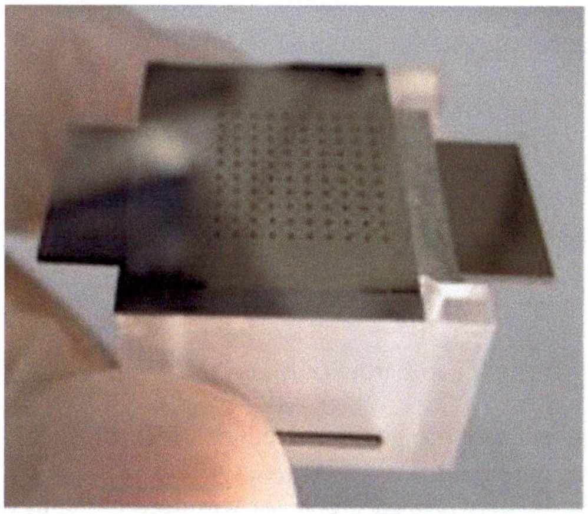

Figure 6: JPL MEP thruster

Assuming 7 thrusters each providing 100 μN of thrust in the same direction, the delta-V manoeuvre can be achieved in 2.17 years. The fuel required for this would be 930 g, which is outside the current capacity of the thrusters of 10g each, however JPL intends to increase the throughput capacity of the thrusters. It is therefore assumed that it will be possible to do so.

NEO In-situ visit

As the current technology is stretched past its capacity with the delta-V manoeuvre to the observation orbit, it is not considered feasible to achieve the additional 4000 m/s required for the rendezvous.

Propulsion System Configuration

The propulsion system will consist of the VACCO Hybrid module and 7 JPL MEP thrusters for delta-V and AOCS manoeuvres. The VACCO module and the 7 MEP thrusters will all provide thrust in the same direction. The total wet mass of the propulsion system is 2799.41g. The budget assumes that the VACCO propulsion module is fully fuelled, but that the MEP thrusters are only fuelled with the required propellant load. 2g of propellant has been added for the MEP thrusters for AOCS.

IV. COMMUNICATIONS

The key driver for the communications system for SOFIA is up to 1000 bits of telemetry per Cubesat over the lifetime of the mission. Before the telemetry rate is reviewed the basic range requirements could be determined from the orbital geometry. As defined above it is proposed to deploy 360 cubesats equally spaced 1° around the Sun at a distance of 0.9AU with communications relayed from satellite to satellite. This leads to secondary requirements on the communications subsystem.

1. Nominal intra-satellite range ~2.35Mkm
2. Nominal intra-satellite Off boresight angle = 0.5°
3. 1 failed Cubesat intra-satellite range ~4.7Mkm
4. 1 failed Cubesat, Off boresight angle, 1°
5. Closest range to Earth ~15Mkm
6. Boresight Range to Earth ~65Mkm

Given the extreme ranges involved in order to enable successful communications between Earth and every satellite in the network, each cubesat will need to act as a data relay satellite so that Earth based communication only occurs between Earth and the nearest of the cubesats. The commands and telemetry are then routed round the network to required destination.

Looking into Data Handling it is recommended that the CCSDS File Delivery Protocol (CFDP) be employed to enable this functionality. This protocol is specifically tuned for long-delay and disrupted communication links between a network of relay nodes, which each cubesat would form a part of. So far some American missions (LRO, Messenger & JWST) are known to use CFDP however no known European mission has yet employed this protocol. It is known that SciSys in conjunction with Keltik and Trinity College Dublin (TCD), have been conducting just such a study for ESOC. It is thought that the soon to be implemented European Data Relay Satellites (EDRS) must also employ a similar receive and re-transmit data protocol.

The telemetry requirement above is based on a fixed number of photos being transmitted back to Earth that allow ground based analysis and detection of the target NEOs. Even at 1kb the data volume is minimal in terms of subsystem design driver. It is likely that any housekeeping telemetry will greatly exceed the max 1kb science return per cubesat.

Assuming 24 hour ground station coverage, each cubesat in the network could have a dedicated communications window of 240 seconds each day. Baselining a telemetry data rate of 50bps, would equate to a total daily downlink data volume per cubesat of 12kbits (1.5kBytes).

Given the proposed 3-axis stabilised configuration and the limited communication window every cubesat would have to be vritually autonomous. It is therefore expected that whilst in nominal operational mode each cubesat would downlink mostly science data and some mode / positioning data. When in safe mode, science data downlink would cease with the downlink mainly composed of AOCS and equipment health data.

The orbital geometry leads to a simple design that allows simultaneous sun pointing, communications and science operations. Two configuration options are immediately obvious;

1) A single antenna on one side of the cubesat. Cubesat rotates about Z-axis to point at +X and –X cubesats in turn.

- This requires Macro AOCS control

- Only one link possible at a time.

2) Two antennae on opposite sides of the cube. One always pointed at +X cubesat, the other pointed at –X cubesat.

- Requires only position stabilisation control

- Constant 2 way link with cubesat network possible.

Due to the necessity for both cubesats to be pointed at each other in order to successfully communicate, option 1 is not recommended. Option 2, which uses two fixed antennae on opposite sides of the cubesat is recommended. This allows for potentially constant intra-network communications and a fully 3-axis stabilised design greatly simplifying the AOCS steering / pointing requirements.

The proposed cubesat configuration easily allows for intra-satellite communications. However as depicted in Figure 7, the Earth direction at closest approach is not within the field of view of either antenna. It is therefore proposed that Earth communications do not use the nearest cubesat but rather the next cubesat in the chain. From Figure 8 we can see that the Earth is directly on boresight of one of the X-face mounted antennae on the trailing cubesat. The significant increase in range, from 2.35Mkm to 65Mkm, will in the next section, be shown to be offset by use of the large ground station antenna, making the requirement for a dedicated Earth communication antenna obsolete.

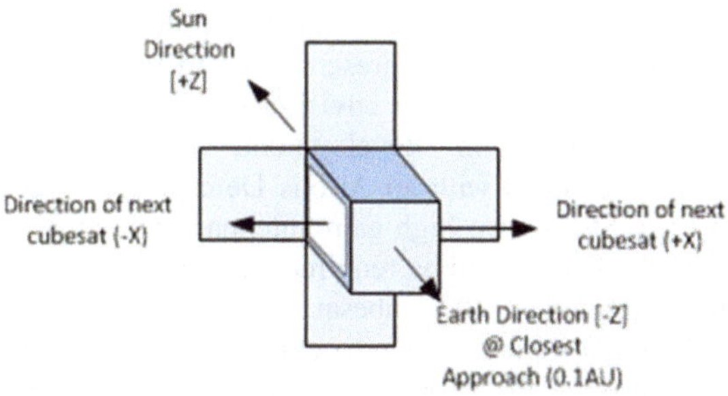

Figure 7: Simplified Cubesat geometry

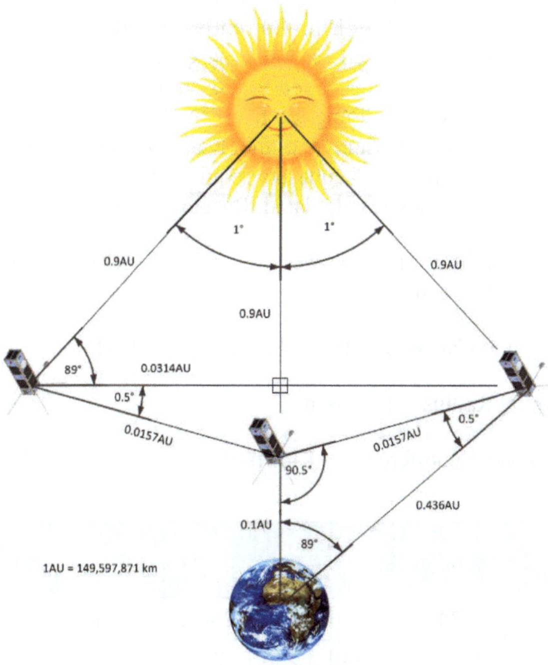

Figure 8: Orbital geometry (not to scale)

Link Budget

Taking current hardware technology out of the equation the necessary spacecraft equivalent isotropically radiated power (EIRP) can be determined in order to close the RF link. Once the necessary EIRP has been determined a simple trade-off follows in order to define the required antenna and power amplifier size in order to produce the required EIRP.

Baseline assumptions:

1) Data Rate (raw) = 50bps

2) Modulation = PCM(NRZ-L)/PSK/PM (according to ECSS Std for data rates <60kbps)

3) Coding = Turbo Code Rate 1/4, Frame Size 8920

4) Intra-satellite range (1 failed satellite) ~4.7Mkm

5) Cubesat to Earth Range 65Mkm

6) TM Margin = 3dB

7) Receiver Carrier Tracking Threshold = -150dBm

8) Receiver PLL = 7Hz

9) Fairbanks 13m G/S (X-band baselined)

The links are in large part driven by the receiver carrier tracking threshold, more so than the data rate which is unusual, but then so are the distances involved with this study. Nominal intra cubesat link: Irrespective of frequency used, a Satellite EIRP of 42.12dBW is required to close the link. Non nominal (one failed cubesat) intra-cubesat link: irrespective of frequency used, a Satellite EIRP of 48.14dBW is required to close the link. Cubesat to Earth downlink: irrespective of frequency used, a Satellite EIRP of 35dBW is required to close the link.

As expected the analysis concludes that the intra-satellite link is clearly the design driver, irrespective of whether there is a failed satellite in the chain or not. Two options therefore exist for the ground station. Either a reduction in ground station dish diameter from 13m to ~7.5m would be possible, opening up the number of available ground stations, with an expected corresponding reduction in costs; Or an increase of the Earth link telemetry rate to ~10kbps. This would be especially beneficial where no failed satellite exists in the chain. The communications subsystem should therefore be sized for 1 failed cubesat intra-satellite link where the required EIRP is equal to 48.14dBW i.e. worst case.

Figure 9 below has been created to demonstrate how the antenna diameter sizes vs amplifier power for the three main frequency bands utilised by spacecraft. The EIRP is fixed at 48.14dBW.

• S-Band (Deep Space): 2290 - 2300MHz

• X-Band (Deep Space): 8400 - 8450MHz

• Ka-Band (Deep Space): 31800 - 32300MHz

This clearly shows that for a fixed transmit power, the higher the frequency, the smaller the required antenna diameter for a given EIRP.

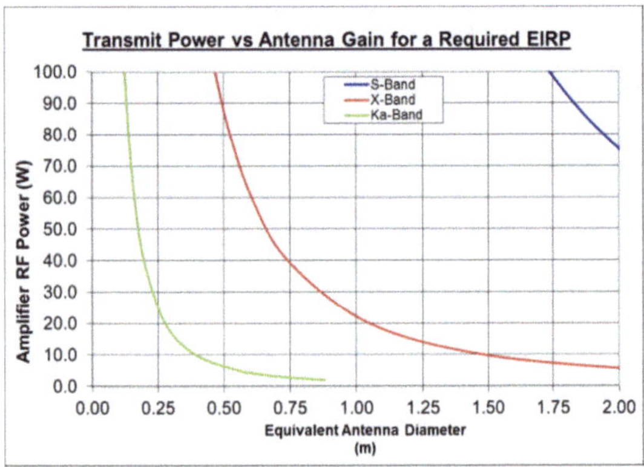

Figure 9: EIRP Comparison. Amplifier Pwr vs Antenna Diameter by Frequency Band

From Figure 9 it becomes apparent that only Ka-Band stands out as potentially feasible for this mission given the limited physical size of the satellite. It is plausible that up to a 30cm diameter planar array antenna could be utilised (fold out panels), however this would likely be quite heavy and would still require a ~17W Transmitter. At 20cm diameter ~40W would be required. At 10cm diameter ~160W would be required.

Hardware Selection

Cubesat communication subsystems are still in their relative infancy. Up until recently most cubesats have relied on UHF / VHF for low Earth orbit communications. In the past couple of years S-band systems have been demonstrated, first transmit only but more recently in receive as well.

• X-band transmitters are now beginning to be demonstrated but with low transmit powers.

• Syrlinks EWC 27 HDR X Band Transmitter (up to 2W).

• IRIS DSN Compatible Small Satellite Navigation and Communications Transponder (0.25W)

• Cubesat size X-band GaN SSPA's are in development and ground test (Airbus DS & JPL). Technology being developed allows adaption to work at Ka-Band. Potentially capable of 17.5W RF output.

Only known X-band receiver in development is IRIS Transponder mentioned above with Rx sensitivity of -130dBm. Additional development would be required to increase the threshold sensitivity to ~-150dBm). Ka-Band transmitter (IRIS ka-band development) and reflecta-array (35dBi) have been developed for a JPL cubesat (ISARA). No known Ka-band receivers in development.

Despite the large amount of investment currently being provided by NASA into cubesat subsystem development, the technology does not at present exist to perform the SOFIA mission as currently envisioned. Using the best known technology in development, IRIS X-band Transponder coupled with an Airbus Defense and Space GaN 17.5W SSPA, two high gain antennae both equal to ~1.2m in diameter would be required, which is unfeasible for the current 1 x 3U cubesat baseline. Non flight inflatable antennae are known to exist (GATR inflatable HGA's), however these are mass intensive (~26kg for a 1.2m dish) and will not meet the launch envelope and mass requirements. Whilst the transmit side of the RF link is not possible without the move to ka-band, the receive side also needs development to increase the sensitivity of the receiver to ~-150dBm. Currently the IRIS receiver development is focused on Earth based communications out to Mars (2.6AU max) and based on the use of a 70m Ground station. Baselining the IRIS Transceiver (both the existing X-band variant and the Ka-band development variant) the number of cubesats required for a successful chain network can be calculated. Baseline assumptions;

• Transmitter RF Power = 17.5W (Portsmouth development SSPA)

• Receiver Carrier Tracking Threshold = -130dBm

• Orbital Radius = 135Mkm

• Orbital Circumference = 846Mkm

	10 cm	20 cm	30 cm	40 cm	50 cm	60 cm
X	39,717	9,953	4,418	2,482	1,589	1,106
Ka	10,575	2,699	1,238	727	492	367

Figure 10: Number of Required Satellites in Chain (1 failed scenario assumed)

The numbers specified in Figure 10, in all instances exceed the number of cubesats allocated to the baseline mission. What makes the scenario worse is that this was evaluated using technology that is not yet flight rated (Airbus DS SSPA and Antenna). The analysis has identified that direct to Earth communications were not the subsystem design driver. As such this section will investigate the possibility of a mission where every cubesat communicates directly with Earth rather than to the next cubesat in the chain. The number of cubsats in this scenario is not limited by the communications subsystem but rather by the Field of View of the payload.

Baselining a 15m Ground Station (ESA's Korou) and the IRIS X-Band Transceiver operating at 2W, coupled with an Airbus DS SSPA as per Section 6.1 both the following scenarios are analysed;

- 15m Ground station

- 35m Ground station

For the 15m ground station although it is possible to establish a downlink (125bps increasing as the range decreases by $1/R^2$) from 1.9AU it is not possible to successfully uplink a telecommand unless the Earth to Satellite range is <0.84AU. Given that the worst case Earth to Cubesat range is to be ~1.9AU this provides an incomplete communications solution.

For the 35m ground station a downlink with increased telemetry data rate (up to 4kbps) is possible from 1.9AU generally increasing as the range decreases by $1/R^2$. A standard ESA uplink data rate of 2ks/s is also possible. Given the limited availability of the 35m ground stations it is proposed to only use the 35m dishes for telecommanding and baseline the more plentiful 15m dishes for the downlink. Although not analysed in any detail here it should be noted that direct to Earth communications will be not be possible during super conjunction (up to a SES angle of ~5°) and will be degraded by between 4dB at 5° SES to 15dB during inferior conjuction.

The alternative design detailed above utilises two technologies that are not flight capable:

1). SSPA

- 17.5W RF output not currently met, further development required.

- Frequency range of phase 1 development model at wrong frequency range. 8.5GHz required.

- Dimensions exceed cubesat standard with one dimension at 160mm.

2). Antenna

Currently baselined is a custom microstrip phased array based on the design produced by Airbus Defence and Space Antenna group for the APPS program. This was a 10cm x 10cm single panel array, weighing 156g per panel and requiring 240mW of DC power. It should be noted this was a design only, no hardware was manufactured and is such considered a new development. Each SOFIA panel, of which there will be 5, will be the equivalent of four APPS panel arrays. The SOFIA Antenna baselined is to be 20 x 60cm on the back of the solar array panel. Predicted performance is 32.6dBi and is focused in the plane of the ecliptic.

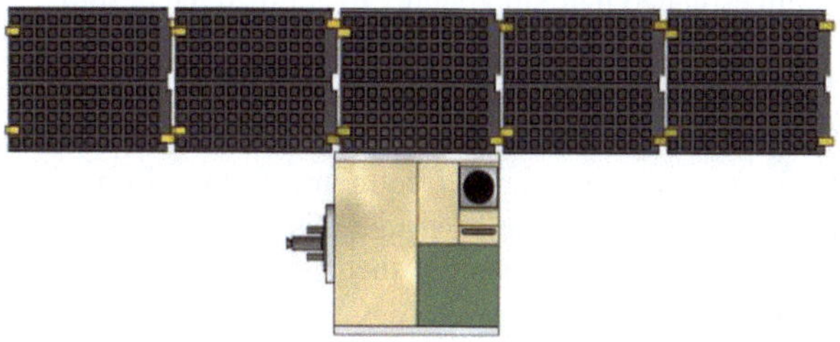

Figure 11: Baseline Configuration Showing Antenna

V. THERMAL DESIGN

The sun illumination angles based on the orbit of the mission, suggest that the PX, MX, PZ & MZ faces should ideally all be completely isolated with MLI, as they will be all be illuminated at various different times during the mission, as shown below. The PY and MY surfaces would then be the optimum candidates for radiators for any items which are expected to dissipate throughout the mission (eg. the PCDU and the OBC).

However, during the Earth communication phase the total internal dissipation of the spacecraft is estimated at around 60W. This requires approximately 0.16m² of radiator area to dissipate, but each side of the cube has an area of only 0.04m², so 4 sides would be needed as radiators to achieve an acceptable steady state temperature.

For the very short communications phases (eg. <10 mins), the communications units may stay within limits simply due to their own inertia. For the longer communications phases, it is worth noting that the MX face will be in shadow for the vast majority of communications windows. Unfortunately the MX face is directly exposed to the sun during closest-Earth communication window, and would then receive approximately 68W of solar flux. And of course, this face will face the sun during all nominal NEO observation windows, so this is unlikely to be an acceptable solution.

Very approximately, if the mass of the whole spacecraft is ~10kg, and the average thermal capacity is taken as 900J/kg/K (which is the appropriate value for aluminium), then an 60W internal power dissipation would cause an increase of about 30°C an hour (assuming no heat loss, which is very conservative). In principle, if the spacecraft is kept cold immediately prior to communication, (by switching most units off), and if the communication units are sufficiently well coupled to the rest of the spacecraft bulk mass, then the thermal inertia may be sufficient to keep the units below their limit during the communication window. This would have to be investigated more thoroughly to confirm the design.

VI. AOCS

The following activities during which the AOCS has to fulfil specific performance requirements to help achieve the mission goals have been identified from the mission concept. Note that, though in theory every activity could be represented by a separate AOCS mode, to optimise the overall mission concept and the AOCS more than one activity might be performed in the same AOCS mode.

Activity	Description
P/L Survey	Search for new NEOs and measuring of their orbital parameters using the P/L camera. For visibility reasons, the survey has to be performed in the direction away from the Sun
Comms	Establishing a COM link to Earth
Transition	Transition from one pointing direction to another
Orbit/Position Determination	Detection of the position of the S/C
Propulsion	Delta-V manoeuvres
Wheels Unloading	Unloading of the reaction wheels
Charging	Charging the battery
Detumbling	Detumbling after separation or in case of an accident
Safe	Safe mode

Figure 12: Activity description

The most challenging identified requirements for SOFIA's AOCS are as follows:

- Absolute Knowledge Error (AKE): < 2 arcsec
- Absolute Pointing Error (APE): < 1 deg (during delta-v activity)
- Pointing Stability: < 2 deg/s
- Maximum slew rate: > 2 deg/s

Obviously, the main design drivers are the AKE during P/L operation and the APE during propulsion usage (high torque). Note that the stability requirement for the survey is not particularly strict because a star tracker is going to be used as P/L camera.

The accuracy requirements for the determination of the S/C position relatively to the sun are assumed to be of the same magnitude as the accuracy requirements for the determination of the NEO's position relatively to the S/C.

To fulfil the challenging AKE requirement a star tracker is needed. This sensor also has the advantage that it provides a 3-axis estimate using only a single sensor.

To detect the current S/C position a set of sensors is needed. A feasible configuration for a CubeSat is as follows:

1. Medium precision sun sensor (~0.2 deg): The attitude of the S/C is already known because of the star tracker. So the sun sensor can be used to detect in which direction relatively to the S/C the Sun is. This reduces the possible positions of the S/C to a single line.

2. COMS delay to ground: The exact delay of a COMS signal from Earth to the S/C and back defines a shell of possible S/C positions around Earth.

From the two intersection points of these geometric shapes, the right one can be selected by knowledge of previous S/C positions and orbit propagation methods. Several measurements of the S/C position over time are then used to determine its orbital parameters. No analysis about the accuracy of this procedure has been performed yet. However, it is assumed to be sufficient.

The APE requirement is much less challenging than the AKE one and the torques – when no propulsion is active –

are assumed to be small (mainly SRP). Therefore relatively weak and in-precise actuators would be sufficient. The main torque experienced by the S/C when doing observations (solar panel array facing the sun, P/L sensor facing anti-sun direction) is the Solar Radiation Pressure (SRP).

The S/C will have to regularly change its attitude by large angles to either point the solar panel array towards the Sun or the antenna array (rear of solar panel) towards the Earth. To not waste energy (i.e. fuel) on these repetitive manoeuvres, reaction wheels are the way forward for closed-loop control of the S/C attitude during most activities. The only exceptions are: delta-v manoeuvres using the high thrust system (ADN) and the 'safe mode'.

The ADN propulsion module includes its own closed-loop attitude control system (4 thrusters). This module is not going to be analysed here. It is assumed that as long as the alignment/positioning requirements of the module relatively to S/C CoM are fulfilled, the attitude control result will be sufficient.

Reaction Wheel (RW) unloading will be performed in the following ways:

• X RW: Change of the y-angle (pitch) of the S/C during P/L operation to invert the torque induced by SRP.

• Z RW: Mostly the same as for the x RW + plus some torque from thrusters (because z cannot be unloaded separately from x).

• Y RW: Thrusters.

The largest momentum input will be about the x-axis because of the placement of the antenna/solar panel. Usage of the pitch angle of the S/C to unload the reaction wheel about this axis is possible since there is no requirement on the pitch angle during P/L operations. This strategy leads to a reduction in fuel mass needed for attitude control by nearly 95 percent.

The following COTS CubeSat bus has been selected to be used (in a tailored version):

• Blue Canyon Technologies: XB1

This module fills 1 CubeSat unit and contains the following AOCS HW satisfying the HW requirements defined earlier:

• 2 Star Trackers (BCT: Nano Star Tracker)

• 3 Reaction Wheels (BCT: Micro Reaction Wheel)

• IMU

• Magnetometer

• Magnetorquers

• GPS

This module was chosen over the XACT module (1/2 CubeSat units) because of its additional bus functionalities. These include:

• OBC

• EPS (including battery)

• Control for P/L, Propulsion, Thermal and Solar Panels

• COMS module

Note that the star tracker of this module will be used as P/L camera.

This module could be used of the shelf – ignoring radiation issues (see corresponding chapter). However, for mass/power/volume optimisation it is assumed to be used in a tailored version removing the following components:

• 1 Star Tracker (1 single star tracker satisfies the requirements)

• Magnetorquers (cannot be used in any phase of the mission)

• COMS module (wrong band for this mission)

Other components like the GPS chip and the magnetometer could be removed but their presumed mass/power/volume needs do not seem to justify the effort. There is detailed data available only about the star tracker; therefore only that device is removed from the mass and power budget (assuming only removing 85 percent of mass/power of the stand-alone version).

The following sun sensor has been selected:

• Sinclair Interplanetary: SS-411

With the following performance:

• Accuracy +/- 0.1 over +/- 70 FoV

Two sun sensors are included in the baseline to be able to deal with different attitudes the S/C has when communicating with Earth.

The following thrusters are needed by the propulsion subsystem:

• 1 Vacco: Hybrid ADN Delta-V /RCS System (high thrust)

• 7 JPL: Microfluidic Electrospray Propulsion (MEP) Thrusters (low thrust)

The configuration is shown in Figure 12. During high thrust manoeuvres (moon swing-by) the ADN unit is used. During low thrust manoeuvres the MEP thrusters are used. The configuration shown here is sufficient for unloading the RWs assuming that the following assumptions are fulfilled:

• The S/C is in nominal attitude most of the day: Solar panel facing sun, star tracker pointing in anti-sun direction (+/- 20 deg).

• For the majority of the day the pitch angle can be freely chosen.

With these assumptions the pitch angle can be used to unload the x RW before it is saturated (happening in >= 12 hours). The x-axis has the highest disturbance torque input (due to the non-symmetric shape of the S/C assembly) and there is no controllability about this axis by MEP thrusters in this configuration. The ADN module's ACS thrusters are not assumed to be used because its fuel is reserved for the attitude control during high thrust delta-v manoeuvres. If these assumptions are not fulfilled, two MEP thrusters will have to be added to allow the unloading of the x-RW and the fuel mass has to be re-calculated (increase by more than factor 15). The current configuration provides at least the following torque capability (+/-) about the y- and z-axes: 1.5e-05 Nm. This is nearly two orders of magnitude larger than the largest disturbance torque, which is sufficient for wheel unloading. Note: It would also be sufficient for direct control (e.g. for a safe-mode).

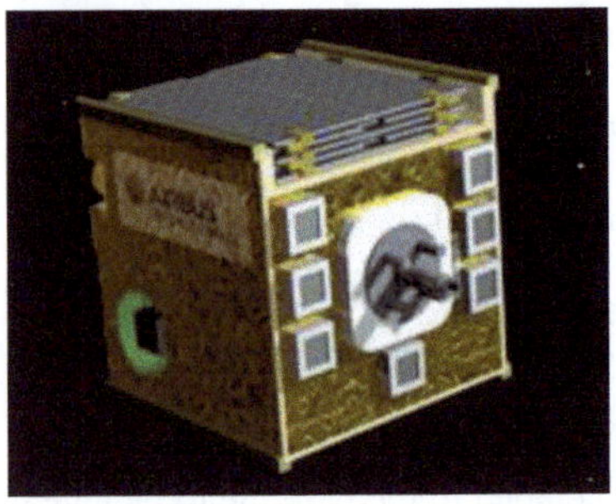

Figure 13: Thruster configuration

VII. POWER SYSTEM DESIGN

Based on our analysis:

• The average power consumption is about 32W per hour along the whole SOFIA mission, with peaks and lower power requirements spread depending on the active specific mode;

• The peak power consumption is about 64W and it is required in correspondence with autonomous communications with Ground. This is mainly driven by the fact that, due to the big distance from Earth, it is required to use an SSPA for providing enough RF power for transmitting the whole set of scientific and/or telemetry data. An additional requirement is that during communication with Ground SOFIA will be Earth-pointing

SOFIA's power system has been designed to be simple and capable to meet the power requirements described in the previous section. For this reason, it includes the following components:

• A large Solar Array, capable to generate enough power to sustain nominal operations and to recharge the batteries,

• A set of Batteries, to support power peak requirements and to provide the required energy during periods in which the solar array is not generating power, i.e. eclipse periods along the transfer orbit, and when communicating with Ground implies to lose the Sun-pointing attitude and therefore a reduced amount of solar power can be collected,

• An Electronic Power System (EPS), which enable power conditioning and distribution between all the subsystems.

In practice, taking into account that SOFIA is a 2U side x 2U side Cubesat, a couple of 2U solar panels has been considered for each side. Each side then can ideally generate up to 10.4W (using Clyde Space product catalog). Therefore, in order to take into account the required average power of about 32W and the need for battery recharge and margins on solar cells failure, a Solar Array composed by five (5) solar panels (each solar panel large as a SOFIA side) has been assumed. The maximum power generation is then estimated as about 52W.

The sizing case for the battery resulted to be when SOFIA is communicating to Ground while in its operational case. In fact, as worst case, SOFIA is Earth-pointing and the Solar Array is not "facing" the Sun. In this condition, the battery has to provide up to 64W per hour. Taking into account that a 25Whr battery module is already included in the XB1 AOCS component, two 30Whr battery modules supplied from Clyde Space have been selected.

VIII. ACCOMMODATION/STRUCTURE

A key advantage of the cubesat technology is the modular approach that enables off the shelf technology to be used, therefore reducing the amount of costly bespoke design and development. The design of the SC shall utilise as much existing cubesat technology as possible. Some bespoke structure will be required to enable accommodation. The structure of the SC shall accommodate the following key hardware items;

- Magnetometer
- X-Band Communication System
- Antenna
- Solar Arrays
- Batteries
- ADN Thruster
- MEP
- Propellant for MEP
- AOCS Unit
- Sun Sensors

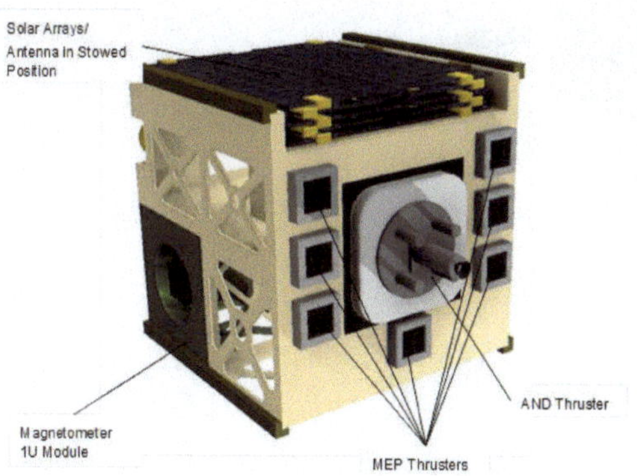

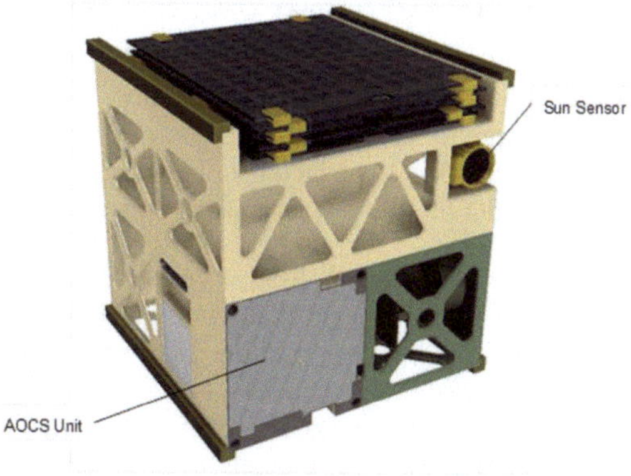

Figure 14: SOFIA LH (top), RH (bottom) view.

The SC will be deployed in a pod. For deployment four rails are required on the edges. The solar arrays will be required to be recessed into the body of the SC. Hold down mechanisms will be required for launch. After launch the arrays will deploy in two stages, see Figure 15. This will be achieved using standard spring driven deployment technology and off the shelf hold down mechanisms.

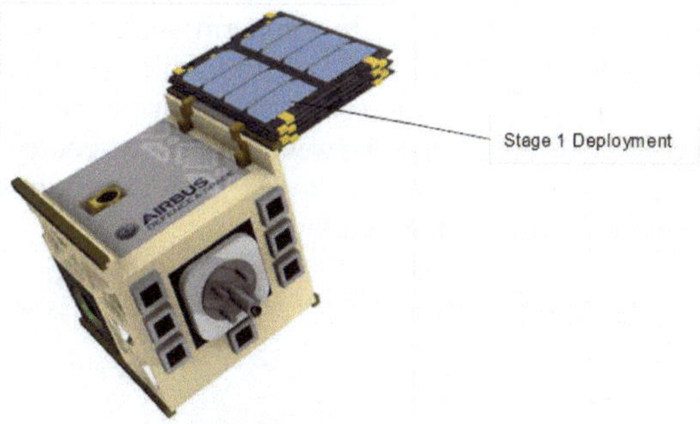

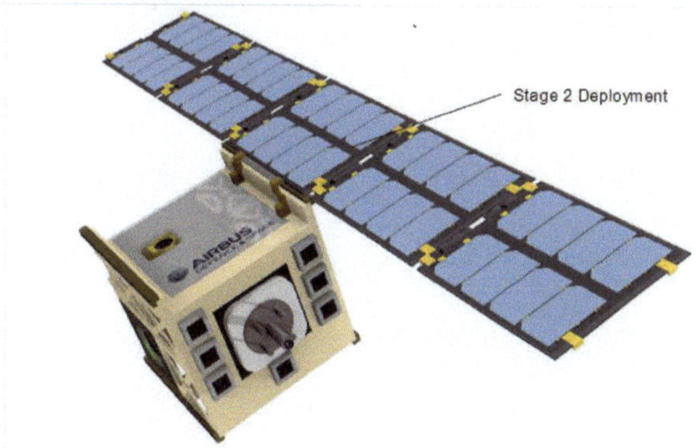

Figure 15: SOFIA solar array deployment

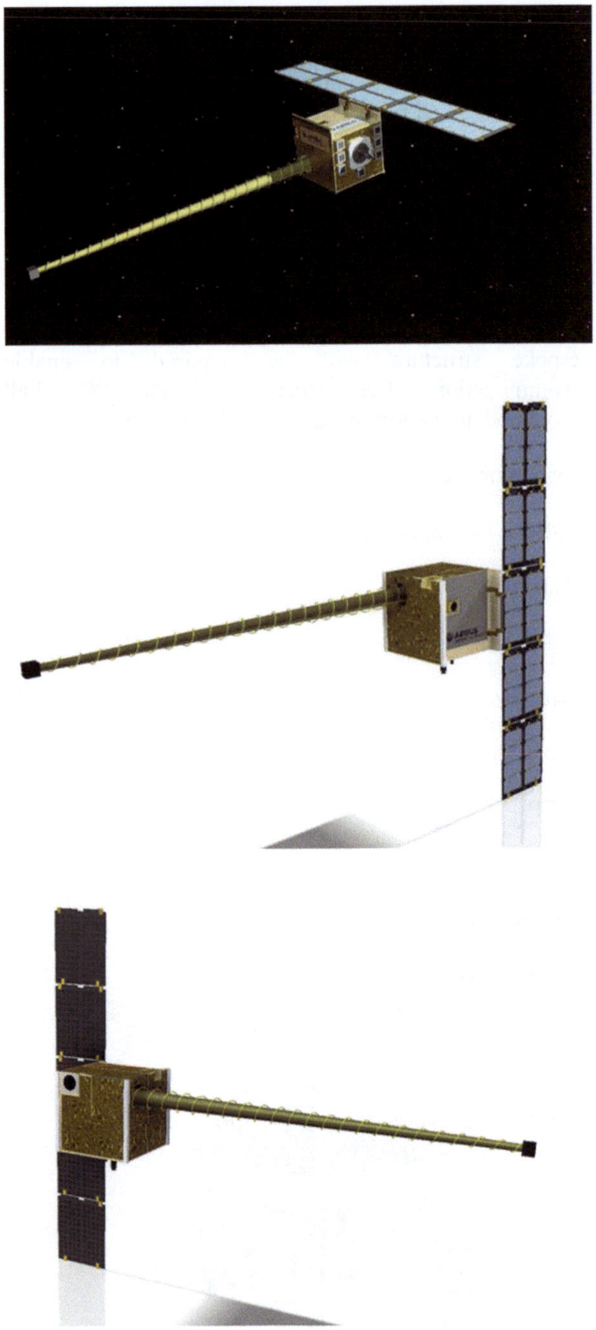

A compact magnetometer will be included as part of the payload. The concept will be similar to previous Cubesat missions where a compact Magnetometer is deployed using a Stacer spring producing a 1m long boom. This will help predict hazardous space weather events.

Initial concept is to launch from a customised dispenser similar to the XPOD system that has flight heritage. The structure is held in a dispenser at the guide rail hard points and along the guide rails themselves. Upon lid release a spring pushes the structure out, sliding at the guide rail to dispenser interfaces. The Launch pusher will need to allow for propulsion system thrusters emerging from base of the unit. The pusher interface points will be at the base of the guide-rails (hard-points) for the structure.

VIII. SUMMARY

With this study we tried to assess whether a large formation of Cubesats could perform an extensive survey and in-situ characterization of NEOs, even those that are affected by the synodic problem. Although the in-situ characterization is far beyond the current technological capabilities we have shown that a 10Kg spacecraft not only could travel to 1.9 AU but it is capable of transmitting a large portion of science data directly back to Earth, without the need of the initially planned ring formation. We established that communications, propulsion and AOCS technologies either exist or are in advanced development statge, to perform such a task. Radiation is still a a major problem and to allow such a mission to survive the hostile space environment lots of development will be needed. However we have secured Airbus Defence and Space funds to take this interplanetary Cubesat to the next level.

BIS-RS-2014-47

The North Star Rocket Family

C.J. Verberne
Nammo Raufoss AS
Enggata 37, NO-2830 Raufoss, Norway; +47 91555817
cj.verberne@nammo.com

ABSTRACT

The North Star Rocket Family is a family of rockets based on hybrid rocket motors clustered together to form 2-stage sounding rockets for scientific research. Two sizes of 2-stage sounding rockets are planned and together these "launcher elements" form the basis for a small launcher. The initial version of the small launcher will be able to serve the market for dedicated launches of 20-25kg into a 250-350km SSO orbit. This market will be extremely cost sensitive and is seen as the low end of the space market. It is crucial for the concept that the component cost is kept low enough. "Space qualified" components are not likely to fit into this model.

Commonality between the sounding rocket market and the small launcher segment are the basis for obtaining economy-of-scale. The selection of components is tailored towards "industry quality" components, characterized by manufacturing processes which provide reliability through robust processes and sufficient quantities. The intended launcher will strive to be the smallest possible to do the job, and size alone will already keep the cost down. Only if the launcher is small enough can the propulsion system also be used as the propulsion system for the sounding rocket market, which is urgently looking for new motors. The propulsion system constitutes the single largest sub-system of a launcher and has therefore been the primary focus of the work.

To reach our goal, disruptive thinking is needed to assemble a new rocket based on a novel propulsion system. But the success also builds on the modularity aspects of the hybrid propulsion, the ease of upgrading it and the inherent safe behavior. The North Star project is the only project around which attacks the launch cost from below, i.e. by developing an efficient, safe and cost effective propulsion system first. The first firings of the scaled-up hybrid motors have been successful.

KEYWORDS: Launch System, Micro-Launcher, Sounding Rocket, Low Cost, Hybrid Propulsion

INTRODUCTION

In Norway, Andøya Space Center (ASC) operates a launch site for sounding rockets and scientific balloons. The site has been operative since 1962 and has launched around 1500 sounding rockets (see Figure 1 and Figure 2). It has a close cooperation with ESA, NASA, DLR and JAXA on scientific sounding rocket campaigns. Andøya Space Center offers launch services capable of rockets up to 20 tons. ASC can construct their own payloads and offers space education through their subsidiary NAROM (Norwegian Centre for Space-related Education). Andøya Space Center is located at 2 degrees north of the Arctic Circle in northern Norway, making it ideally situated for scientific research in the auroral oval (see Figure 3). The supply of adequate rocket motors for these scientific experiments has become increasingly more difficult.

Figure 1 The launch of a LOX/HTPB hybrid rocket from Andøya[1] (Picture: ASC)

S. Hatton (ed.), *Proceedings of the 12th Reinventing Space Conference*, DOI 10.1007/978-3-319-34024-1_19

Figure 2 Sounding rocket launch from Andøya Space Center in Northern Norway (Picture: ASC)

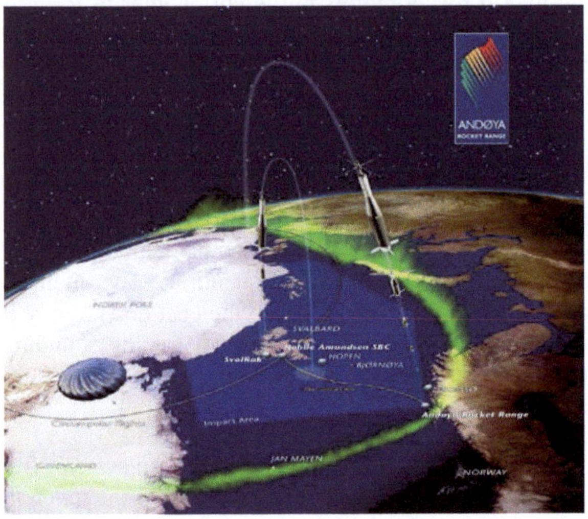

Figure 3 Andøya Space Center located north of the Polar Circle (Source: ASC)

Andøya Space Center does not manufacture rocket motors and is dependent on their customers to provide their own rocket motors or acquire rocket motors on the open market. Increasing difficulties in acquiring suitable rocket motors for their customers has made them approach the rocket motor manufacturer Nammo Raufoss (Nammo) for a solution to this problem. Recent successes with hybrid propulsion made by Nammo facilitated an opportunity to develop environmentally friendly, flexible and safe rocket motors.

In 2008, the rocket range and Nammo conceived the vision to cooperate on the realization of a family of rockets based on hybrid rocket propulsion. They nurtured a plan to develop and introduce environmentally friendly, flexible and safe rocket motors to the market. The family of rockets created was called the North Star Rocket Family.

The North Star Rocket Family is based on a modular concept existing of hybrid rocket motors clustered together to form a 2-stage sounding rocket for scientific research. There are plans for two sizes of these 2-stage sounding rockets, see Figure 4. Ultimately, these "launcher elements", will form the basis of a small launcher (Figure 5). If the scaling of the technology proves to be efficient, the cost can be kept to a minimum.

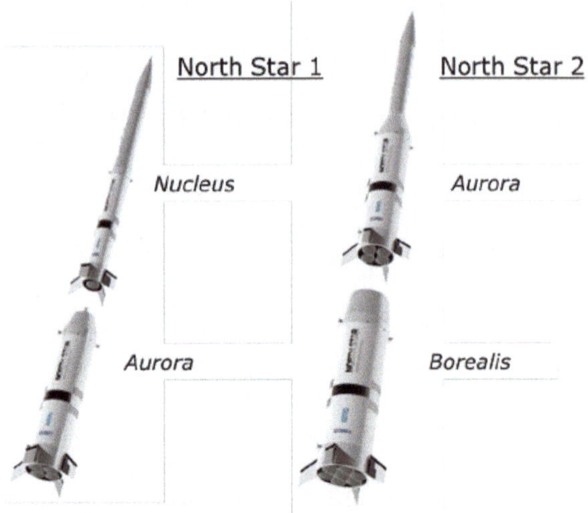

Figure 4 The two 2-stage sounding rockets from the North Star Rocket Family (Source: ASC)

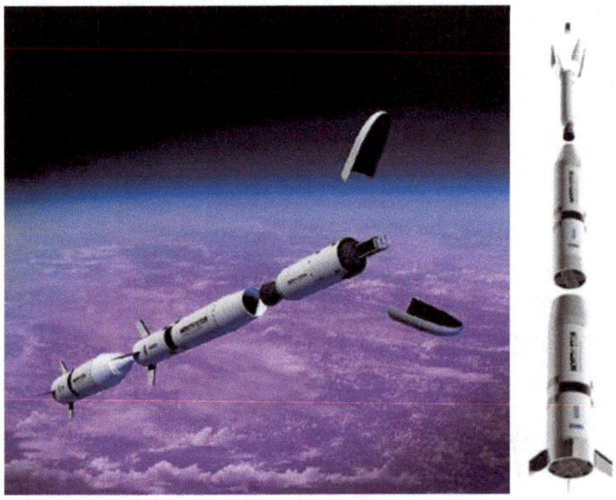

Figure 5 Artist impression of the North Star Launch Vehicle (NSLV) concept (Source: ASC)

The propulsion system of the North Star Rocket Family is based on Nammo's hybrid technology. With the experience from all three readily available green oxidizers; liquid oxygen (LOX), nitrous oxide (N_2O) and hydrogen peroxide (H_2O_2), Nammo concluded that the alternative based on hydrogen peroxide was the most promising alternative for a low cost launcher. A series of maturation programs were conducted[2] with support from the Norwegian Space Agency and ESA. SAAB Dynamics (SAAB) in Sweden contributed with their H_2O_2 experience to the team. SAAB has over 50 years of experience in handling H_2O_2 for military torpedoes, and also supplied the catalysts for these programs. The small scale firings exhibited consistent efficiencies above 95-98%, removing the supposed Achilles heel of hybrid rocket engines. It has long been known that the theoretical performance of hybrid rocket motors can surpass the performance of a solid rocket motor, but with the combustion efficiencies falling behind, it had been hard to prove their excellence through real firings. With the efficiencies obtained in these small scale firings, Nammo has opened the door for a new look at implementing hybrid rockets into real applications.

After successful demonstration of the potential of Nammo's hybrid rocket technology, Norway provided funding through ESA's Future Launcher Preparatory Program (FLPP) to demonstrate the up-scaling to a useable size. The initial application is to use this up-scaled engine as the building block for a small sounding rocket for Andøya Space Center. This project has now progressed from a period of additional small scale testing[3] and initial design work on a conceptual micro-launcher, into a phase of ground testing of the rocket motor concept at a larger scale. To get to this point, Nammo had to invest in a large scale test facility for hybrid rocket motor testing. This newly erected Green Propulsion test stand is capable of ground testing of hybrid rocket engines up to 500kN. The FLPP hybrid demonstrator activities correlate well with the need for new sounding rocket motors for Andøya Rocket Range in northern Norway.

There are at the moment more sounding rocket launches than dedicated micro-launcher campaigns. This is because of all the piggy-back solutions offered. The latter launches are either heavily sponsored or deliver their payloads in rather unfavorable orbits as they are just secondary payloads. Another issue coming up is the conjunction of space which might lead to restrictions on at which altitude these low cost satellites can be released. With their number rapidly increasing, they become a real threat on their way down for important installations as the International Space Station (ISS) and other satellite services depending on relatively low altitude orbits, i.e. 350km approx. Actually, the problem is more that it takes them too long, up to several decades, to come down from their typical release orbits (500-600km) which increases the probability of a collision. Most of these low-cost satellites have stopped working long before that.

It is this market for dedicated launches of nano-satellites and CubeSats and delivering them to their desired altitudes and orbit when *they* want (and not their host), which is the motivation for taking the hybrid sounding rocket ambition also into the micro-launcher world.

OPERATIONAL ADVANTAGES

The H2O2 based hybrid technology offers more than just the thrust to reach an altitude. The commonly reported advantages of hybrid propulsion are also valid for Nammo's hybrid technology and Nammo has already demonstrated these benefits with their designs:

- a Green Propulsion alternative
- start/stop capabilities
- throttleability
- non-hazardous materials
- non-toxic
- thrust profile tailored to the mission
- low cost

If these properties are combined with thrust vector capabilities, these sounding rockets based on hybrid rocket motors provide the scientist with a rocket which can be steered into the area of interest and stay there for a prolonged period in time (Figure 6). Atmospheric conditions and seasonal effects require tailoring of the rocket to allow it to reach the ideal altitude. The North Star sounding rocket will offer the scientist such flexibility. The traditional solid propellant based sounding rockets cannot adapt to such variations and often overshoot the area of interest if they are not loaded down with additional lumped mass.

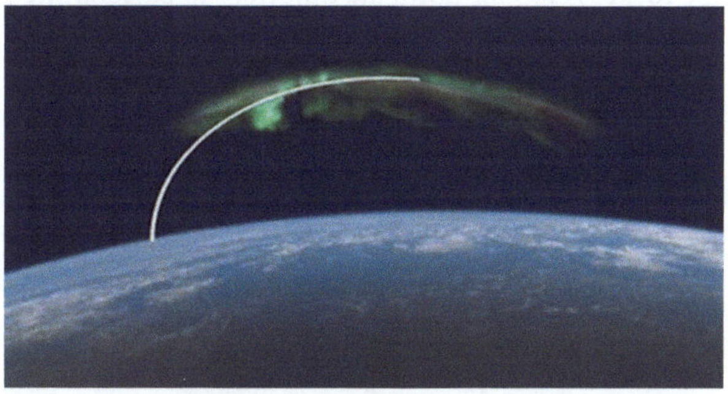

Figure 6 Precise flight control adapts the attitude of the hybrid rocket to the phenomenon of interest
(Source: ASC)

THE OPERATIONAL CONCEPT

The operational concept behind the North Star Rocket family has not been fully developed yet, but there are considerable pieces in place already. Andøya Space Center is a highly capable rocket range for sounding rockets up to the sizes of the North Star 1 and 2. Launching a micro-launcher of the same size will be new for ASC but represents a manageable step forward, mostly involving controlling a different trajectory and managing the third stage which will be responsible for the orbit insertion. Launching into orbit from Andøya has been an ambition which has become more and more pronounced.

In an established North Star Launch Vehicle (NSLV) operation, the hybrid rocket engines and other propulsion components will be manufactured by Nammo in Raufoss, Norway. Because of the nature of the hybrid rockets and by clustering of these motors, no big investments are needed in new manufacturing facilities. Not even for the largest hybrid rocket motor of 450kN approx. The fuel grains will be relatively small when compared with solid rocket motor substitutes and can be manufactured in Nammo's existing manufacturing facilities. Hybrid rockets are by definition inert, so they are easily transported from Nammo to the launch site by standard shipping methods.

No big investments are needed in launch infrastructure either. The already available U3 launch rail at Andøya Space Center supports a maximum lift-off mass of 20 tons. The target Gross Lift-Off Mass (GLOM) of the NSLV will be in the range of 15 tons or less. The rocket motor launcher elements will be assembled into a sounding rocket (or micro-launcher) at Andøya, where the payload will be prepared as well, Figure 7 and Figure 8. For work directly on sensitive payloads, clean-room facilities will be established. ASC has all the facilities available to support a launch campaign, including tracking radars and telemetry.

In support of a future micro-launcher campaign, after the rocket is too far down the trajectory to be controlled from Andøya, Norway has other existing infrastructure at the satellite station SVALSAT, Longyearbyen, Spitsbergen, which will be utilized for orbital insertion. In this way, the complete flight from launch to orbital insertion can be controlled from Norwegian territory.

Figure 7 Sounding rocket assembly at Andøya

Figure 8 Payload assembly at Andøya

The capability of launching micro-launchers from Andøya will provide Andøya Space Center with a new business area offering dedicated CubeSat and nanosat launches and will provide launch services for In-Flight Experiments. To be able to offer the complete range of services, a third stage to inject the payload into orbit will be needed in addition to the two stages of the largest North Star 2 sounding rocket.

LOW COST APPROACH

If the hybrid technology as demonstrated by Nammo keeps scaling-up as planned and the anticipated cost levels can be realized, a dedicated low cost micro-launcher service will be within reach. It is crucial for this concept that the component cost is kept at the cost levels acceptable to the sounding rocket market. This is where the economy-of-scale can be achieved while the dedicated micro-launcher market has not taken off yet.

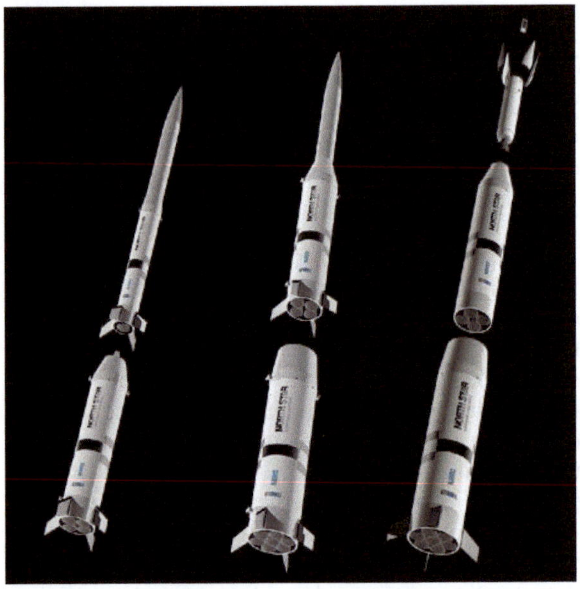

Figure 9 A modular concept is adopted and engines and complete stages are reused (Source: ASC)

Commonality between the sounding rocket market and the small launcher segment are the basis for obtaining the economy-of-scale (Figure 9). The selection of components is tailored towards "industry quality" components, characterized by manufacturing processes which provide reliability through robust processes and sufficient quantities. The launcher design will be the smallest possible to do the job, and size alone will already keep the cost down. Only if the launcher is small enough can the propulsion system also be used as the propulsion system for the sounding rocket market, a market which is also urgently looking for new motors. The propulsion system constitutes the single largest sub-system of a launcher and has therefore been the primary focus of the work.

To realize this plan, not only the cost of the launcher elements should be kept affordable, also the non-recurring cost during development and the launch campaign(s) itself have to be optimized. Substantial up-front financing has not been identified yet, simply because the targeted customers do not have that kind of money available. But it should be mentioned that the concept has been well received by the community.

To keep the cost under control, an approach has been adopted which takes the project forward in smaller steps, with each step being affordable, with each step making the project move closer to the final goal and with each step resulting in a new building block which can be used for the next step.

The project has therefore adapted to the following philosophy in order to move forward:

- Primary goal is to offer sounding rocket launch services
- Secondary goal is to develop a nano-sat launch service if economically viable
- Incremental progress through a spiral development concept
- Each step has a clear goal and a well-defined outcome
- For each step, the confidence in the concept will grow
- Modularity is the basis for keeping the cost down
- Modularity combined with the flexibility and robustness of the hybrid engines allows for future upgrades

The concept has been compared with several other initiatives based on air-launched concepts and disposable micro-launch initiatives based on solid propellants. These projects all appear to be aiming at payloads of around 100kg and upwards, while the North Star Launch Vehicle is aiming at the lower end payload range of 20-25kg. The mass of 20-25kg is equivalent to a stack of 8-12 CubeSats with dispensing units and support equipment. For this market, low cost is the far most important property of the solution.

GREEN PROPULSION TEST STAND

In 2012, Nammo recognized the single most important hurdle to pass before one could start the development of the hybrid rocket engines for the North Star Family: these new rocket motors needed to be tested in a test stand capable of handling large thrust levels, long burn times and should be equipped with an installation which can handle large quantities of hydrogen peroxide. Such a facility did not exist anywhere in Europe.

In support of the North Star Rocket Family development and the FLPP hybrid demonstrator program, Nammo made the decision to invest in such a facility themselves[4]. Large amounts of H2O2 needed to be handled to support long burn times and high mass flow rates. To support a test campaign, also a long term storage facility for large quantities of H2O2 needed to be established.

The facility was built at the Nammo Test Centre and took 18 months to complete. The Nammo Test Centre is already a well-equipped test center for the development and qualification of solid rocket motors for tactical missiles. It has a complete range of environmental test cells including shock and vibration testing to support the development and the qualification of products and components for both military and aerospace customers. Nammo's portfolio of Ariane 5 products have been tested and qualified at this test center. Together with the existing capabilities of the Nammo Test Center, the new Green Propulsion Rocket Test Stand will be a unique installation capable of supporting many future rocket development programs.

In April 2014 the facility was approved and certified by the authorities for the use of H2O2 and is now operative. Our partner SAAB Dynamics assisted in the design and construction of all H2O2 related equipment. The test cell consists of a reinforced building (Figure 10) containing the fortified test bench which is holding down the test object during firings. On the other side of a strong wall, the long term storage and handling facilities of the H2O2 have been established. Half of the building is taken up by the long term storage and handling equipment for the H2O2. A separate building contains the control room for data acquisition and controlling the test procedure (Figure 11).

Figure 10 The new test cell for large scale hybrid rocket motor firings (500kN)

The test stand is dimensioned for hybrid rocket motor firings up to 500kN of thrust and a long term storage capacity of 20.000 liter of H2O2. The capacity of the facility is large enough to support the development of the largest booster of the North Star 2 sounding rocket which will also be the first stage of the North Star Launch Vehicle.

As an example of the amount of propellants consumed during a single firing, the FLPP hybrid demonstrator will consume around 1080kg of oxidizer and 152kg of fuel in order to demonstrate a hybrid engine with a thrust of 112kN during a 25-28 seconds burn.

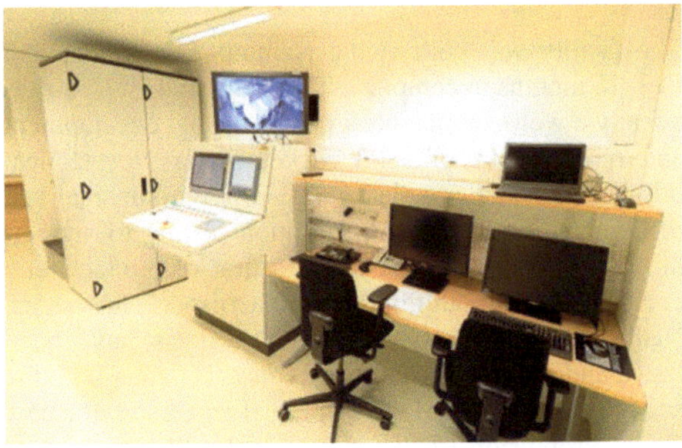

Figure 11 The control room

In support of large testing campaigns, two 10.000 liter H2O2 storage tanks were installed. The tanks are manufactured in pure aluminum, providing a stable storage period of years for the 87.5% H2O2. The tanks have been mounted in a raised position above a pool of water (see Figure 12). The water basin is for safety. In case a leakage or spill occurs, the water will dilute the H2O2 quickly into a harmless concentration.

Figure 12 Both tanks in place just above the water basin below floor level

When H2O2 is delivered from the factory, the filling of H2O2 will be done from a standard road delivery on a special lorry. In Figure 13 one can see the truck under the delivery of H2O2 to Nammo. By having the possibility to use readily available propellant ingredients available in large quantities, the cost of the North Star concept can be kept low.

Figure 13 Filling of the tanks from standard truck

MODULAR MOTOR DEVELOPMENT

The North Star Rocket family is based on the communality of motors. The initial step is to up-scale the technology from the laboratory scale motors into the first, and smallest, usable size of the hybrid rocket motor. This will require a scaling factor of 20 when compared with the lab test motors. This first practical motor will be the rocket motor for the second stage of North Star 1 and will be an important building block of the concept. It is called the Unitary Motor. After successful up-scaling to the Unitary Motor size, the booster stage of North Star 1 will be a clustering of 4 Unitary Motors. Clustering of motors (see Figure 14) will keep the cost down and the development risk of the hybrid rocket motors manageable.

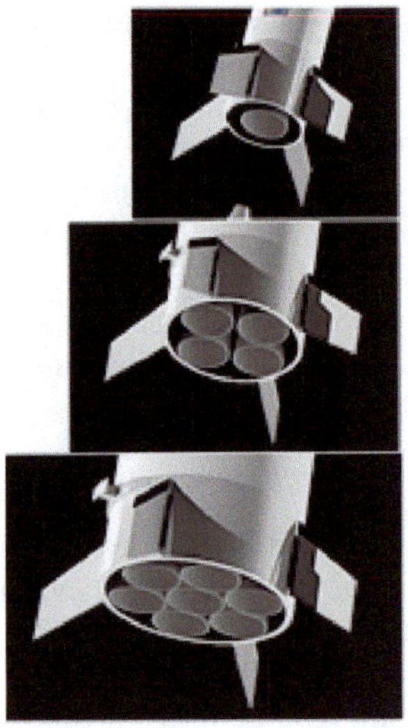

Figure 14 Clustering of motors

The performance of the 4 Unitary Motors clustered together, is the configuration which will be demonstrated in the FLPP-project[5]. This low-risk approach allows for the realization of the North Star 1 without further scaling of the hybrid rocket motor technology. In Figure 15 the correlation between the different sizes of engines is visualized. The clustering of 4 identical motors will increase the production rate of each of the components needed, which in turn assures a more favorable cost and quality on all parts being used.

After sufficient funding has been identified, this process will be repeated for the large booster needed for the first stage of North Star 2. A new slightly larger Unitary Motor is foreseen for this step.

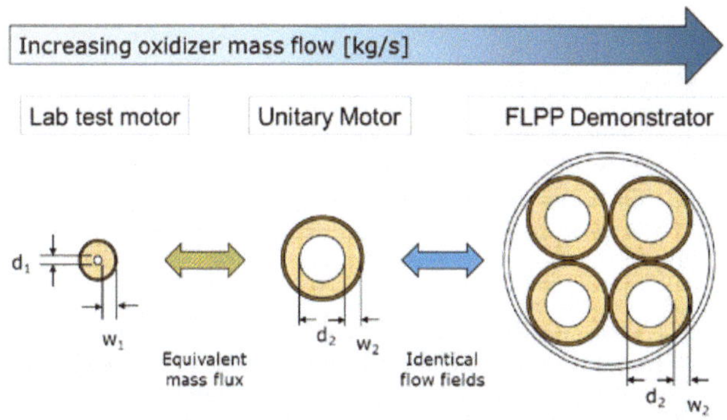

Figure 15 Correlation between the small and large scale motors

Design data for the Unitary Motor is given in Table 1. These are the target values of the performance of the Unitary Motor based on the available background information and the design data and experience gained from the small scale firings. Delivered performance data are sea level performance data including nozzle and combustion (in-)efficiencies. These values should not be compared with theoretical data and expansion to vacuum conditions.

Table 1 Design data for the Unitary Motor

Thrust	28 kN	Total propellant mass	308 kg
Total impulse	700 kNs	Total oxidizer mass	270 kg
Burn time	25 s	Total fuel mass	38 kg
Chamber pressure	35 bar	Liquid oxidizer volume	196 L
Optimal OF (CEA)	7.1	Solid fuel volume	39.6 L
Delivered Isp (at sealevel)	232.01 s	Propellant mass flow	12.3 kg/s
Delivered c*	1542.4 m/s	Oxidizer mass flow	10.8 kg/s
Delivered Cf (at sealevel)	1.4751	Fuel mass flow	1.5 kg/s

To give an impression of the size of the Unitary motor, one can see one of the fuel grains in Figure 16. They will be machined into their final shape depending on the results from the development tests and the intended purpose of the test. This is a good example of the versatility and robustness of the hybrid propulsion. Try to do this with a solid rocket motor grain. The motor case housing used during initial testing can be adapted to different grain lengths. The outer diameter of the breadboard Unitary Motor is slightly oversized to provide a margin for longer firings or higher regression rates, but under a nominal test run there should be some remaining fuel after shut-off.

Figure 16 The fuel grain for the Unitary Motor

FIRING RESULTS

The initial motor firings of the Unitary Motor have started and two motors have been fired by Oct. 2014. The oxidizer feed system was tested first separately to demonstrate that a full burn duration of 25 seconds with an oxidizer mass flow of 10.8 kg/s could be achieved. During these firings the catalyst was tested as well.

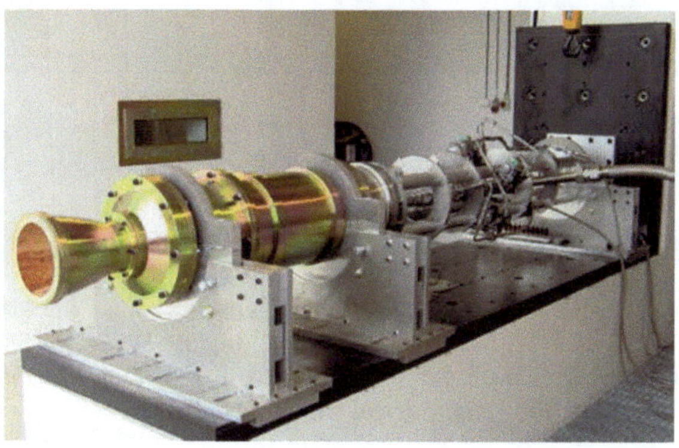

Figure 17 The breadboard Unitary Motor on the test stand just before firing

After successful demonstration of the capacity of the oxidizer feed system and by showing full decomposition of the H_2O_2 for the full duration of the firing, the test program of the Unitary Motor could start.

The first two firings were complete hybrid rocket motor firings (see Figure 17) and the duration of burn was gradually increased, the first with a burn time of 11 sec, the second 16 sec. Ignition occurred instantly in both cases and also termination of the motor functioned as commanded. Both firings reached the desired thrust level of 28kN and the efficiency which was demonstrated at the smaller scale was successfully repeated at the 20x larger scale. C* and Isp values were also according to the predictions. These results are a big step forward for the North Star Concept, as both thrust level and efficiency at the larger scale were seen as the most challenging aspects of the hybrid propulsion technology. Figure 18 shows the Unitary motor during its first firing. Water vapor is created initially at start-up and creates some fog under the extreme humid outdoor conditions under this firing.

Figure 18 The Unitary Motor during test firing

Figure 19 The plume just after start-up (top) and during the remainder of the burn (bottom)

With a plume consisting of only H2O and CO2 a small cloud is created at start-up which is quickly replaced by a completely smokeless behavior for the remainder of the burn as shown in Figure 19. Indirectly, this is an excellent proof of the burning efficiency of this hybrid rocket motor design.

Testing will continue with full duration firings and the design will be optimized based on the demonstrated performance and the condition of the hardware. Flight weight Unitary Motors are expected to be manufactured in spring 2015.

ACKNOWLEDGMENTS

The European Space Agency (ESA) who is funding the FLPP Hybrid Demonstration activities at Nammo Raufoss. The Norwegian Space Centre who is supporting the FLPP program with their national contribution to the program. Innovation Norway for their support to the Green Propulsion Test Stand.

Special thanks go to Kolbjørn Dahle and Trond Abrahamsen from Andøya Rocket Range for their continuous support and providing the artwork of the North Star initiative.

REFERENCES

1 Jan-Erik Rønningen: "The HTR Program – An Overview of the Rocket System and Program Achievements", *18th ESA Symposium on European Rocket and Balloon Programmes and Related Research*, Visby, Sweden, 3-7 June 2007.

2 Jan-Erik Rønningen and Johanne Husdal: "Nammo Hybrid Rocket Propulsion TRL Improvement Program", *AIAA 48th Joint Propulsion Conference*, Atlanta, Georgia, USA, August 2012.

3 Jan-Erik Rønningen and Johanne Husdal: "Test Results from Small-Scale Hybrid Rocket Testing", *Space Propulsion Conference 2012*, Bordeaux, France, May 2012.

4 Onno Verberne, Adrien J. Boiron and Terje Vesterås: "Green Propulsion Rocket Test Stand", *Space Propulsion Conference 2014*, Cologne, Germany, May 2014.

5 Onno Verberne, Adrien J. Boiron and Terje Vesterås: "FLPP Large Scale Hybrid Rocket Demonstrator", *Space Propulsion 2014*, Cologne, Germany, May 2014

MIX
Papier aus verantwortungsvollen Quellen
Paper from responsible sources
FSC® C105338

If you have any concerns about our products,
you can contact us on
ProductSafety@springernature.com

In case Publisher is established outside the EU,
the EU authorized representative is:
Springer Nature Customer Service Center GmbH
Europaplatz 3, 69115 Heidelberg, Germany

Printed by Libri Plureos GmbH
in Hamburg, Germany